DATA MINING IN AGRICULTURE

Springer Optimization and Its Applications

VOLUME 34

Managing Editor
Panos M. Pardalos (University of Florida)

Editor—Combinatorial Optimization
Ding-Zhu Du (University of Texas at Dallas)

Advisory Board
J. Birge (University of Chicago)
C.A. Floudas (Princeton University)
F. Giannessi (University of Pisa)
H.D. Sherali (Virginia Polytechnic and State University)
T. Terlaky (McMaster University)
Y. Ye (Stanford University)

Aims and Scope
Optimization has been expanding in all directions at an astonishing rate during the last few decades. New algorithmic and theoretical techniques have been developed, the diffusion into other disciplines has proceeded at a rapid pace, and our knowledge of all aspects of the field has grown even more profound. At the same time, one of the most striking trends in optimization is the constantly increasing emphasis on the interdisciplinary nature of the field. Optimization has been a basic tool in all areas of applied mathematics, engineering, medicine, economics and other sciences.

The *Springer Optimization and Its Applications* series publishes undergraduate and graduate textbooks, monographs and state-of-the-art expository works that focus on algorithms for solving optimization problems and also study applications involving such problems. Some of the topics covered include nonlinear optimization (convex and nonconvex), network flow problems, stochastic optimization, optimal control, discrete optimization, multiobjective programming, description of software packages, approximation techniques and heuristic approaches.

For other titles published in this series, go to
www.springer.com/series/7393

DATA MINING IN AGRICULTURE

By

ANTONIO MUCHERINO
University of Florida, Gainesville, FL, USA

PETRAQ J. PAPAJORGJI
University of Florida, Gainesville, FL, USA

PANOS M. PARDALOS
University of Florida, Gainesville, FL, USA

 Springer

Antonio Mucherino
Institute of Food & Agricultural
Information Technology Office
University of Florida
P.O. Box 110350
Gainesville, FL 32611
USA
amucherino@ufl.edu

Petraq J. Papajorgji
Institute of Food & Agricultural
Information Technology Office
University of Florida
P.O. Box 110350
Gainesville, FL 32611
USA
petraq@ifas.ufl.edu

Panos M. Pardalos
Department of Industrial & Systems Engineering
University of Florida
303 Weil Hall
Gainesville, FL 32611-6595
USA
pardalos@ise.ufl.edu

ISSN 1931-6828
ISBN 978-1-4614-2935-7   e-ISBN 978-0-387-88615-2
DOI 10.1007/978-0-387-88615-2
Springer Dordrecht Heidelberg London New York

Printed on acid-free paper

Springer is part of Springer Science+Business Media (www.springer.com)

Dedicated to Sonia
who supported me morally
during the preparation of this book.

To the memory of my parents
Eleni and Jorgji Papajorgji
who taught me not to betray my principles
even in tough times.

Dedicated to my father and mother
Miltiades and Kalypso Pardalos
for teaching me to love nature
and to grow my own garden.

Preface

Data mining is the process of finding useful patterns or correlations among data. These patterns, associations, or relationships between data can provide information about a specific problem being studied, and information can then be used for improving the knowledge on the problem. Data mining techniques are widely used in various sectors of the economy. Initially they were used by large companies to analyze consumer data from different perspectives. Data was then analyzed and useful information was extracted with the goal of increasing profitability.

The idea of using information hidden in relationships among data inspired researchers in agricultural fields to apply these techniques for predicting future trends of agricultural processes. For example, data collected during wine fermentation can be used to predict the outcome of the fermentation while still in the early days of this process. In the same way, soil water parameters for a certain soil type can be estimated knowing the behavior of similar soil types.

The principles used by some data mining techniques are not new. In ancient Rome, the famous orator Cicero used to say *pares cum paribus facillime congregantur* (*birds of a feather flock together* or literally *equals with equals easily associate*). This old principle is successfully applied to classify unknown samples based on known classification of their neighbors. Before writing this book, we thoroughly researched applications of data mining techniques in the fields of agriculture and environmental studies. We found papers describing systems developed to classify apples, separating good apples from bad ones on a conveyor belt. We found literature describing a system that classifies chicken breast quality, and others describing systems able to predict climate forecasting and soil classification, and so forth. All these systems use various data mining techniques.

Therefore, given the scientific interest and the positive results obtained using the data mining techniques, we thought that it was time to provide future specialists in agriculture and environment-related fields with a textbook that will explain basic techniques and recent developments in data mining. Our goal is to provide students and researchers with a book that is easy to read and understand. The task was challenging. Some of the data mining techniques can be transformed into optimization problems, and their solutions can be obtained using appropriate optimization methods. Although this transformation helps finding a solution to the problem, it makes the presentation difficult to understand by students that do not have a strong mathematical background.

The clarity of the presentation was the major obstacle that we worked hard to overcome. Thus, whenever possible, examples in Euclidean space are provided and corresponding figures are shown to help understand the topic. We make abundant use of MATLAB® to create examples and the corresponding figures that visualize the solution. Besides, each technique presented is ranked using a well-known publication on the relevance of data mining techniques. For each technique, the reader will find published examples of its use by researchers around the world and simple examples that will help in its understanding. We made serious efforts to shed light on when to use the method and the quality of the expected results. An entire chapter is dedicated to the validation of the techniques presented in the book, and examples in MATLAB are used again to help the presentation. Another chapter discusses the potential implementation of data mining techniques in a parallel computing environment; practical applications often require high-speed computing environments. Finally, one appendix is devoted to the MATLAB environment and another one is dedicated to the implementation of one of the presented data mining techniques in C programming language.

It is our hope that readers will find this book to be of use. We are very thankful to our students that helped us shape this course. As always, their comments were useful and appropriate and helped us create a consistent course. We thank Vianney Houles, Guillermo Baigorria, Erhun Kundakcioglu, Sepehr M. Nasseri, Neng Fan, and Sonia Cafieri for reading all the material and for finding subtle inconsistencies. Last but certainly not least, we thank Vera Tomaino for reading the entire book very carefully and for working all exercises. Her input was very useful to us.

Finally, we thank Springer for trusting and giving us another opportunity to work with them.

Gainesville, Florida   *Antonio Mucherino*
January 2009   *Petraq J. Papajorgji*
                                                                       Panos M. Pardalos

Contents

List of Figures

Chapter 1
Introduction to Data Mining

1.1 Why data mining?

There is a growing amount of data available from many resources that can be used effectively in many areas of human activity. The Human Genome Project, for instance, provided researchers all over the world with a large set of data containing valuable information that needs to be discovered. The code that codifies life has been read, but it is not yet known how life works. It is desirable to know the relationships among the genes and how they interact. For instance, the genome of food such as tomato is studied with the aim of genetically improving its characteristics. Therefore, complex analyses need to be performed to discover the valuable information hidden in this ocean of data. Another important set of data is created by Web pages and documents on the Internet. Discovering patterns in the chaotic interconnections of Web pages helps in finding useful relationships for Web searching purposes. In general, many sets of data from different sources are currently available to all scientists.

Sensors capturing images or sounds are used in agricultural and industrial sectors for monitoring or for performing different tasks. In order to extract only the useful information, these data need to be analyzed. Collections of images of apples can be used to select good apples for marketing purposes; sets of sounds recorded from animals can reveal the presence of diseases or bad environmental conditions. Computational techniques can be designed to perform these tasks and to substitute for human ability. They will perform these tasks in an efficient way and even in an environment harmful to humans.

The computational techniques we will discuss in this book try to mimic the human ability to solve a specific problem. Since such techniques are specific for certain kinds of tasks, the hope is to develop techniques able to perform even better than humans. Whereas an experienced farmer can personally monitor the sounds generated by animals to discover the presence of diseases, there are other tasks humans can perform only with great difficulties. As an example, human experts can check apples in a conveyor belt to separate good apples from bad ones. The percentage of removed bad apples (the ones removed from the conveyor) is a function of the speed of the

A. Mucherino et al., *Data Mining in Agriculture*, Springer Optimization and Its Applications 34,
DOI: 10.1007/978-0-387-88615-2_1,
© Springer Science + Business Media, LLC 2009

conveyor and the amount of human attention dedicated to the task. It is proved that it is rather difficult for the human brain to be focused on a particular subject for a long time, thus inducing distraction. Unlike humans, computerized systems implementing computational techniques to solve a particular problem do not have these kinds of problems as they are immune to distraction. Furthermore, there are tasks humans cannot perform at all, such as the task of locating all the interactions among all the genome genes or finding patterns in the World Wide Web. Therefore, researchers are trying to develop specialized techniques to successfully address these issues.

Data mining is designed to address problems such as the ones mentioned above. Techniques used in data mining can be divided in two big groups. The first group contains techniques that are represented by a set of instructions or sub-tasks to carry out in order to perform a certain task. In this view, a technique can be seen as a sort of recipe to follow, which must be clear and unambiguous for the executor. If the task is to "cook pasta with tomatoes," the recipe may be: heat water to the boiling point and then throw the pasta in and check whether the pasta has reached the point of being *al dente*; drain the pasta and add preheated tomato sauce and cheese. Even a novice chef would be able to achieve the result following this recipe.

Moreover, note that another way to learn how to cook pasta is to use previous cooking experience and try to generalize this experience and find a solution for the current problem. This is the philosophy the second group of data mining techniques follows. A technique, in this case, does not provide a recipe for performing a task, but it rather provides the instructions for *learning* in some way how to perform the task. As a newborn baby learns how to speak by acquiring stimuli from the environment, a computational technique must be "taught" how to perform its duties. Although learning is a natural process for humans, it is not the case for computerized systems designed to replace humans in performing certain tasks. In the case of the novice chef, he has all the needed ingredients (pasta, water, tomato sauce, cheese) at the start, but he does not know how to obtain the final product. In this case, he does not have the recipe. However, he has the capability of learning from the experience, and after a certain number of trials he will be able to transform the initial ingredients into a delicious tomato pasta dish and be able to write his own recipe.

In this book we will present a number of techniques for *data mining* (or *knowledge discovery*). They can be divided in two subgroups as discussed above. For instance, the k-nearest neighbor method (Chapter 4) provides a set of instructions for classification purposes, and hence it belongs to the first group. Neural networks (Chapter 5) and support vector machines (Chapter 6), instead, follow particular methods for learning how to classify data.

Let us consider the following example. A laboratory is performing blood analysis on sick and healthy patients. The goal is to correlate patients' illness to blood measurements. Why? If we were able to find a subgroup of blood measurement values corresponding to sick patients, we would predict the illness of future patients by checking whether their blood measurements fall in the found subgroup. In other words, the correlation between blood measurements and patient's conditions is not known and the goal is to find out the nature of this relationship. Available data accumulated in the past can be used to solve the problem. The laboratory may perform blood analysis and then check a patient's conditions in a different way that is totally reliable (and probably expensive and invasive). When a reasonable amount of data is collected, it is then possible to use the accumulated knowledge for classifying patients on the basis of their illness. In this process two sets of data are identified: input data (i.e., blood measurements) and a set of corresponding outputs (patient illnesses). Data mining techniques such as k-nearest neighbor (which follows a list of instructions for classifying the patients) or neural networks (which are able to learn how to classify the patients) can be used for this purpose.

Unfortunately, all needed data may not always be available. As an example, let us consider that only blood measurements are available and there is no information about patients. In this case, solving the problem becomes more difficult because only input data are available. However, what can be done is to partition inputs into clusters. Each cluster can be built so that it contains similar data, and the hope is that each cluster would represent the expected outputs. The techniques belonging to this group are referred to as *clustering techniques* or as *unsupervised classification techniques*, because the couples of corresponding inputs/outputs are actually absent. In the cases this information is available, *classification techniques* are instead used.

Data mining techniques can be therefore grouped in two different ways. They can be clustering or classification techniques. Furthermore, some of them provide a list of instructions for clustering or classification purposes, whereas others learn from the available data how to perform classifications. Note that clustering techniques cannot learn from data, because, as explained earlier, only a part of the data is available. In classification techniques, the categories in which the data are grouped are referred to as *classes*. Similarly, in clustering techniques, such categories are referred to as *clusters*. The object contained in the set of data, i.e., blood measurements, apples, sounds, etc., are referred to as *samples*. Section 1.2.1 provides an overview of data mining techniques.

Based on what is presented above, the following can be a good definition of data mining or knowledge discovery:

Data mining is a nontrivial extraction of previously unknown, potentially useful and reliable patterns from a set of data. It is the process of analyzing data from different perspectives and summarizing it into useful information.

1.2 Data mining techniques

1.2.1 A brief overview

Many data mining techniques have been developed over the years. Some of them are conceptually very simple, and some others are more complex and may lead to the formulation of a global optimization problem (see Section 1.4). In data mining, the goal is to split data in different categories, each of them representing some feature the data may have. Following the examples provided in Section 1.1, the data provided by the blood laboratory must be classified into two categories, one containing the blood measurements of healthy patients and the other one containing the blood measurements of sick patients. Similarly, apples must be grouped as bad and good apples for marketing purposes. The problem is slightly more complicated when using, for instance, data mining for recognizing animal sounds. One solution can be to partition recorded sounds into two categories, in which one category contains the sounds to be recognized and the other category contains the sounds of no interest. However, sounds that may reveal signs of diseases in animals can be separated from other sounds the animals can generate and from noises of the surrounding environment. If more than two categories are considered, then sounds signaling signs of diseases in animals can be more accurately identified, as in the application described in Section 5.4.1.

Let us refer again to the example of the blood analysis for shedding some more light on the data mining techniques discussed in this book. Once blood analysis data are collected, the aim is to divide these data into two categories representing sick and healthy patients. Thus, a new patient is considered sick or healthy based on the fact that his blood values fall in the first (sick) or the second (healthy) category. The decision whether a patient is sick or healthy can be made using a classification or clustering technique. In the case that for every blood analysis, in a given set of blood measurements data, it is known whether the patient is sick or healthy, then the set of data is referred to as a *training set*. In fact, data mining techniques can exploit this set for classifying a patient based on his blood values. In this case, classification techniques such as k-nearest neighbor, artificial neural network and support vector machines can be successfully used.

Unfortunately, in some applications available data are limited. As an example, blood measurements data may be available, but no information about a patient's conditions may be provided. In these cases, the goal is to find in the data inherent patterns that would allow their partitioning in clusters. If a clustering technique finds a partition of the data in two clusters, then one of them should correspond to sick patients and the other to healthy patients. Clustering techniques include the k-means method (with all its variants) and biclustering methods.

Statistical methods such as principal component analysis and regression techniques are commonly used as simple methods for finding patterns in sets of data. Statistical methods can also be used coupled with the above-mentioned data mining techniques.

There are different surveys of data mining techniques in the literature. Some of them are [17, 46, 72, 116, 136, 239]. A graphic representation of the classification of data mining techniques discussed in this book is given in Figure 1.1.

Fundamental for the success of a data mining technique is the ability to group available data in disjoint categories, where each category contains data with similar properties. The similarity between different samples is usually measured using a distance function, and similar samples should belong to the same class or cluster. Therefore, the success of a data mining technique depends on the adequate definition of a suitable distance between data samples. If the blood data pertain to the glucose level and the related disease is diabetes, then the distance between two blood values

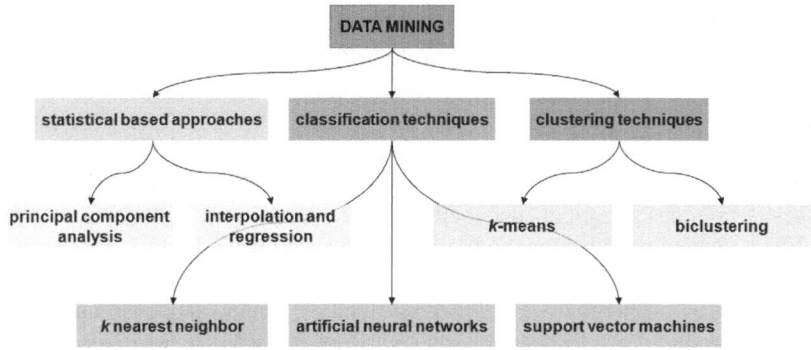

Fig. 1.1 A schematic representation of the classification of the data mining techniques discussed in this book.

is simply the difference in glucose levels. In the case that more complex analysis needs to be performed, then more complex variables may be needed for representing a blood test. Consequently, the distance between two blood tests cannot always be defined as the simple difference between two real numbers, but more complex functions need to be used. The definition of a suitable distance function depends on the representation of these samples. Section 1.2.2 provides a wide discussion on the different data representations that can be used.

Clustering techniques are divided in *hierarchical* and *partitioning*. The hierarchical clustering approach builds a tree of clusters. The root of this tree can be a cluster containing all the data. Then, branch by branch, the initial big cluster is split in sub-clusters, until a partition having the desired number of clusters is reached. In this case, the hierarchical clustering is referred to as *divisive*. Moreover, the root of the tree can also consist of a set of clusters, in which each cluster contains one and only one sample. Then, branch by branch, these clusters are merged together to form bigger clusters, until the desired number of clusters is obtained. In this case, the hierarchical clustering is referred to as *agglomerative*. In this book, we will not consider hierarchical techniques.

The partition technique referred to as k-means and many of its variants will be discussed in Chapter 3. The k value refers to the number of clusters in which the data are partitioned. Clusters are represented by their centers. The basic idea is that each sample should be closer to the center of its own cluster. If this is not verified, then the partition is modified, until each sample is closer to the center of the cluster it belongs to. The distance function between samples plays an important role, since a sample can migrate from a cluster to another one based on the values provided by the distance function.

Among the partitioning techniques for clustering are also the recently proposed methods for biclustering (Chapter 7). Such methods are able to partition the data simultaneously on two dimensions. While standard clustering techniques consider only the samples and look for a suitable partition, biclustering partitions simultaneously the set of samples, and the set of attributes used for representing them, in biclusters. First, biclustering was introduced as clustering technique. Later, methods have been developed for exploiting training sets for obtaining partitions in biclusters. Therefore, biclustering methods can be used for both clustering and classification purposes.

In this book, the following classification techniques will be described: the k-nearest neighbor method, the artificial neural networks and the support vector machines. A brief description of such methods is presented in the following.

The k-nearest neighbor method is a classification method and is presented in Chapter 4. In this approach, the k value has a meaning different from the one in the k-means algorithm that we will explain soon. A training set containing known samples is required. All the samples which are not contained in the training set are referred to as *unknown samples*, because their classification is not known. The aim is to classify such unknown samples by using information provided by the samples in the training set. Intuitively, an unknown sample should have a classification close to the one its neighbors in the training set have. Therefore, each unknown sample can be classified accordingly to the classification of its neighbors. The k value defines the number of nearest known samples considered during the classification.

Artificial neural networks can also be used for data classification (Chapter 5). This approach tries to mimic the way the human brain works and they try to "learn" how to classify data using knowledge embedded in training sets. A neural network is a set of virtual neurons connected by weighted links. Each neuron performs very easy tasks, but the network can perform complex tasks when all its neurons work together. Commonly, the neurons in networks are organized in layers, and these kinds of networks are referred to as *multilayer perceptrons*. Such networks are composed by layers of neurons: the input layer, one or more "hidden" layers and finally the output layer. A signal fed to the network propagates through the network from the input to the output layer. A training set is used for setting the network parameters so that a predetermined output is obtained when a certain input signal is provided. The hope is that the network is able to generalize from the samples in the training set and to provide good classification accuracy.

Support vector machines are discussed in Chapter 6. This is a technique for data classification. Its basic idea is inspired by the classification of samples into two different classes by a linear classifier. The method though can be extended and used for classifying data in more than two classes. This is achieved by using more than one support vector machine organized in a tree-like structure, since each of them is able to distinguish between two classes only. The case where data are not linearly separable can also be considered. Kernel functions are used to transform the original space in another one where classes are linearly separable.

1.2.2 Data representation

The representation of the data plays an important role in selecting the appropriate data mining technique to use. In the example of the blood analysis, the data can be represented as real numbers. Usually one variable does not suffice for representing a sample, and hence vectors or matrices of variables need to be used. For instance, an apple can be represented by a digital image portraying the fruit. A digital image is a matrix of pixels with a certain color. In this case, the image of the apple is represented as a matrix of real numbers. A sound can instead be represented as a set of consecutive audio signals. In this case the data are represented as vectors of real numbers. The length of the representing vector is important as longer vectors represent the sound more accurately. Other representations can make use of graphs or networks, as is the case of the financial application discussed in Section 1.3.3.

Some of the data mining techniques use distances between samples for partitioning or classifying data. Computing the distance between two samples means computing the distance between two vectors or two matrices of variables representing the samples. An efficient representation of the data impacts the definition of a good distance function. Even in the cases where data mining techniques do not use the distance function (such is the case of artificial neural networks), data representation is important as it helps the technique to better perform the task.

In order to understand the importance of data representation, let us consider as an example the different ways a DNA (deoxyribonucleic acid) sequence can be represented. The DNA contains the genetic instructions used in the development and the functioning of all living organisms. It consists of two strands that wrap around each other. Chemical bonds hold together the two strands. Each strand contains a sequence of 4 bases and each base has a complementary base. This means that one strand can be rebuilt by using the information located on the other one. Only one sequence of bases is therefore sufficient for representing a DNA molecule. One of the possible representations can be the sequence of initials of the name of the bases: A for adenine, C for cytosine, G for guanine and T for thymine. On a computer, a character is represented using the ASCII code, an 8-bit code. However, as pointed out in [49], there are more efficient representations. Four names or initials can be coded by 4 integer numbers, for instance 0 for adenine, 1 for cytosine, 2 for guanine and 3 for thymine. These numbers can be represented on computers using a 2-bit code: 00, 01, 10, 11. This code is certainly more efficient than the ASCII code, since it needs one fourth of the variables for representing the same data. Figure 1.2 gives a schematic comparison of the possible representations for the DNA molecules.

In living organisms a DNA molecule can be divided into genes. Genes contain the information for coding *proteins*. Proteins have been studied for many years because of their high importance in biology, and finding out the secrets they still hide is one of the major challenges in modern biology. Because of its relevance, this topic is largely treated in the specialized literature. There is a considerable amount of work dedicated to the protein representation and its conformations. In January 2009, Google Scholar provided more than 6000 papers on "protein folding" published during 2008, and already about 300 papers published in 2009. Just to quote one of them, the work in [115] presents the recent progress for uncovering the secrets of protein folding.

Even though protein molecules are not specifically studied in agricultural-related fields, we decided to discuss here the different ways a protein conformation can be modeled. This is a very interesting example, because it shows how a single object,

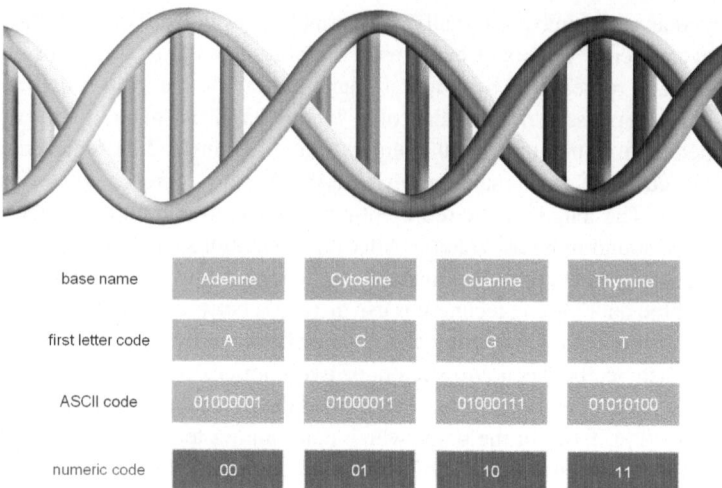

base name	Adenine	Cytosine	Guanine	Thymine
first letter code	A	C	G	T
ASCII code	01000001	01000011	01000111	01010100
numeric code	00	01	10	11

Fig. 1.2 The codes that can be used for representing a DNA sequence.

the protein, can be modeled in different ways. The model to be used can then be chosen on the basis of the experiments to be performed. In the following, only the spatial conformations that proteins can assume are taken into consideration, leaving out protein chemical and physical features.

Proteins are formed by other smaller molecules called *amino acids*. There are only 20 different amino acids that are involved in the protein synthesis, and therefore proteins can be built by using 20 different molecular bricks only. Each amino acid has a common part and a part that characterizes each of them, which is called *side chain*. The amino acids forming a protein are bonded chemically to each other through the atoms of their common parts. Therefore, a protein can be seen as a chain of amino acids: the sequence of atoms contained in the common parts form the so-called *backbone* of the protein, where the side chains of all the amino acids are attached.

Among the atoms contained in the common part of each amino acid, more importance is given to the carbon atom usually labeled with the symbol C_α. In some models presented in the literature [38, 172, 175], this atom has been used alone for representing an entire amino acid in a protein. Then, in this case, protein conformations are represented through the spatial coordinates of n atoms, each of them representing an amino acid. It is clear that these models give a very simplified representation of a protein conformation. In fact, information about the side chains are not included at all, and therefore the model cannot discriminate among the 20 amino acids. However, this representation is able to *trace* the protein backbone.

More accurate representations of the protein backbones can be obtained if more atoms are considered. If three particular atoms from the common part of each amino acid are considered (two carbon atoms C_α and C and a nitrogen N), then this information is sufficient for rebuilding the whole backbone of the protein. Therefore, a protein backbone can be represented precisely by a sequence of $3n$ atomic coordinates, where n is the number of amino acids.

This representation is however not much used, because there is another representation of the protein backbones which is much more efficient. A torsion angle can be computed among four consecutive atoms of the sequence of atoms N, C_α and C representing a protein backbone. Then, a corresponding sequence of $3n - 3$ torsion angles can be computed. This other sequence can be used for representing the protein backbone as well, because the torsion angles can be computed from the atomic coordinates, and vice versa. The representation which is based on the torsion angles is more efficient, because the protein backbone is represented by using less information. Indeed, a sequence of $3n$ atoms is a sequence of $9n$ coordinates, whereas a sequence of $3n - 3$ angles is just a sequence of $3n - 3$ real numbers.

In the applications, the representation based on the sequence of torsion angles is further simplified. The sequence of atoms on the backbone is a continuous repetition of the atoms N, C_α and C. Each quadruplet defining a torsion angle contains two atoms of the same kind that belong to two bonded amino acids. Then, the torsion angles can be divided in 3 groups, depending on the kind of atom that appears twice. Torsion angles of the same group are usually denoted by the same symbol: the most used symbols are Φ, Ψ and ω. Statistical analysis on the torsion angle ω proved that its value is rather constant. For this reason, often all the torsion angles ω are not considered as variables, so that only $2n - 2$ real numbers are needed for representing a protein backbone by the sequence of torsion angles Φ and Ψ. One of the most successful methods for the prediction of protein conformations, ASTROFOLD, uses this efficient representation [130, 131].

Depending on the problem that is under study, different representations of the protein backbones can be convenient. In the problem studied in [138, 139, 140, 141, 152, 153], for instance, the distances between the atoms of each quadruplet that can be defined on the protein backbone are known. This information is used for computing the cosine of the torsion angle among the atoms of each of such quadruplets. Thus, if the cosine of a torsion angle is known, the torsion angle can have only two possible values. If all these values are preliminarily computed, then the sequence of torsion angles Φ and Ψ can be substituted by a sequence of binary variables that can have two possible values only, 0 and 1. In this representation, $2n - 2$ variables are still needed for representing the protein backbone, but the variables are not real numbers anymore, but rather binary variables.

The representation of entire protein conformations is more complex. The full-atom representation consists in the spatial coordinates of all the atoms of the protein. Even though some of the atoms can be omitted because their coordinates can be computed from others, the full-atom representation still remains too much complex, especially for large proteins. Another possibility is to represent the protein backbone with the Φ and Ψ torsion angles, and to represent each side chain through suitable torsion angles χ that can be defined on each side chain. A protein molecule can contain 20 different amino acids, and therefore 20 different sets of torsion angles χ need to be defined, each of them tailored to the different shape of each side chain.

Figure 1.3 shows three possible representations of myoglobin, a very important protein. On the left, the full-atom representation of the protein is shown: atoms having a different color or gray scale refer to different kinds of atoms. In the middle,

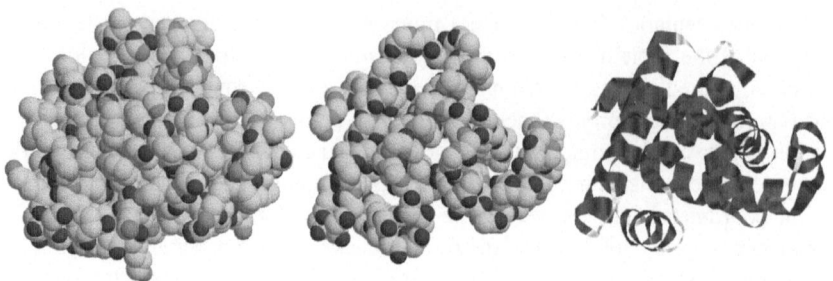

Fig. 1.3 Three representations for protein molecules. From left to right: the full-atom representation of the whole protein, the representation of the atoms of the backbone only, and the representation through the torsion angles Φ and Ψ.

the same representation is presented, where all the atoms related to the side chains are omitted. The figure gives an idea on how many atoms more are needed to be considered when the information about the side chains is also included. Finally, on the right, the path followed by the protein backbone is shown, which can be identified through the sequence of torsion angles Φ and Ψ. Note that we did not include the representation of the protein backbone as a sequence of binary variables, because it would just be a sequence of numbers 0 and 1. The conformation of the protein in Figure 1.3 has been downloaded from the Protein Data Bank (PDB) [18, 186], a public Web database of protein conformations.

Depending on the problem to be solved, a representation can be more convenient than others. For instance, in [175], the protein backbones are represented by the trace of the C_α carbon atoms, because the considered model is based on the relative distances between such C_α atoms. The model is used for simulating protein conformations. In [131], the sequence of torsion angles is instead used, because the aim is to predict the conformation of proteins starting from their chemical composition. The complexity of the problem needs a representation where the maximum amount of information is stored by using the minimum number of variables. Finally, in [139], the molecular distance geometry problem is to be solved. In this case, some of the distances between the atoms of the protein backbone are known, and the coordinates of such atoms must be computed. By using the information on the distances, the representation can be simplified to a sequence of binary variables. In this way, the complexity of the problem decreases, and it can then be solved efficiently. Protein molecules have been studied also by using data mining techniques. Recent papers on this topic are, for instance, [47, 107, 242].

1.3 General applications of data mining

In this section, some general application of data mining is presented, with the aim of showing the applicability of data mining techniques in many research fields. An overview of the applications in agriculture discussed in this book is given in Section 1.5.

1.3.1 Data mining for studying brain dynamics

Data mining techniques are successfully applied in the field of medicine. Some recent works include, for instance, the detection of cancers from proteomic profiles [149], the prediction of breast cancer survivability [56], the control of infections in hospitals [27] and the analysis of diseases such as bronchopulmonary dysplasia [199]. In this section we will focus instead on another disease, epilepsy, and on a recently proposed data mining technique for studying this disease [20, 31].

Epilepsy is a disorder of the central nervous system that affects about 1% of the population of the world. The rapid development of synchronous neuronal firing in persons affected by this disease induces seizures, which can strongly affect their quality of life. Seizure symptoms include the known uncontrollable shaking, accompanied by loss of awareness, hallucinations and other sensory disturbances. As a consequence, persons affected by epilepsy can have issues in social life and career opportunities, low self-esteem, restricted driving privileges, etc. Epilepsy is mainly treated with anti-epileptic drugs, which unfortunately do not work in about 30% of the patients diagnosed with this disease. In such cases, the seizure could be cured by surgery, but not all the patients can be cured in this way. The main problem is that the procedure cannot be performed on brain regions that are essential for the normal functioning of the patient. In order to check the eligibility for surgery, electroencephalographic analysis is performed on the patient's brain.

Since not all the patients can be treated by surgery and since surgery is a very invasive procedure, especially if we know that the procedure is performed on the brain, there have been other attempts to control epileptic seizures. These attempts have to do with the electronic stimulations of the brain. One of these is the chronic vagus nerve stimulation. A device can be inplanted subcutaneously in the left side of the chest for electric stimulations of the cervical vagus nerve. Such device is programmed to deliver electrical stimulation with a certain intensity, duration, pulse width, and frequency. This method for controlling epileptic seizures has been successfully applied, and patients had the possibility to benefit from it, after that the device has been tuned. Each patient has to be stimulated in his own way, and therefore the stimulation parameters need to be tuned in newly implanted patients. This process is very important, because the device must be personalized for the patient's needs.

Unfortunately, the only way for tuning the device is currently a trial-and-error procedure. Once the device has been implanted, it is tuned on initial parameters, and patient reports help in modifying such parameters until the ones that better fit the patient are found. The problem is that the patient, during this process, may still continue experiencing seizures because the parameter values are not good for him, or he may not tolerate some other parameter values. Then, locating the optimal parameters more rapidly would save money due to fewer doctor visits, and would help the patient at the same time. Data from electroencephalography have been collected from epileptic patients and they have been analyzed by data mining techniques, in order to predict the efficacy of the numerous combinations of stimulation parameters. In these studies, support vector machines (Chapter 6) have been used in the experiments presented in [20], whereas a biclustering approach (Chapter 7) has been used in [31]. The results of the analysis suggest that patterns can be extracted from electroencephalographic measures that can be used as markers of the optimal stimulation parameters.

1.3.2 Data mining in telecommunications

The telecommunication field has some interesting applications of data mining. In fact, as pointed out in [197], the data generated in the telecommunications field has reached unmanageable limits of information, and data mining techniques have showed their advantages in helping to manage this information and transforming it into useful knowledge. In the quoted paper, a real-time data mining method is proposed for analyzing telecommunications data.

An interesting application in this field consists of the detection of the users that potentially will perform fraudulent activities against telecommunication companies. Million of dollars are lost every year by telecommunication companies because of frauds. Therefore, the detection of users that can have a fraudulent behavior is useful for the companies in order to monitor and avoid such activities. The hope is to identify the fraudulent users as soon as possible, starting from the time they subscribe.

The studies that are the focus of this section are related to a telecommunication company and details can be found in [69]. The aim of the studies is to develop a system for identifying fraudulent users at the time of applications. In this example, a neural network approach is used (see Chapter 5). The data used for training the neural network are collected from different databases managed by the company. The data consist of information regarding each single user and the classification of the user's behavior as fraudulent or not. For each user, information such as name, address, data of birth, ID number, etc., are collected. The classification of the user's behavior is performed by an expert by checking his payment history. Once the neural network is trained, it is supposed to do this job on new users, whose payment history is not available yet.

The personal information that each user provides when he subscribes can contain clues about his future behavior. If a user has the same name and ID number of another user in the database which already had a fraudulent behavior, then there is a high probability that this behavior will be repeated again. In the specific case discussed in [69], a public database is available where insolvency situations mostly related to banks and stores are registered. Therefore, the user's behavior can be checked also in other situations beyond the ones related to the telecommunication company itself. Users having the same address can also behave in similar ways. Moreover, when the application for a new phone line is filled, the new user is asked to provide an existing phone number as reference. The new and the existing phone lines have high probabilities to be classified in the same way. By using this information, a particular kind of fraudulent behavior can be detected. Before that the telecommunication company finds out that a particular line is related to a fraud and it blocks such line, the fraudster can apply for a new phone line under another name but providing the old line during the application. This could be repeated in a sort of chain, if the line provided in the application is not verified.

The user's behaviors can be classified as fraudulent or not. This is a simplified classification in 2 classes only. In general, each subscriber can be classified in more than 2 classes when he applies for a new phone line. In the first class, the most fraudulent users can be cataloged: they do not pay bills or their debt/payment ratio is very high and they have suspicious activities related to long distance calls. The *otherwise* fraudulent users are instead those that have a sudden change in their calling behavior which generates an abnormal increase of the bill amount. Users having two or more unpaid bills and having a debt less than 10 times their monthly bill are classified as insolvent. Finally, users who paid all the bills or with one unpaid bill only can be classified as normal.

The neural network used in these studies is a multilayer perceptron in which the neurons are organized on three layers (see Section 5.1). The 22 neurons on the input layer correspond to the 22 pieces of information collected from the user during the application. The 2 neurons on the output layer allow the network to distinguish only between two classes: fraudsters and non-fraudsters. The internal layer, the hidden layer, contains 10 neurons. The data obtained from the databases of the telecommunication company and successively classified by an expert are divided in a training set, a validation set and a testing set. In this way, it is possible to control if the network is correctly learning how to classify the data during the training phase using the validation set. After this process, the network can then be tested on known data, the ones in the testing set. For more details about validation techniques, refer to Chapter 8.

1.3.3 Mining market data

Data mining applied to finance is also referred to as *financial data mining*. Some of the most recent papers on this topic are [240], in which a new adaptive neural network is proposed for studying financial problems, and [247], in which stock market tendency is studied by using a support vector machine approach. In fact, in finance, one of the most important problems is to study the behavior of the market. The large number of stock markets provides a considerable amount of data every day in the United States only. These data can be visualized and analyzed by experts. However, the quantity of data allows the visualization of small parts of all the available data per time and the expert's work can be difficult. Automated techniques for extracting useful information from these data are therefore needed. Data mining techniques can help solve the problem, as in the application presented in [25].

Recently, stock markets are represented as networks (or graphs). As discussed in Section 1.2.2, the success of a data mining method strongly depends on the data representation used. In this approach, a network connecting different nodes representing different stocks seems to be the optimal choice. The network representation of a set of data is currently widely used in finance, and also in other applied fields. In this example, each node of the network represents a stock and two nodes are linked in the network if their marketing price is similar over a certain period of time. Such network can be studied with the purpose of revealing the trends that can take place in the stock market.

Given a certain set of marketing data, a network can be associated to it. In the network, stocks having similar behaviors are connected by links. Grouping together stocks with similar market properties is useful for studying the market trends. Clustering techniques can be used for this purpose. However, in this case, the problem is different from the usual. Section 1.2.1 introduces clustering techniques as techniques for grouping data in different clusters. In this case, there is only one complex variable, the network, and its nodes have to be partitioned. Similar nodes can be grouped in the same cluster, which defines a sort of sub-network of the original one. In such sub-networks, nodes are connected to each other, because they are similar. These kinds of networks are called *cliques* in graph theory. Thus, this clustering problem can be seen as the problem of finding a clique partition of the original network. Such problem is considered challenging because the number of clusters and the similarity criterion are usually not known a priori.

Recently, in [10], the food market in the United States has been analyzed by using this approach. The food market in United States is one of the largest in the world, since it is a major exporter and significant consumer of food products. For instance, the agricultural exports in the US were about $68 billion for the year 2006. The food sector in the US includes retailers, wholesalers and all food services that link the farmers to the consumers. In general, the food market industry in the US has a significant global impact and it provides a representative sample for food economic studies.

In [10], the food market of the US has been represented by a network and its trends have been analyzed by looking for a clique partition of such network. An optimization problem has been formulated for this purpose, and it has been solved by using the software CPLEX9 [114]. The obtained cliques showed the markets with a high correlation. For instance, the clustering showed that *beverages*, *grocery stores*, and *packaged foods* markets have significantly high market capitalization. This can also help in predicting the behaviors of different stock markets. Indeed, if some market in a clique is known, then the trend of other markets in the same clique has to be similar to the known one.

1.4 Data mining and optimization

Optimization is strongly present in our everyday life. For instance, every morning we follow the shortest path which leads to our office. If we were farmers, we would want to minimize the expenses while trying to maximize the profits. We are not the only ones which try to optimize things, since there are many optimization processes in nature. Molecules, such as proteins, assume their equilibrium conformations when their energy is minimum. As we try in the morning to minimize our travel time, rays of light do the same by following the shortest paths during their travel. In all these cases, there is something, called *objective*, which has to be minimized or maximized, in other words *optimized*. Objectives can be the length of paths which lead from home to the office, the total expenses in a farm, the total profit in a farm, the energy in a molecule, the length of paths followed by a ray of light, etc. The objectives depend on certain characteristics of the system which are called *variables*. In these cases, variables can be the set of roads on which we drive, the set of things we need to buy for the farm, the set of farm products we expect to sell, the positions of the atoms in a molecule, the set of light paths. Sometimes these variables are not free to have any possible value. For instance, if there are roads closed in our home city, we need to avoid driving on these roads, even though they may decrease the travel time. Therefore, the set of roads we can drive on is restricted, in other words the variables are *constrained*. The process of identifying objective, variables, and constraints for a given problem is known as *modeling* of the *optimization problem*.

Data mining techniques seek the best classification or clustering partition of a set of data. Among all the possible classifications or partitions, the best one, the optimum one, is searched. Indeed, many of the data mining techniques we will discuss in this book lead to the formulation of an optimization problem. For instance, k-means algorithms (see Chapter 3) try to minimize an error function which depends on the possible partitions of the data in clusters. The error function is the objective in this case, and the partitions represent its independent variables, which are not constrained. A neural network (see Chapter 5) and a support vector machine (see Chapter 6) lead also to an optimization problem. In these two cases, the optimization problem has to be solved in order to teach the neural network or the support vector machine how to classify sets of data, by defining certain parameters. The objective is the error which occurs by classifying data with a given set of parameters, corresponding to the variables of the objective. Such variables are constrained in the support vector machine approach.

From a mathematical point of view, optimization is the minimization or maximization of a function (the objective) subject to constraints on its variables. x is usually used for indicating the vector of independent variables, $f(x)$ is the objective function, and functions c_k represent the constraints. Since minimizing $f(x)$ is equivalent to maximizing $-f(x)$, the general optimization problem may be formulated as follows:

$$\min_x f(x)$$

subject to

$$c_i(x) = 0 \quad \forall i$$
$$c_j(x) \leq 0 \, \forall j.$$

Functions c_i and c_j represent the equality and inequality constraints, respectively. They may not be present in some formulations, and in that case the optimization problem is unconstrained. There is not only one way for solving these problems, but rather a collection of algorithms, which can be chosen on the basis of the particular needs. Properties of the objective function, or of the constraints, can determine the choice of one algorithm or another. A large variety of optimization methods and algorithms for optimization can be found in [76, 184].

Methods for optimization are mainly divided into *deterministic* or *exact* methods and *meta-heuristic* methods. Deterministic methods are based on mathematical theories. If some hypotheses are met, they guarantee that the solution can be found. Meta-heuristics instead are based on probabilistic mechanisms and there are only probabilities that the solutions can be found. Deterministic methods can usually be applied to a certain subset of optimization problems only, whereas meta-heuristics are more flexible. The implementation of meta-heuristic methods is also easier in general, and the basic ideas behind these methods are usually simple. For this reason, meta-heuristic methods are widely applied in many research fields. Due to their simplicity and flexibility, meta-heuristic methods are the choice of many researchers who are not experts in computer science and numerical analysis. Even though one cannot be sure if the solution found by applying a meta-heuristic method is correct or not, often such solutions are good approximations of the real one. In general, easier methods might provide a solution with a lower accuracy. However, researchers commonly use such methods. They first seek to find out the method which is the best fit for their problem. This decision may result in trading off the quality of the solution with speed or ease of implementation. For high-quality solutions, modeling issues may usually become more complex, requiring additional programming skills and powerful computational environments [174].

Once a global optimization problem has been formulated, the usual approach is to attempt to solve it by using one of the many methods for optimization. The choice of the method that fits the structure of the problem is very important. An analysis of the complexity of the model is required and the expected quality of the solution needs to be determined. The complexity of the problem can be derived from the data structures used, and from the mathematical expression of the objective function and the constraints. If the objective function is linear, or convex quadratic, and the problem has box, linear or convex quadratic constraints, then the optimization problem can be solved efficiently by particular methods, which are tailored to the objective function and constraints [33, 76, 100]. For instance, the optimization problem arising when training support vector machines has a convex quadratic function and linear constraints (see Chapter 6 for details). Methods for solving these particular kinds of problems include the active set methods and the interior point methods [33, 100]. However, there are methods tailored to the support vector machines for solving such quadratic optimization problems, and hence the general methods are often not used. If the objective function and the constraints are instead nonlinear without any restriction, then more general approaches must be used. For differentiable functions, whose gradient vector can be computed, deterministic methods can be used. As already pointed out, these methods are able to guarantee that the solution can be found if certain hypotheses are met. Functions that are twice differentiable with a computable Hessian matrix can be locally approximated by a quadratic function. Typical examples of methods which exploit the quadratic approximation of a differential function are the trust region algorithms [40]. Other deterministic approaches include for instance the branch and bound methods [1, 2, 5].

Meta-heuristic methods are often used in applied fields such as agriculture because they are, in general, easier to implement and more flexible. The ideas behind the most used meta-heuristics for global optimization follow. Most of them took inspiration from animal behavior or natural phenomena and try to reproduce such processes on computers. In the simulated annealing algorithm, for instance, the temperature of a given system is slowly decreased in order to obtain a crystalline structure, which corresponds to the optimal solution of an optimization problem [128]. More details about this optimization technique are given is Section 1.4.1. Genetic algorithms [88] mimic the evolution of a population of chromosomes that can procreate child chromosomes, which can undergo genetic mutations. Harmony search [82] is inspired by jazz music improvisation, and it seeks the optimal value of an optimization problem the same way musicians look for perfect harmonies. Many meta-heuristic methods took inspiration from animal behavior. Swarm intelligence can be defined as the collective intelligence that emerges from a group of simple entities, such as ant colonies, flocks of birds, termites, swarm of bees, and schools of fish [148]. Ant colony optimization [64] algorithms simulate the behavior of a colony of ants finding and conserving food supplies, whereas particle swarm optimization [126] simulates the motion of a large number of insects or other organisms. Finally, the recently proposed monkey search [173] is inspired by the behavior of a monkey climbing trees in its search for food supplies.

It is worth noting that hybrid methods which are in part deterministic and in part meta-heuristic have been developed with the aim of combining their qualities [190]. Moreover, optimization problems that would require the use of complex methods are sometimes reformulated, so that an easier and more effective method for optimization can be used. To reformulate an optimization problem means to transform the original problem into another problem that is equivalent or similar to the original one, and that is easier to manage. A lot of research is devoted to suitable reformulations of difficult global optimization problems [151, 213].

In this section, we referred only to optimization problems with a single objective function. However, there are several application in which there is not only one function to be optimized, but rather a small set of functions. These problems are referred to as *multi-objective optimization problems*. Let us consider again the problem of a farmer who tries to maximize his profits while the expenses must be as small as possible. In this example, there are in fact two objectives: the profits (to be maximized) and the expenses (to be minimized). In these situations, the easiest strategy is to combine the two objectives in order to obtain a unique objective function, so that the multi-objective optimization problem is reformulated as an optimization problem having only one objective function. As for example, if $f(x)$ represents the profit, and $g(x)$ are the expenses, then a maximization problem with objective function $\alpha_1 f(x) - \alpha_2 g(x)$ would be a possible reformulation of the original problem, where α_1 and α_2 are two real and positive constants. The reader is referred to [162, 178, 194] for recent surveys on methods for solving multi-objective optimization problems.

1.4.1 The simulated annealing algorithm

In this section, we give some more details about one of the easiest methods for optimization, the simulated annealing (SA) [128]. It is a meta-heuristic method, which is inspired by a physical process. Since it is very easy to implement, it can be used to perform the first experiments on a given optimization problem. Because of its simplicity, the solutions provided by SA might lack a high accuracy, especially on more complex problems. Depending on the problem at hand, the solutions found by SA can be either considered as accurate enough, or just an initial approximation of the solutions that can be found later by more complex and more accurate methods.

SA is a meta-heuristic method for optimization, and therefore it is based on a probabilistic mechanism. It is based on an analogy with the annealing physical process, in which the temperature of a given system is decreased slowly, in order to obtain a crystalline structure. As an example, let us consider a simple glass of water. If the system "glass of water" is kept to the normal temperature of 20°C, then the molecules of water in the glass are free to move. That is why the water is a liquid at this temperature. However, if we put the glass of water in the cooler, then the temperature of the glass of water decreases slowly to 0°C. The more the temperature is lowered, the less are the molecules free to move. When the temperature reaches and passes 0°C, the glass contains an ice piece having the same shape of the glass. The molecules of water in the glass cannot move so freely anymore, because they are now organized in a crystalline structure.

This physical process is simulated for solving a given optimization problem. The variables of the objective function play the role of the molecules of water. They are free to move when the temperature is high. Their mobility is simulated by applying suitable perturbations to the variables. When the temperature decreases, the variables are less free to change their values. This is monitored through the corresponding objective function value: the lower is the temperature, the less variability is allowed on the objective function values. The hope is that, when the temperature approaches to zero, the variables of the problem contain values which represent a good approximation of the solution.

The basic SA algorithm can be described by two nested loops. At the start, random and feasible values are assigned to the variables, defining the initial approximation to the solution $X^{(0)}$. The inner loop generates at each iteration a new candidate approximation to the solution, by applying random perturbations to the previous one. The new approximation is accepted or rejected, by using a random mechanism based on an acceptance function, whose value depends on the temperature parameter. The lower is the temperature, the smaller is the number of accepted approximations. The outer loop controls the decrease of the temperature parameter, i.e., defines the so-called cooling schedule.

It follows that SA is built up from three basic components: next candidate generation, acceptance strategy and cooling schedule. To generate the next candidate approximation to the solution, totally random or customized perturbations can be applied. The acceptance strategy usually used is based on the Metropolis acceptance function [164]. If $X^{(k)}$ is the approximation of the solution at a step k of the SA and $\hat{X}$ is a new candidate approximation, then $\hat{X}$ is accepted if

$$A(X^{(k)}, \hat{X}, t^{(k)}) = \min\left\{1, e^{-\frac{f(\hat{X}) - f(X^{(k)})}{t^{(k)}}}\right\} > p,$$

```
t = t0
maxout = maximum allowed number of outer iterations
nsteps = number of steps at constant temperature
X = random starting solution
nout = 0
while (f(X) not stable and nout ≤ maxout)
   nout = nout + 1
   for k = 1, nsteps
      X(k) = random perturbation on X
      p = uniform random number in (0,1)
      if (A(X,X(k),t)) > p) then
         X = X(k)
      end if
   end for
   t = γ t, γ < 1
end while
```

Fig. 1.4 The simulated annealing algorithm.

where f is the objective function to be minimized, $t^{(k)}$ is the temperature value at step k and p is a random number from the uniform distribution in $(0, 1)$. The candidate approximation can be accepted even if it does not increase the value of f, depending on $t^{(k)}$ and p. At high temperatures, many candidate approximations can be accepted, but, as the temperature decreases, the number of candidate approximations decreases, in analogy with the physical process of annealing. The cooling strategy has an important role in SA. The temperature must be decreased very slowly to avoid trapping into local optima that are far from the global one. This reflects the behavior of the physical annealing, in which a fast temperature decrease leads to a polycrystalline or amorphous state. Figure 1.4 gives a sketch of the SA algorithm.

1.5 Data mining and agriculture

Data mining is widely applied to agricultural problems. For instance, the prediction of wine fermentation problems can be performed by using a k-means approach (Section 3.5.1). Knowing in advance that the wine fermentation process could get stuck or be slow can help the enologist to correct it and ensure a good fermentation process. Weather forecasts can be improved using a k-nearest neighbor approach (Section 4.4.1), where it is assumed that the climate during a certain year is similar to the one recorded in the past. The same data mining technique can also be used for estimating soil water parameters (Section 4.4.2).

Apples and other fruits are widely analyzed in agriculture before marketing. Apples running on conveyors can be checked by humans and the bad apples (the ones presenting defects) can be removed. The same task can be efficiently performed by a recognition system based on the k-means method (Section 3.5.2). In this approach, digital pictures of the fruit are taken. However, some defect can be internal and not visible at the exterior. The approach discussed in Section 5.4.2 uses X-ray images for checking the apple watercore. It is based on an artificial neural network which learns from a training set how to classify the X-ray images. Neural networks are also used for classifying sounds from animals such as pigs for checking the presence of diseases (Section 5.4.1). Support vector machines can be used for recognizing animal sounds as well, such as sounds from birds (Section 6.5.1). Besides the scientific interest in the classification of such sounds, there are practical applications related to these kinds of studies. For instance, collisions between aircraft and birds can cause damage to the vehicle and the bird's death. Then, the recognition of a bird by its sounds is helpful.

Other applications of data mining techniques include the detection of meat and bone meal in feedstuffs destined to farm animals (Section 6.5.2), the control of chicken breast quality (Section 2.3.1), and the analysis of the effects of energy use in agriculture (Section 2.3.2). An interesting recent review of data mining techniques and applications to agriculture can be found in [48].

1.6 General structure of the book

In this book, we will discuss several data mining techniques and we will provide many applications in the agricultural field. Chapter 2 presents simple and common statistical methods which can be used as a data mining technique itself or combined with more complex techniques. The statistical based methods presented are principal component analysis, interpolation and regression. Chapters 3 to 7 present widely used data mining techniques. Chapter 3 is devoted to the k-means methods and to many of its variants. Chapter 4 focuses on the k-nearest neighbor approach. In this chapter, many strategies for reducing the training sets used in the k-nearest neighbor approach are presented. Chapter 5 is dedicated to artificial neural networks, and hence to the training, pruning and testing process of a neural network. Chapter 6 is on support vector machines. This technique is introduced as a simple linear classifier able to discriminate between two classes only. Then it is extended to the general case when the classes are more than two and they are not linearly separable. Finally, Chapter 7 is focused on biclustering techniques. Biclustering has been recently proposed and it is very efficient in some kind of applications. There are no applications in agriculture yet which use this method. However, a chapter in this book is devoted to it for completeness, and an application in the field of biology is presented.

Chapters have a common structure. The first sections are dedicated to the data mining techniques. Basic ideas are given, as well as variants and improvements of the technique proposed over time. Several applications in agriculture of the data mining technique are then provided, and a couple of applications per chapter are presented in detail. Our aim is to give the reader the instruments for applying the data mining techniques for his purposes. For this reason, experiments in MATLAB® and/or applications of freeware software for data mining are discussed in each chapter. The simplicity behind the k-means and the k-nearest neighbor allows one to implement them by using little code. Codes in MATLAB are provided for both techniques. They are very simple and may not work in some kinds of situations. Our aim is to keep the simplicity, however the reader could even modify such codes for solving particular problems. Artificial neural network and support vector machines are much more complex. Therefore, various software implementing such techniques are presented and examples on how to use them are discussed. At the end of each chapter, a section devoted to exercises is given. The solutions of such exercises can be found in Chapter 10.

All the data mining techniques can be validated by using validation techniques. A review of the most common validation techniques is provided in Chapter 8. Then, for some of the data mining techniques discussed in the previous chapters, examples of applications of the validation techniques are provided. The last chapter of the book, Chapter 9, focuses on the implementation of data mining techniques in a parallel environment. The parallel version of some of the data mining techniques discussed in the book are given.

This book provides two appendices. Appendix A gives some details about the MATLAB environment. The reader who is interested in MATLAB can also find a lot of textbooks in literature. Therefore, only the basic concepts needed for understanding the several examples in MATLAB given in this book are discussed. Appendix B presents an entire application in C programming language. The implemented algorithm is the k-means algorithm. The aim of this appendix is to provide to the reader the instruments for programming personal applications when software performing the desired tasks does not exist or is not available. The k-means algorithm has been chosen because it is one of the simplest algorithms in data mining.

Chapter 2
Statistical Based Approaches

2.1 Principal component analysis

Principal component analysis (PCA) is a method used to reduce the dimension of a given set of data while retaining the variability present in the set. Each set of data contains information represented through vectors of single variables (that usually have real, integer or binary values). For instance, a geometric point in the three-dimensional space can be represented through a vector having three variables, each one associated to one of the three coordinate axes x, y and z. In general, a sample can be represented by a vector formed by a certain number of variables. Such number of variables defines the length of the vectors contained in the set, and hence the dimension of the set. Moreover, for each variable, a certain range of variability can be defined, which determines the interval of values that the single variable can take. For instance, if the set of data contains three-dimensional points delimited into a cube having side 1 and centered in $(0, 0, 0)$, then the three variables representing the Cartesian coordinates are bounded to have values in $\left[-\frac{1}{2}, \frac{1}{2}\right]$. This interval defines the range of variability of the three variables. The aim of PCA is to find hidden patterns amongst the data and transform the original data in such a way that emphasizes their similarities and differences. Once the patterns are found, the data can be represented as components ordered by their relevance and it is possible then to discard components of low level of relevance without loss of important information.

PCA is able to reduce the dimension of a set of data if the original variables used for representing the data are correlated. In order to clarify this concept, let us consider again the example of the three variables representing the three coordinates of points in a cube. If, for instance, all the points in the set lie on a suitable plane, then the three variables are correlated. PCA applied to this particular problem transforms the three variables in a way that one of them has a null variability. In the new transformed space, the points can therefore be represented by two variables only, and hence in a space having a dimension less with respect to the original one. The information regarding the third dimension (the discarded dimension) is irrelevant, because the

A. Mucherino et al., *Data Mining in Agriculture*, Springer Optimization and Its Applications 34,
DOI: 10.1007/978-0-387-88615-2_2,
© Springer Science + Business Media, LLC 2009

points actually lie on a two-dimensional space. This is a very simplified situation. The following examples introduce the PCA method in more detail.

Let us suppose that the considered set of data contains the points having as coordinates

$$(-2, -1), \quad (-1, 0), \quad (0, 1), \quad (1, 2), \quad (2, 3)$$

in a two-dimensional space. The values of x vary in the interval $[-2, 2]$, and the values of y vary in the interval $[-1, 3]$. These two intervals show the variability of the variables x and y. As it is easy to note, these two variables are correlated. Indeed, the x coordinates increase in value when the y coordinates increase, and vice versa: a straight line passes through them. Hence, one of the two coordinates can be obtained if the other one is known. The idea behind PCA is to transform these variables in a way that they become uncorrelated. Doing so, the dimension of the set of data can be reduced if only the variables having the larger variability are considered and all the others are discarded. The variables with larger variability are here called *principal components*. They are usually sorted by their variability, so that only the *first* principal components can be used for representing the data. Note that there are cases where a low order principal component exhibiting low variance within the ensemble does not necessarily imply that it is unimportant in regression models [13].

In this example, the following transformation can be applied (see Figure 2.1). The straight line passing through all the points of the set and the x axis of the Cartesian system form a certain angle. All the points can be rotated so that they change their configuration from the one in Figure 2.1(a) to the one in Figure 2.1(b). As the figure shows, the transformation brings all the points on the x axis. Therefore, they all have zero as y coordinate. After the transformation, the points are represented by two new variables $\hat{x}$ and $\hat{y}$, where $\hat{x}$ has a variability similar to the one x has, and $\hat{y}$ has a null variability. The variable $\hat{y}$ can then be discarded, so that the dimension of the set of points decreases to 1. The original points in the two-dimensional space can be actually represented in a one-dimensional space without losing any information. After the transformation is applied to the original set of points, the points are represented with vectors having a shorter length. In this example, they were represented in a two-dimensional space before, and they are represented in a one-dimensional space now. The values of the variables used for the representation are completely different. However, the distances between these points is preserved. This is very important. Indeed, distances are usually used for evaluating the similarities and the differences among the data. Note that Figure 2.1(a) and 2.1(b) have two different scales, and therefore the distances between the points in Figure 2.1(b) look shorter but they are actually the same.

Let us suppose now that the considered set of data contains points that are not perfectly aligned. Let us suppose that the coordinates of the points are

$$(-2.1, -1), \quad (-1, 0), \quad (0, 1), \quad (1, 2), \quad (2, 3.2).$$

Figure 2.2(a) shows that there is no straight line passing through these points as in the previous example. However, a similar kind of transformation of the data can be

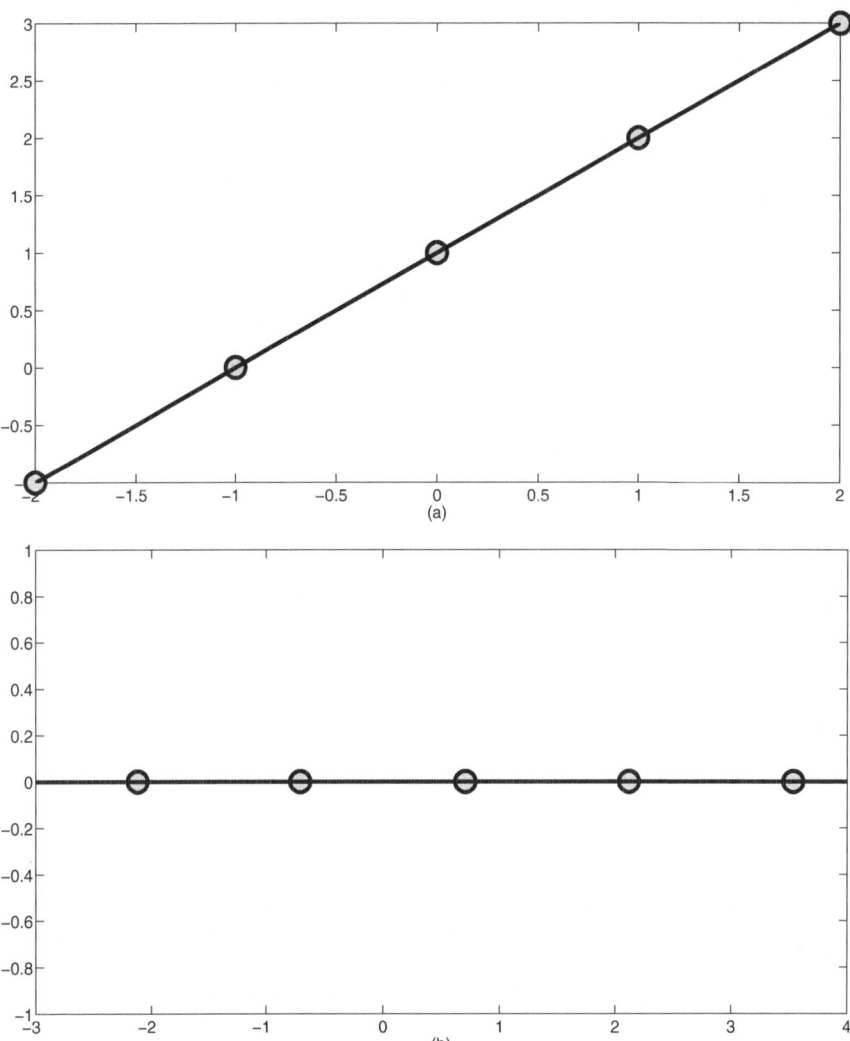

Fig. 2.1 A possible transformation on aligned points: (a) the points are in their original locations; (b) the points are rotated so that the variability of their y component is zero.

performed. The linear regression function defined by these points and the x axis form a certain angle, and all the points can be rotated by this angle (see Section 2.2 for details on regression functions). The result is the set of points shown in Figure 2.2(b). As the points are not aligned, not all of them lie on the x axis as in the previous case. However, the new variables $\hat{x}$ and $\hat{y}$ obtained after the transformation have interesting properties with respect to the original ones x and y. The variable x has values ranging in the interval $[-2.1, 2]$, and the new variable $\hat{x}$ has a similar variability, as Figure 2.2 shows. The variable y can have instead values in the interval $[-1, 3.2]$, whereas

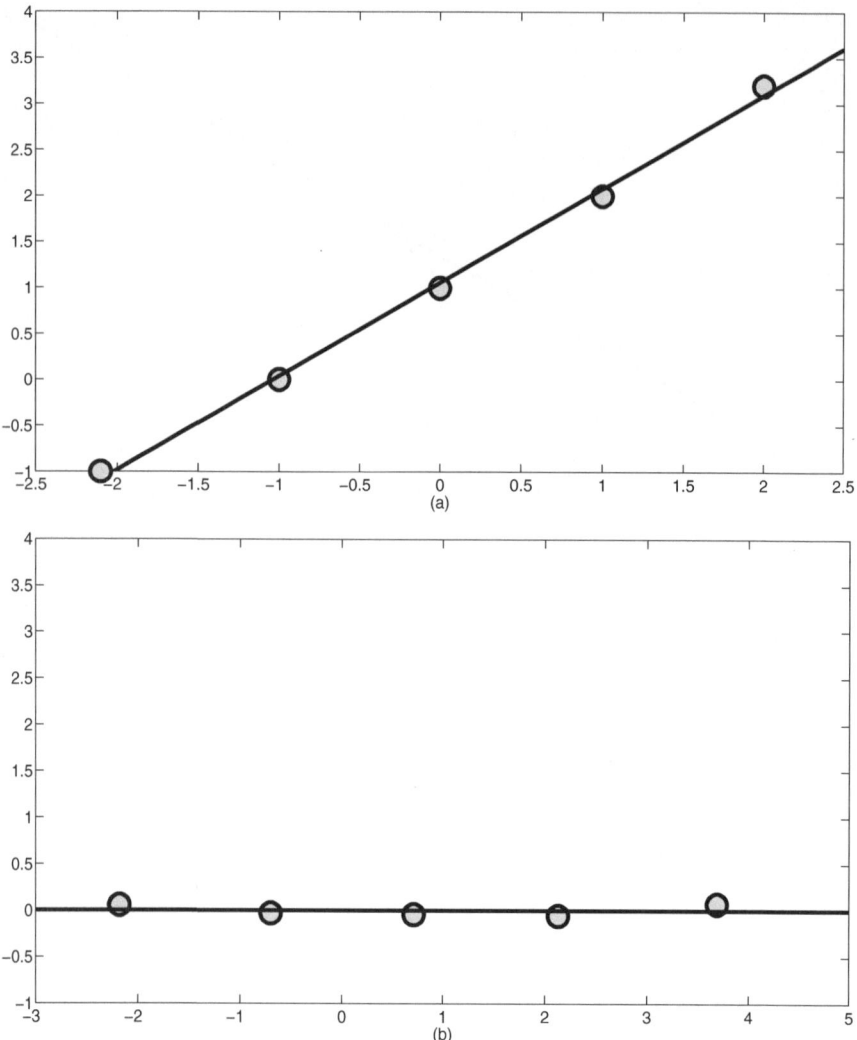

Fig. 2.2 A possible transformation on quasi-aligned points: (a) the points are in their original locations; (b) the points after the transformation.

the corresponding variable $\hat{y}$ has almost a null variability. In this second example, $\hat{y}$ has a certain variability, but it is very small. It can then be discarded in order to decrease the dimension of the set of data. Since its variability is small, the loss of information is small as well. For instance, the distances of the points in the new space are different, but the introduced error is small.

In general, PCA can be applied for reducing the dimension of a set of data, where samples are represented by using m-dimensional vectors. Reducing the dimension of the set means to find a representation of the same samples in a lower-dimension space, where vectors have a number of components smaller than m. In other words, the PCA method applied to these components finds a set of principal components that are able to represent the same sample by using shorter vectors.

The following introduces and motivates the PCA method [120]. Its basic idea is quite simple, however it requires a little knowledge on eigenvalues and eigenvectors of a matrix (see Glossary for the definitions) for understanding it. This topic can be difficult for the readers who do not have a mathematical background. The reader can therefore continue reading at the end of this section, where a practical example is provided. What is needed to know is that PCA computes the kth principal component as a linear combination of the original variables, where the coefficients used in the linear combination come from the elements of the kth eigenvector of a covariance matrix. The eigenvectors of the covariance matrix are sorted in ascending order by the value of the corresponding eigenvalues. Even though the names "covariance matrix," "eigenvalue" and "eigenvector" may seem related to very difficult mathematical concepts, they can be easily computed with software for mathematical computations, such as MATLAB®. The reader can refer to the example at the end of this section and to the exercises at the end of this chapter for learning how to apply PCA to simple examples by using MATLAB.

In order to find the first principal components, the variables x_i in the generic sample $x = \{x_1, x_2, \ldots, x_m\}$ of the set of data need to be transformed so that they become uncorrelated. Let us consider a linear combination of all the variables:

$$\alpha_1^T x = \alpha_{11} x_1 + \alpha_{12} x_2 + \cdots + \alpha_{1m} x_m = \sum_{i=1}^{m} \alpha_{1i} x_i, \qquad (2.1)$$

where α_1 is the vector containing all the linear coefficients and α_1^T is its transposed vector. The variability of a variable can be monitored using the so-called *covariance matrix* Σ, whose element (i, j) represents the covariance between the ith and j^{th} elements of x when $i \neq j$, and the variance of the ith element when $i = j$. The real covariance matrix is not known in applications, and an approximation of this matrix can be computed using the samples x of the set of data. It can be proved that the variability (or *variance*) of $\alpha_1^T x$ can be expressed as

$$\alpha_1^T \Sigma \alpha_1. \qquad (2.2)$$

In order to find the linear transformation of the variables x_i maximizing its variance or variability, the quantity (2.2) needs to be maximized. Since there are infinite coefficient vectors α_1 that are solutions to this problem and one unique solution is searched, the vector α_1 is normalized. The quantity (2.2) can be therefore maximized subject to the constraint $\alpha_1^T \alpha = 1$. This is a simple optimization problem. Indeed, it does not require a computational method (see Section 1.4) to be solved, but it can be solved analytically. The constraint on the coefficient vector α_1 can be considered as a penalty term in the objective function:

$$\alpha_1^T \Sigma \alpha_1 + \lambda(\alpha_1^T \alpha_1),$$

where λ determines the trade-off between constraint satisfaction and maximization of the variance. The derivative with respect to α_1 of this function helps locating the function stationary points. The stationary points of the function include their minimum and maximum points. Such stationary points are the ones satisfying the equation:

$$\Sigma \alpha_1 - \lambda \alpha_1 = 0.$$

This equation can be equivalently written as

$$(\Sigma - \lambda I_m) \alpha_1 = 0, \tag{2.3}$$

where I_m is the square identity matrix of dimension m. The equation (2.3) corresponds to the definition of eigenvalue and eigenvector of a matrix. In this case, the matrix is represented by Σ, the eigenvalue is represented by λ and the eigenvector by α_1. For this reason, the problem of finding the first principal component becomes the problem of finding the eigenvalues and eigenvectors of the matrix Σ. All the eigenvectors related to Σ are stationary points of the considered objective function. However, only the vector α_1 maximizing the $\alpha_1^T x$ variance is searched. The matrix Σ has in fact m eigenvectors and m eigenvalues, and each corresponding couple (λ, α_1) satisfies equation (2.3). Then, the variance

$$\alpha_1^T (\Sigma \alpha_1) = \alpha_1^T (\lambda \alpha_1) = \lambda (\alpha_1^T \alpha_1) = \lambda$$

equals the eigenvalue related to α_1. The first principal component is therefore defined as the variable $\alpha_1^T x$, where α_1 is the eigenvector related to the larger eigenvalue λ of Σ and x is the vector of the original variables. In general, the kth principal component of x is $\alpha_k^T x$, where λ_k is the kth largest eigenvalue of Σ and α_k is the corresponding eigenvector. The demonstration for $k > 1$ is provided in [120].

Let us consider the set of points

$$(-2.1, -1), \quad (-1, 0), \quad (0, 1), \quad (1, 2), \quad (2, 3.2)$$

again and let us apply the PCA method. According to the definition, the covariance matrix related to these points is

$$\Sigma = \begin{pmatrix} 2.6020 & 2.6510 \\ 2.6510 & 2.7080 \end{pmatrix}.$$

The eigenvectors related to Σ are

$$\alpha_1 = (-0.7141, 0.7000), \quad \alpha_2 = (0.7000, 0.7141)$$

and the corresponding eigenvalues are

$$\lambda_1 = 0.0035, \quad \lambda_2 = 5.3065.$$

One of the eigenvalues is very small, and this means that the corresponding transformed variable has a small variability. Indeed, if the transformed variables $\alpha_1^T x$ and

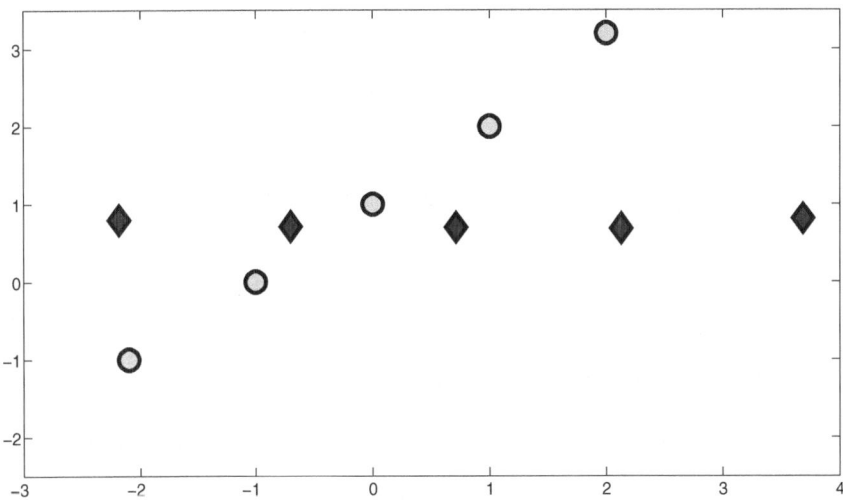

Fig. 2.3 A transformation on a set of points obtained by applying PCA. The circles indicate the original set of points.

$\alpha_2^T x$ are computed, then it is possible to see that the first one (corresponding to a small eigenvalue) has variability equal to $\Delta x = 0.01$, whereas the other one has variability $\Delta y = 5.87$. Figure 2.3 shows the set of points before and after the transformation. The computation of the covariance matrix of a given set of data and the computation of the eigenvalues and eigenvectors of a given matrix can be performed by using software such as MATLAB (see Section 2.4).

As a conclusion, we can say that initially we had points on a two-dimensional Cartesian system. Each point had two coordinates x and y in the system, but no information was provided regarding how each point was related to each other. After the transformation, we have points expressed in terms of eigenvectors. As eigenvectors are orthogonal, they define the Cartesian system in which they are now expressed. Therefore, we are still considering the same exact points, even though they are represented in a different system. This new system helps in finding out how the points are related.

Finally, it is worth noting that PCA has an interesting property that allows one to have an estimation of the information loss when discarding eigenvectors of low value. To better explain this property, let us consider generic data points $X = \{x_1, x_2, ...x_N\}$ where each $x_i \in \Re^n$. After applying PCA, a subspace $Y = span\{u_1, u_2, ..., u_m\}$ is obtained, where each $u_i \in \Re^m$ with $m \leq n$. According to [211]:

$$\sum_{j=1}^{N} d^2(x_j, Y) = \sum_{i=m+1}^{n} \lambda_i$$

where d represents the distance between the generic point x_j and Y. It follows from the formula that the sum of squared distances of the points x_j to the subspace Y is equal to the sum of discarded eigenvectors. λ_i represents the error of approximation by the subspace Y. This error is small if the sum of discarded eigenvalues is small and, therefore, the impact of discarded eigenvectors is also small.

2.2 Interpolation and regression

In this section, interpolation and regression techniques for data mining are introduced step by step through several examples. The aim is to model a given set of data with a suitable mathematical function. The sets of data obtained in real applications usually contain a discrete number of samples which describe a certain process or phenomenon. By applying interpolation or regression techniques, the hope is to find a function that is able to describe this phenomenon or process in general.

Let us suppose that the quantity of water y in a certain soil is monitored during time x. Experimental analysis can be used for obtaining y at different times x, so that a set of points (x', y') can be defined. As always in real life applications, the number of experiments is discrete and limited, whereas a general function able to relate each time x to a water level y is searched. Finding this function by using the data available (the x' and y' pairs in this case) means to find a model which is able to provide the correct water level y for any time x. In this simple example, the points (x', y') belong to a two-dimensional space, and hence all the functions defined in $\Re$ and having values in $\Re$ can be a good model for the process under study.

In general in mathematics, given an independent variable x, a function f provides a value for the corresponding dependent variable $y = f(x)$. The functions that are the focus of this section must obey the following properties. Given a known x', they must be able to provide the corresponding y' or a good approximation of y'. Moreover, they must be able to generalize: given an x which is unknown (no pairs (x, y) are contained in the set of data), the value of y provided by the function must be an estimation of the behavior of the modeled process. The final goal is to find the general rule that relates x and y. For example, let us suppose that water levels y in a given soil are measured every hour x for 10 successive hours. The 10 pairs (x', y') containing the details of these measurements represent the available set of data. An interpolation or regression function modeling this set of data is required to provide a good estimation of the water levels y even for times x that are not included in the data. If this aim is reached, no more measurements are needed, but the process can be monitored using the obtained model.

Let $\{(x_1, y_1), (x_2, y_2), \ldots, (x_{n+1}, y_{n+1})\}$ be a set of points representing a given process to be modeled. This set can be called *training set*, because it can be used for learning how to model the process. Most of the following discussion is limited to functions defined in $\Re$ and having values in $\Re$.

The easiest way for modeling a set of data by a function $f : \Re \longrightarrow \Re$ is the following one. All the points can be simply linked by linear segments. The *join-the-dots* functions are able to model the data with no errors on the known pairs (x_i, y_i). In other words, they are able to provide the exact y_i when they have as input the

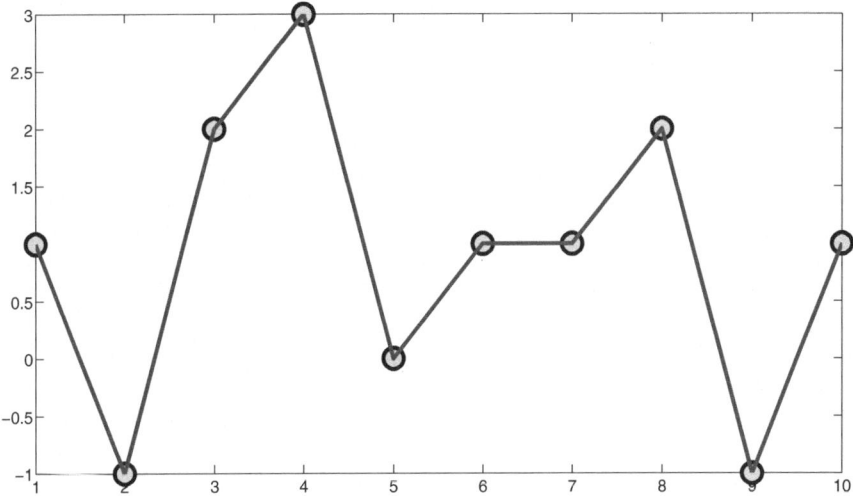

Fig. 2.4 Interpolation of 10 points by a join-the-dots function.

corresponding x_i. The value of the function in points $x \in (x_i, x_{i+1})$ is instead a sort of *mean* between the two known values y_i and y_{i+1} (see Figure 2.4). The join-the-dots functions are very easy to define, but they usually do not provide an accurate model. Moreover, the join-the-dots functions are not smooth, and they are not differentiable in all the points of the training set. These are properties that might be useful when the model is successively used.

Smoother functions that can be used as models are the polynomials. Polynomials having degree 1 are straight lines in the two-dimensional space, and polynomials having degree 2 are parabolas in the two-dimensional space. If it is required that each y_i must correspond to $p(x_i)$, i.e., the graphic of the polynomial p must *pass* through the known points, then the degree of the polynomial plays a crucial role. In fact, two points suffice for defining a straight line. If there is a third point in the training set which is not aligned with the first two, then the straight line is not sufficient, and a polynomial having degree 2 is needed. In general, a degree equal to n is needed for defining a polynomial p such that $p(x_i) = y_i$ for each $i = 1, 2, \ldots, n+1$. If the set of points satisfies particular properties, then a smaller degree could suffice. For instance, if n points are perfectly aligned, a polynomial having degree 1 is sufficient. These polynomials are called *interpolating* polynomials.

Let us introduce a simple rule for building interpolating polynomials. The general formula of a straight line in a two-dimensional space is

$$y = ax + b,$$

where $x \in \Re$ is the independent variable, where $y \in \Re$ is the dependent variable and where a and b are two real constants, the coefficients. A line on a plane can be unequivocally identified by the values given to the two coefficients a and b. A generic straight line can also be expressed as

$$y = a_0 + a_1(x - x_1),\tag{2.4}$$

where a_0, a_1 and x_1 are real constants. It is very easy to show that these two equations are equivalent if $a = a_1$ and $b = a_0 - a_1 x_1$. Note that x_1 is associated to $y = a_0$ by the equation. Therefore, if a line passing through a point (x_1, y_1) is searched, then one of its equations is (2.4) where $a_0 = y_1$. a_1 can have any value, and each of them defines one of the infinite lines passing through (x_1, y_1). The passage through (x_1, y_1) is guaranteed because $a_1(x - x_1)$ is zero when $x = x_1$ and then $y = a_0$.

Defining a model by using a set with only one point does not have any practical meaning. Let us suppose then that there is another point (x_2, y_2) in the training set. There are infinite straight lines passing through (x_1, y_1), and if the passage through (x_2, y_2) is also required, one of these infinite lines has to be chosen. As previously noticed, a_1 can be any real number in (2.4), for guaranteeing the passage from the point (x_1, y_1). Let us now define a_1 as follows:

$$a_1 = \frac{y_2 - y_1}{x_2 - x_1}.\tag{2.5}$$

The line (2.4) having as coefficients $a_0 = y_1$ and a_1 as defined in (2.5) passes through both (x_1, y_1) and (x_2, y_2). Indeed, if $x = x_2$, it follows that

$$y = a_0 + a_1(x_2 - x_1) = y_1 + \frac{y_2 - y_1}{x_2 - x_1}(x_2 - x_1) = y_2,$$

and then the line passes through (x_2, y_2) as well.

Supposing that there is a third point (x_3, y_3) in the training set, then a straight line is not sufficient anymore, unless the three points are aligned. The following polynomial having degree 2 can be used

$$y = a_0 + a_1(x - x_1) + a_2(x - x_1)(x - x_2)$$

for modeling the set of data. In general, the Newton polynomial of degree n

$$y = a_0 + \sum_{i=1}^{n} a_i \prod_{j=1}^{i}(x - x_j)$$

can be used for modeling sets of data represented through $n + 1$ points in a two-dimensional space. The coefficients a_i can be substituted with the so-called Newton's *divided differences*:

$$y = f(x_1) + \sum_{i=2}^{n+1} f[x_1, \ldots, x_i] \prod_{j=1}^{i-1}(x - x_j).$$

The divided differences can be defined iteratively by the following formula:

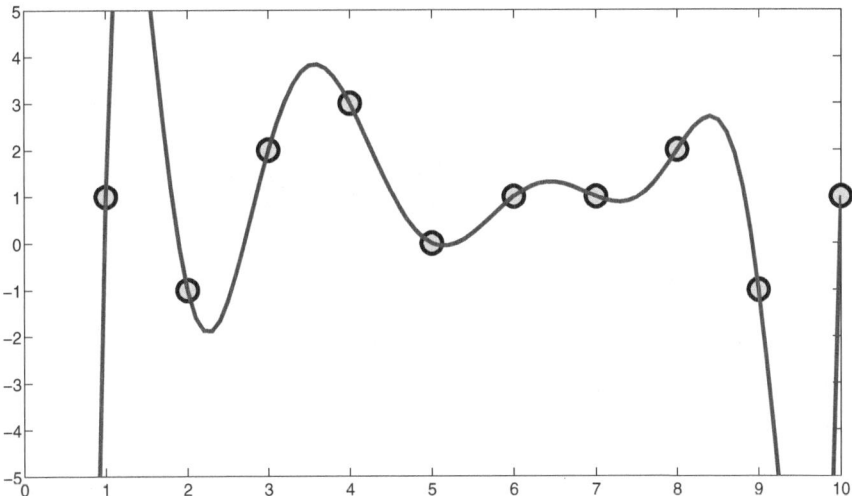

Fig. 2.5 Interpolation of 10 points by the Newton polynomial.

$$f[x_1, x_2, \ldots, x_{n+1}] = \frac{f[x_2, x_3, \ldots, x_{n+1}] - f[x_1, x_2, \ldots, x_n]}{x_{n+1} - x_1}$$

where

$$f[x_1, x_2] = \frac{y_2 - y_1}{x_2 - x_1},$$

which corresponds to the a_1 coefficient used before.

Figure 2.5 shows a polynomial having degree 9 interpolating 10 points in the two-dimensional space. The figure shows that the polynomial has high oscillations, especially in the interval $[1, 2]$ of the x axis. In fact, the greater is the polynomial degree, the more are the polynomial oscillations. For this reason, when there are many points to consider, the oscillations of the polynomial can be much higher. This could not model the points in the correct way.

If particular properties about the model are not known, but high oscillations must be avoided, then a *spline* function can be used, instead of a polynomial. A spline is a function defined piecewise by polynomials. It is used for avoiding the phenomenon of the increase of oscillations when the degree of a polynomial increases. Indeed, a spline locally is a polynomial having a low degree, so that its oscillations are low. In its general form a polynomial spline

$$S : [a, b] \longrightarrow \Re$$

consists of polynomial pieces

$$P_i : [t_i, t_{i+1}) \in [a, b] \longrightarrow \Re \quad \forall i \in \{1, 2, \ldots, K\},$$

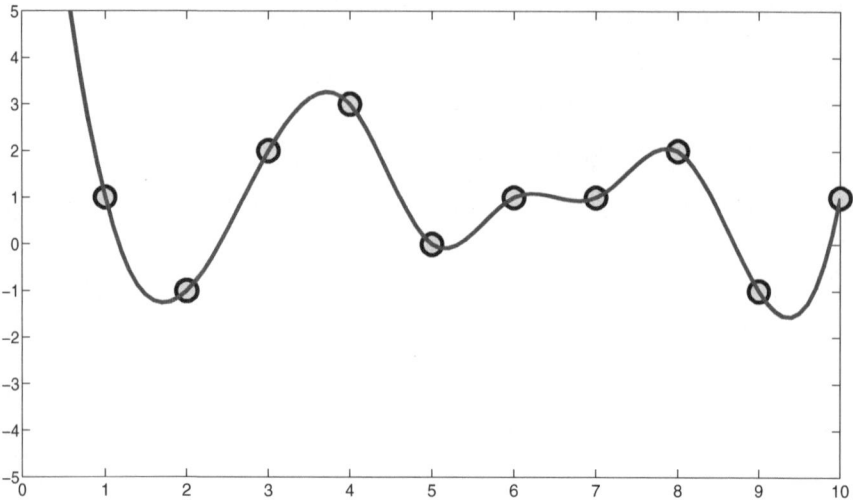

Fig. 2.6 Interpolation of 10 points by a cubic spline.

where $a = t_1 < t_2 < \cdots < t_K < t_{K+1} = b$. Each P_i has a predefined degree. The most used degree is 3, and in this case S is called *cubic spline*. By using a cubic spline, a large number n of points can be interpolated while the oscillations of the values of the function are kept low. Figure 2.6 shows a cubic spline interpolating the same points in Figure 2.5. In the interval $[1, 2]$ of the x axis there are not high oscillations anymore.

There are applications where some information about the process to model is known a priori. Sometimes it might be known that the model has to be linear, or quadratic or other, and hence particular functions need to be used. Let us suppose, for instance, that the model must be linear. As pointed out above, the only way for finding a line passing through more than 2 points is to have all these points aligned. If a polynomial is used for interpolating these non-aligned points, its degree corresponds, in general, to the total number of points minus one. Therefore, if the model must be linear and the points are not aligned, then a function *approximating* these points can be searched, instead of an interpolating function. Functions that approximate a given set of data are called *regression functions*. The main difference between interpolation and regression functions is that the equality $y_i = f(x_i)$, for each pair (x_i, y_i) of the training set, must be satisfied in the first case, whereas $f(x_i)$ must be only an approximation of y_i in the second case.

The easiest regression function is the linear regression. A straight line of equation

$$y = ax + b$$

is considered. For each point (x_i, y_i) of the training set,

$$r_i = y_i - (ax_i + b)$$

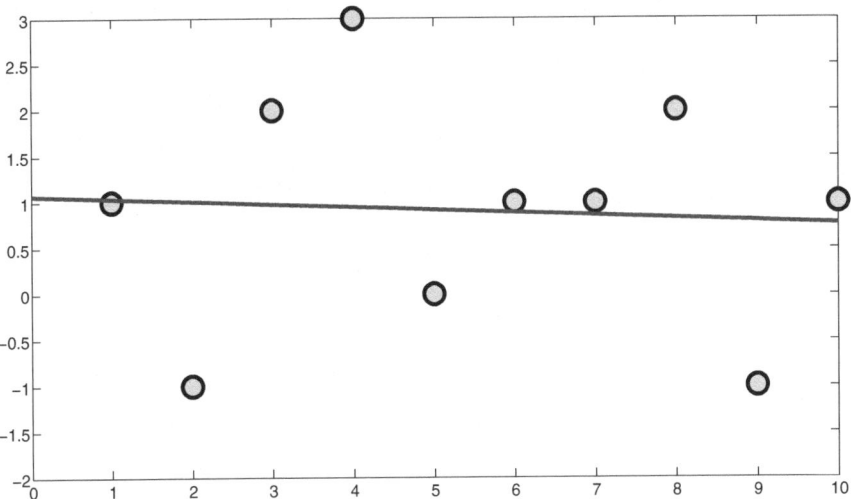

Fig. 2.7 Linear regression of 10 points on a plane.

corresponds to the so-called *residual*. A residual is zero if the point (x_i, y_i) belongs to the straight line, i.e., when the line passes through the point (x_i, y_i). Instead of forcing the residual to be zero for all the points (interpolation), it is minimized (regression). In this way, the straight line that better approximates the points can be found. The problem to be solved can be seen as an optimization problem having as objective function

$$R = \sum_{i=1}^{n+1} r_i^2$$

where $n+1$ is the number of points. A linear regression related to the set of 10 points used in the previous examples is shown in Figure 2.7.

Nonlinear regression models include quadratic regression (see Figure 2.8), and in this case the residual is defined as

$$r_i = y_i - (ax_i^2 + bx_i + c),$$

or logistic regression, where

$$r_i = y_i - \frac{1}{1 + e^{-ax_i}}.$$

The estimation of the parameters of the regression models can be seen as an optimization problem. Different approaches have been developed over the years for performing this estimation. For instance, in [109, 235], surveys on regression techniques based on least-square models are presented.

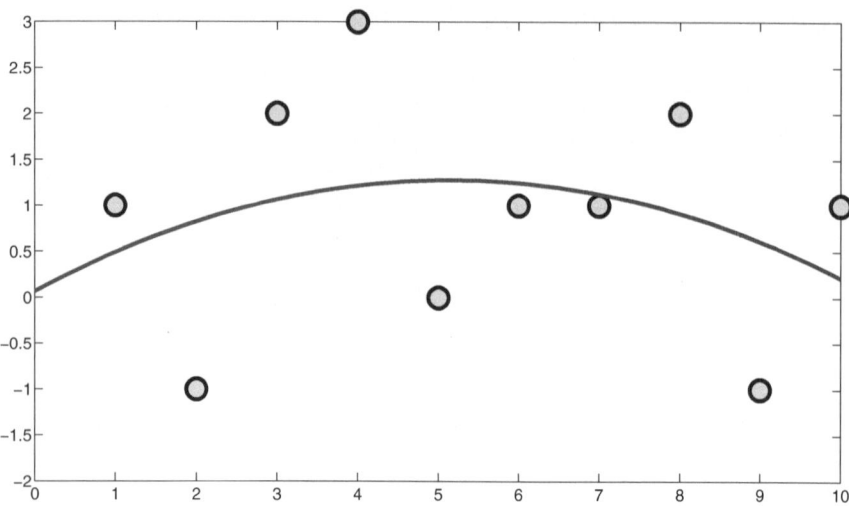

Fig. 2.8 Quadratic regression of 10 points on a plane.

It is important to note that instead of one independent variable x, more variables can be employed for more complex problems. If the generic point $(x_{i1}, x_{i2}, \ldots, x_{ik}, y_i)$ contains k independent variables and one dependent variable y_i, then all the independent variables can be put together in the expression

$$z_i = \beta_0 + \beta_1 x_{i1} + \beta_2 x_{i2} + \cdots + \beta_k x_{ik},$$

where the generic β_i is a real coefficient. Then, the new variable z_i can be used instead of x_i in the previous equations.

2.3 Applications

The PCA method, and interpolation and regression models, are used as techniques for mining data. PCA allows one to represent samples using vectors with a smaller dimension without losing important information on the data. Interpolation and regression techniques allow one to model by simple functions a given set of data and to generalize from it. They can be considered basic *statistical techniques*. They can provide good results in some applications, but they may not be adequate for solving more complex problems. However, some examples can be found in the literature of successful application of these techniques. For instance, PCA is used for studying the star formation history of galaxies [73], and also for analyzing gene expressions in cells [246]. Linear and nonlinear regression is used for the prediction of babies' birth weight among maternal demographic characteristics in [70]. Interpolation is for instance used in [106] for analyzing the human brain.

PCA can be used as a data mining technique itself, but more often it is used for reducing the dimension of a set of data before applying some other data mining technique. Applications in agriculture in which PCA has been used alone are listed below and one of them is presented in detail in Section 2.3.1. Moreover, PCA is also used in some of the applications discussed in other chapters of this book, which are devoted to other data mining techniques. For instance, in Section 3.5.1, PCA is used in conjunction with the k-means algorithm for data partitioning. In this application, the wine fermentation process is studied and the aim is to find clues that reveal bad results at the beginning of the fermentation process [230]. The main technique used is k-means and not PCA. However, PCA helps k-means in partitioning the data in clusters, since it reduces the dimension of the set of data before applying the k-means algorithm.

Some applications of the statistical techniques in agriculture are presented in the next section of this chapter. PCA is for instance applied for characterizing beef meat [58], for analyzing chicken breast quality [156], for locating the origin of potentially toxic elements in soils [23], and for evaluating the impact of irrigation water quality [161]. Interpolation models are used for analyzing climate data [8]. Regression models are used for evaluating soil liquefaction probability [137], for predicting the distributions of New Zealand's freshwater fishes [143], for predicting aroma properties of aged red wines [12], and for monitoring the effects of energy use in agriculture [125]. In [248], nonlinear regression models are used for predicting shrimp growth. The same studies on the shrimps are conducted by using a neural network approach and they show that neural networks perform better.

In the following we will focus on two applications. In Section 2.3.1 we will discuss the application of PCA for controlling the quality of chicken breast meat. In Section 2.3.2 we will present the application of regression models for evaluating the effects of energy use in the agricultural field.

2.3.1 Checking chicken breast quality

Chicken breast meat is widely used as a food resource. After the death of the chicken, the animal has to be deboned. The quality of the meat strongly depends on the post-mortem aging time. Characteristics that are directly related to the physical components of meat products can provide reliable information about meat quality. However, humans go beyond the physical components to describe a wide range of factors involved in mastication and afterfeel/aftertaste sensations, such as appearance, flavor, and texture. All these characteristics can be used for analyzing the variations of physical, color, and sensory properties of chicken breast meat deboned at different times after death.

We will focus in this section on the studies published in [156]. In order to analyze the meat quality and the deboning time, a set of 36 chicken carcasses has been considered and randomly divided into 4 subgroups, each one containing 9 carcasses. These subgroups are designed for different deboning times. Precisely, chickens in the different groups have been deboned after 2, 4, 6 and 24 hours after death. During the period between death and deboning the carcasses have been kept at a 2°C temperature. After deboning, the breasts have been cut in two parts, and each part has been subject to a different set of analysis.

Several parameters have been used for evaluating the meat quality. A colorimeter has been used for measuring the color of both the breasts of the chickens, and a pH meter has been used for measuring their pH levels. The hardness of the meat has been evaluated by using a blade which sheares the meat perpendicularly to the longitudinal orientation of the muscles fibers. Sensory attributes include cardboardy, wet feathers, springiness, cohesiveness, hardness, moisture release, particle size, bolus size, chewiness, and metallic aftertaste-afterfeel. These attributes have been evaluated by 9 expert panelists. The numerical scale for each attribute ranged from 0 to 15.

Before that the PCA method can be applied, the corresponding covariance matrix needs to be created. It provides the variance of each variable and also the variance of each single variable with respect to the other variables. This matrix provides useful information on the nature of the data. Figure 2.9 shows the variance of the considered variables at different deboning times. In general, the considered variables show a steady decrease in value when the deboning time increases. In particular, the pH levels decreased gradually in meats deboned from 2 to 6 hours, while it remained constant when the meat was deboned from 6 to 24 hours. This suggests that complex biochemical reactions are active during postmortem aging. The chicken breast lost redness during time: in particular, its redness decreases while its yellowness increases. The meats deboned at earlier postmortem time require more force to shear and therefore they are less tender. The attributes evaluated by the panelists also decrease gradually. The two sensory flavor attributes (cardboardy and wet feathers), the seven sensory texture attributes (springiness, cohesiveness, hardness, moisture release, particle size, bolus size, and chewiness), and the afterfeel-aftertaste attribute also decreased with the increase of deboning time. In general, these observations suggest that the optimal deboning time for chicken breast meat is after 4 hours from death.

In this application, the covariance matrix includes the 24 variables used for evaluating the chicken meat. One hundred forty-four samples are used and their variance and covariances are computed for generating the covariance matrix. The PCA method just finds the eigenvalues and eigenvectors of the covariance matrix, so that it can locate the principal components. As previously explained in detail, the first few principal components should be able to represent most of the variations in the data. In other words, they should be able to represent the data with minimal loss of information. In this example, the first seven principal components are able to represent about 70% of the total variations on the data. Moreover, the first four principal components represent about 50% of the total variations. In particular, the first principal component takes 23.3% of the variations, the second one 13.6%, the third one 8.8% and finally the fourth one 6.9%. An analysis on the data showed that the first component was mainly defined by the shear force and by the attributes decided by the panelists.

Breast characteristic	Deboning time				
	2h	4h	6h	24h	All together
pH	6.06 ± 0.20	6.02 ± 0.18	5.98 ± 0.18	5.98 ± 0.16	6.01 ± 0.18
Lightness (%)	70.56 ± 7.59	69.68 ± 7.52	70.71 ± 9.29	72.03 ± 10.14	70.74 ± 8.66
Redness (%)	−24.30 ± 16.98	−25.16 ± 15.98	−29.07 ± 15.28	−34.87 ± 14.39	−28.35 ± 16.08
Yellowness (%)	257.53 ± 126.45	267.18 ± 148.95	404.57 ± 372.85	345.38 ± 503.33	318.66 ± 330.18
Cooking yield (%)	73.79 ± 3.00	74.68 ± 2.69	75.14 ± 3.21	75.33 ± 3.03	74.74 ± 3.01
Shear force (kg)	9.40 ± 3.26	7.08 ± 2.83	5.79 ± 1.74	3.90 ± 1.01	6.54 ± 3.10
Brothy	3.58 ± 0.92	3.77 ± 0.67	3.99 ± 0.74	3.73 ± 0.69	3.77 ± 0.77
Chicken-meaty	4.17 ± 0.56	4.22 ± 0.58	4.26 ± 0.48	4.12 ± 0.45	4.19 ± 0.52
Cardboardy	2.87 ± 1.06	2.71 ± 0.95	2.63 ± 0.86	2.39 ± 0.97	2.67 ± 0.97
Wet feathers	2.88 ± 0.97	2.83 ± 0.86	2.77 ± 0.95	2.53 ± 0.88	2.75 ± 0.92
Bloody-serumy	3.30 ± 1.24	3.48 ± 1.17	3.48 ± 1.36	3.12 ± 1.14	3.34 ± 1.23
Sweet	2.21 ± 0.80	2.11 ± 0.94	2.22 ± 0.89	2.40 ± 0.66	2.24 ± 0.79
Salty	2.05 ± 0.74	1.91 ± 0.85	2.02 ± 0.87	2.19 ± 0.73	2.04 ± 0.80
Sour	2.89 ± 0.95	2.71 ± 0.78	2.85 ± 0.85	2.95 ± 0.77	2.85 ± 0.84
Springiness	3.81 ± 1.10	3.87 ± 1.25	3.81 ± 1.24	3.42 ± 1.30	3.73 ± 1.23
Cohesiveness	5.95 ± 1.70	5.63 ± 1.66	5.08 ± 1.51	4.61 ± 1.38	5.32 ± 1.63
Hardness	5.60 ± 1.02	5.45 ± 1.27	5.01 ± 1.11	4.34 ± 1.18	5.10 ± 1.24
Moisture release	3.82 ± 0.82	3.68 ± 0.86	3.69 ± 0.68	3.57 ± 0.87	3.69 ± 0.81
Particle size	3.74 ± 0.76	3.71 ± 1.05	3.36 ± 0.95	3.01 ± 0.93	3.46 ± 0.97
Bolus size	4.16 ± 0.76	3.95 ± 0.99	3.57 ± 0.98	3.32 ± 0.98	3.75 ± 0.98
Chewiness	5.63 ± 1.15	5.18 ± 1.35	4.66 ± 1.28	4.27 ± 0.97	4.96 ± 1.28
Toothpack	3.66 ± 1.00	3.83 ± 1.06	3.68 ± 0.96	3.57 ± 0.94	3.68 ± 0.99
Metallic	3.31 ± 1.17	3.31 ± 1.13	3.06 ± 1.26	3.09 ± 1.28	3.19 ± 1.20
Oily-greasy	1.28 ± 0.92	1.18 ± 0.96	1.28 ± 1.00	1.27 ± 0.90	1.26 ± 0.94

Fig. 2.9 Average and standard deviations for all the parameters used for evaluating the chicken breast quality. Data from [156].

Therefore, these attributes are the most important variables for the evaluation of the chicken breast quality.

2.3.2 Effects of energy use in agriculture

Modern agricultural sector needs an increasing demand of energy resources. Such resources include electricity, fuels, natural gases and coke. Much of this energy is directly used in agriculture for a wide range of purposes. For instance, operating vehicles need fuel or electricity, and irrigation pumps need gas and water. Fertilizers, seeds, pesticides, etc., can also be considered as *indirect* use of energy in agriculture. In general, energy has an important role in the social and economic development of a country. In [125], studies are presented for analyzing the effect of the energy factor on agricultural productivity.

These studies are focused on the agricultural productivity in Turkey. In this country, energy consumption has increased more than 55% during the past three decades. Interesting is to note that, during the same period, the agricultural productivity in Turkey has increased as well. The studies are based on data obtained from the Ministry of Energy of Turkey, and they cover the period 1970–2003. The data have been used for modeling a regression function having the following form:

$$\ln(API(t)) = \alpha_1 + \alpha_2 \ln(EC(t)) + \alpha_3 \ln(AFA(t)) + \epsilon_t.$$

In the formula, $API(t)$ is an index for agricultural productivity (eight products have been used: wheat, barley, sunflower, cotton, sugar, beet, chickpea, tomato and milk), $EC(t)$ is the energy consumption of the agricultural sector, $AFA(t)$ represents gross additional assets, and ϵ_t is a real value denoting possible noise in the data. All these parameters are known year by year, and one can refer to a different year through the variable t. α_i are the coefficients to be found for modeling the data. Note that the equation used is a double logarithmic linear regression.

After the estimation of the coefficients α_i, the results showed that the energy consumption coefficient $EC(t)$ is statistically significant. Its positive sign indicates that agricultural productivity increases with the increase in energy consumption. There is actually a very strong relationship between energy use and agricultural productivity. The energy consumption $EC(t)$ is more sensitive to productivity than the gross additional assets $AFA(t)$. The found coefficient for the gross additional assets is also positive, and therefore an increase in $AFA(t)$ also results in an increase in the productivity.

2.4 Experiments in MATLAB

In this section some few experiences in MATLAB regarding the techniques discussed in this chapter are presented.

```
x = rand(1,10);
y = 2*x + rand(1,10);
A = cov(x,y);
[v,d] = eig(A);
x1 = v(1,2)*x + v(2,2)*y;
y1 = v(1,1)*x + v(2,1)*y;
plot(x,y,'ko','MarkerSize',10,'MarkerEdgeColor','k','MarkerFaceColor',
    [.49 1 .63])
hold on
plot(x1,y1,'kd','MarkerSize',10,'MarkerEdgeColor','k','MarkerFaceColor',
    [.49 0 .63])
var_x1 = max(x1) - min(x1);
var_y1 = max(y1) - min(y1);
```

Fig. 2.10 The PCA method applied in MATLAB to a random set of points lying on the line $y = x$.

An example of the use of the PCA method in MATLAB is given in Figure 2.10. The figure shows the set of instructions in MATLAB for computing the principal components of a random set of two-dimensional points. In the following, all the instructions in Figure 2.10 are commented on step by step. The function rand is used for generating a random vector of points close to the line $y = 2x$. The x coordinates of the points are created by rand, while the y coordinates are obtained by adding a random number to $2x$. In MATLAB all the variables, if it is not differently specified, are matrices of real numbers (for details about MATLAB the reader is referred to Appendix A). Then, x and y are matrices, where one of their dimensions is 1, and this makes them actually vectors. It is very important to keep in mind that MATLAB considers variables as matrices, when functions such as rand need to be used. Indeed, rand takes two input parameters: the number of rows and the number of columns of the random matrix to be generated. If a vector is needed, one of these two parameters has to be 1.

After defining a set of points, the covariance matrix related to the variables used for representing these points, the coordinates x and y, needs to be computed. In MATLAB, the function cov computes the covariance matrix of a given set of variables. The result is stored in A and used as an input parameter for the function eig. eig computes the eigenvalues d and the eigenvectors v of the covariance matrix A. The eigenvectors play the role of the vector α_1 in equation (2.1). They can be used for computing the transformed variables. The two new variables are x1 and y1.

Two calls to the function plot creates Figure 2.11. The figure contains the original set of points and the transformed set of points. The original points are marked by circles. From the figure, it is clear that the variability on one of the transformed variables is very small. The variables var_x1 and var_y1 contain this information. Note that the basic plot function needs two input parameters only: a vector containing the x coordinates and a vector containing the y coordinates of the points to draw. In this case, other optional parameters are also used. For a description of these options refer to Appendix A. They are used for marking each point with a particular marker having a certain color. The vector [.49 1 .63] specifies a particular tonality of green. The instruction hold on is used for letting the different graphs created by plot overlap on each other.

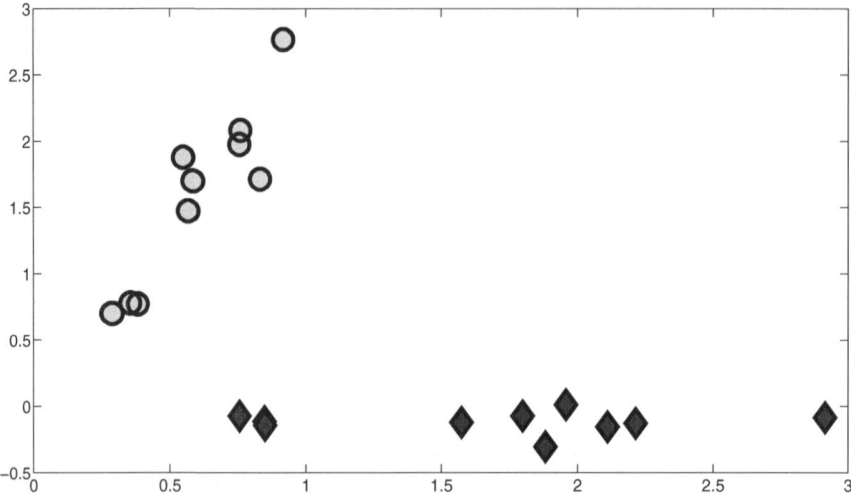

Fig. 2.11 The figure generated if the MATLAB instructions in Figure 2.10 are executed.

Let us generate now in MATLAB interpolating and regression functions. In Figure 2.12 a sequence of MATLAB instruction is shown. The calls to the function plot generate Figure 2.13(a). In this example, a set of 9 points in a two-dimensional space is considered. The 9 points are specified in MATLAB through their x and y coordinates, contained in the vector x and the vector y, respectively. These points are drawn in the figure by using the first call to the function plot. The function plot is then used another time for drawing all the points in the set. This time no options are used, and, by default, the function plot connects the points to draw by a line. What is drawn is therefore the join-the-dots function interpolating the set of points. The polynomial interpolating the points is instead computed by using the function polyfit. The specified degree is 8, since the polynomial passing through 9 points is unique if its degree equals the number of points minus one. The function polyfit needs as input parameters the x and y coordinates of the points to interpolate, and the degree of the polynomial. The output of the function is a vector c containing the coefficients of the polynomial. In order to draw this polynomial, it must be evaluated on a certain number of independent variables, and the couples of independent/dependent

```
x = [-8 -6 -3 -2 1 5 7 9 10];
y = [1 2 2 1 -1 1 0 0 -1];
plot(x,y,'ko','MarkerSize',10,'MarkerEdgeColor','k','MarkerFaceColor',
     [.49 1 .63])
hold on
plot(x,y)
c = polyfit(x,y,8);
xx = -8:0.1:10;
yy = polyval(c,xx);
plot(xx,yy,'r:')
```

Fig. 2.12 A sequence of instructions for drawing interpolating functions in MATLAB.

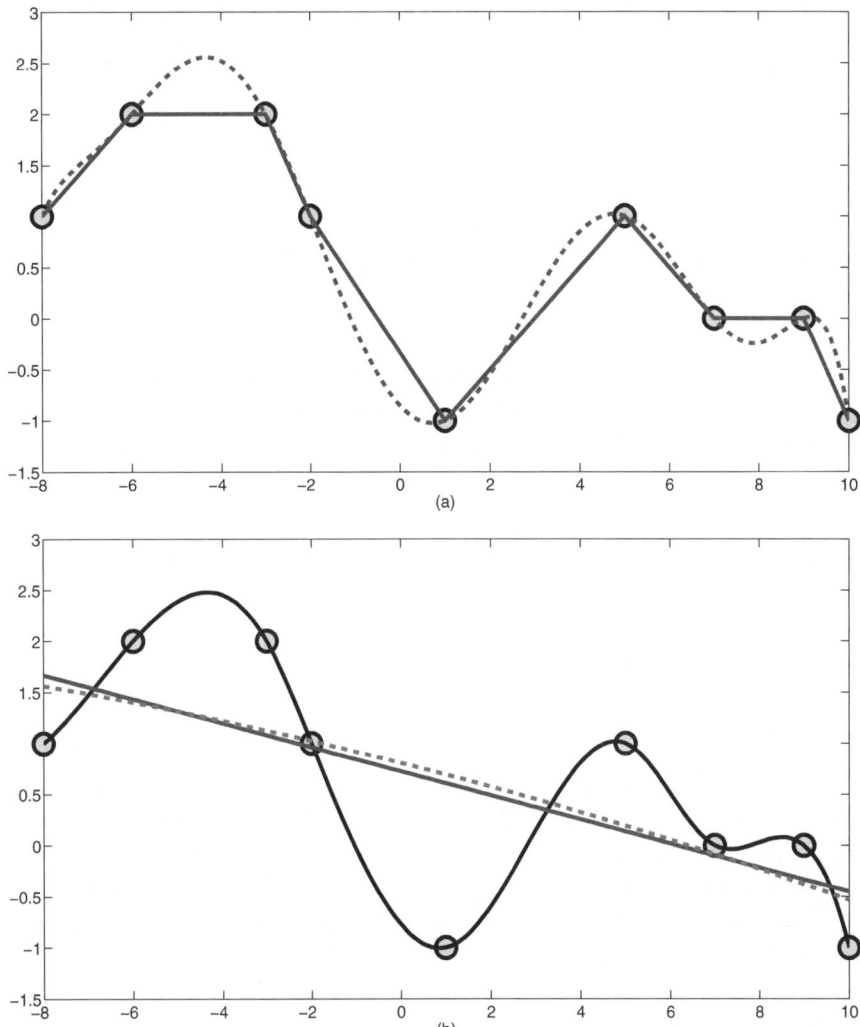

Fig. 2.13 Two figures generated by MATLAB: (a) the instructions in Figure 2.12 are executed; (b) the instructions in Figure 2.14 are executed.

variables can be used to draw the polynomial using the function `plot`. If the used independent variables are sufficiently close to each other, then the figure generated by the function `plot` is a good approximation of the polynomial. The vector `xx` is used for storing the independent variables. It is a vector whose first component is −8 (the smallest value in x), whose last component is 10 (the largest value in x), and such that the difference between any consecutive components in `xx` is 0.1. The function `polyval` can evaluate a polynomial. It takes as input parameters the poly-nomial coefficients and a vector `xx` containing a set of independent variables. The

```
plot(x,y,'ko','MarkerSize',10,'MarkerEdgeColor','k','MarkerFaceColor',
     [.49 1 .63])
hold on
yy = spline(x,y,xx);
plot(xx,yy,'k')
c = polyfit(x,y,1);
yy = polyval(c,xx);
plot(xx,yy)
c = polyfit(x,y,2);
yy = polyval(c,xx);
plot(xx,yy,'m:')
```

Fig. 2.14 A sequence of instructions for drawing interpolating and regression functions in MAT-LAB.

result, the set of corresponding dependent variables, is given in output and stored in yy. The function plot is then called for drawing the points specified in xx and yy. The option 'r:' forces the figure to be in red and drawn with dashed lines.

As discussed above, there are other ways for interpolating or approximating a certain set of points by a function. Suppose that the variables x and y are still in memory as defined in the code in Figure 2.12, then the code in Figure 2.14 generates Figure 2.13(b). The points are drawn another time, by the first call to the function plot. Then, the cubic spline interpolating such points is computed. The function spline evaluates the cubic spline passing through the given points specified in x and y in the independent variables in xx. The corresponding dependent variables are stored in yy. Once again, the function plot is called for drawing the points specified in xx and yy. This time 'k' is used as option, meaning that the figure must be black. After that, the linear regression approximating the points is computed by using the function polyfit. This function has been used before for finding the coefficients of the interpolating polynomial. The only difference stands in the degree of the polynomial: it has to be 1 if the linear regression function is needed. The two coefficients of the linear function are then stored in c, the function polyval is used for evaluating such linear function in a set of points that are utilized by plot. The same procedure is used at the end for drawing the quadratic regression function.

2.5 Exercises

Some exercises related to the principal component analysis, the interpolating functions and the regression functions are presented in this section. All the solutions are reported in Chapter 10.

1. Given the set of points

$$(1, -1), \quad (3, 0), \quad (2, 2),$$

   compute the range of variability of their components.
2. In MATLAB, generate randomly a set of points in a two-dimensional space lying on the line $y = x$. Apply PCA in order to reduce the dimension of the set of points.

3. Compare the original set of points randomly generated in Exercise 2 to the set with reduced dimension obtained by PCA. For this purpose, create a figure in MATLAB that displays the two sets.

4. Given 2 points in a two-dimensional space:

$$(1, 0), \quad (0, -2),$$

compute the equation of the unique line passing through them.

5. Build a figure in MATLAB of the line obtained in Exercise 4.

6. Given 3 non-aligned points in a two-dimensional space:

$$(0, 1), \quad (1, 2), \quad (-1, 3),$$

compute the equation of the unique parabola passing through them.

7. Consider 5 points in a two-dimensional space:

$$(4, 2), (2, 2), (1, 4), (0, 0), (-1, 3).$$

Build a MATLAB figure containing the points and the join-the-dots function interpolating them.

8. Consider the same points of the previous exercise. Build a MATLAB figure containing the points and the quadratic regression approximating such points.

9. Consider 6 points in a two-dimensional space:

$$(1, 2), (2, 3), (1, -1), (-1, 3), (1, -2), (0, -1).$$

Build a MATLAB figure in which the points are represented with their linear and quadratic regression functions.

10. Consider the same set of points of the previous exercise. Suppose that each point (x, y) of the set is approximated with the corresponding point $(x, f(x))$ of the linear regression f obtained in the previous exercise. Compute the mean arithmetic error on the whole set of points using MATLAB.

Chapter 3
Clustering by k-means

3.1 The basic k-means algorithm

Clustering techniques are used for finding suitable groupings of samples belonging to a given set of data. There is no knowledge a priori about these data. Therefore, such set of samples cannot be considered as a training set, and classification techniques cannot be used in this case. The k-means algorithm is one of the most popular algorithms for clustering [103]. It is one of the most used algorithms for data mining, as it has been placed among the top 10 algorithms for data mining in [237].

The k-means algorithm partitions a set of data into a number k of disjoint clusters by looking for inherent patterns in the set. The parameter k is usually much smaller than the dimension of the set of samples, and, in general, it needs to have a predetermined value before using the algorithm. There are cases where the value of k can be derived from the problem studied. For instance, in the example of the blood test analysis (see Section 1.1), the aim is to distinguish between healthy and sick patients. Hence, two different clusters can be defined, and then $k = 2$. In other applications, however, the parameter k may not be defined as easily. In the example of separating good apples from bad ones (see Section 1.1), images of apples need to be analyzed. The set of apple images can be partitioned in different ways. One partition can be obtained by dividing apples into two clusters, one containing apples with defects and another one containing good apples. In this case $k = 2$. However, defective apples can be classified based on the degree of the defect. For instance, if the apples have a defect which is not very visible, then these apples could be sold with a lower price. Therefore, even defective apples can be grouped in different clusters. In this case, k shows the number of defects that are taken into consideration. When there is uncertainty on the value of the parameter k, a set of possible values is considered and the algorithm is carried out for each of the values. The best obtained partition in clusters can then be considered.

Let us suppose that X represents the available set of samples. Each sample can be represented by an m-dimensional vector in the Euclidean space $\Re^m$. For instance, blood analysis can be represented by a vector whose components contain the exper-

A. Mucherino et al., *Data Mining in Agriculture*, Springer Optimization and Its Applications 34, 47
DOI: 10.1007/978-0-387-88615-2_3,

imentally found blood measurements. The image of a fruit can be represented by a matrix of pixels that can be organized row by row in a vector. Moreover, the image of the fruit can be analyzed, and some properties regarding the image can be inserted in a vector that can be used for representing the image. Thus, in the following, $X \equiv \{x_1, x_2, \ldots, x_n\}$ will represent a set of n samples, where the generic sample x_i is an m-dimensional vector.

Given a predetermined k value, the aim of the k-means algorithm is to find a partition of k disjoint clusters of X. If S_j represents one of these clusters, then the following conditions must be satisfied

$$X = \bigcup_{j=1}^{k} S_j, \qquad S_j \cap S_l = \emptyset \quad 1 \le j \ne l \le k.$$

Each cluster is a subset of X and contains samples with some similarity. In this approach, the similarities between samples are measured by metric functions. The distance between two samples provides a measure of similarity: it shows how similar or how different two samples are. In other words, if sample x_1 is closer to x_2 than to x_3, then x_1 is considered to be more similar to x_2 than to x_3.

A representative can be assigned to each cluster. In the k-means approach, the representative of a cluster is defined as the mean of all the samples contained in the cluster. The mean is referred to as the *center* of the cluster, and is calculated by the following formula:

$$c_j = \frac{1}{n(S_j)} \sum_{i=1}^{n(S_j)} x_{j(i)}.$$

In the formula, $n(S_j)$ is the number of samples contained in cluster S_j, and $j(i)$ represents the index of the ith sample in cluster S_j. Then, each $x_{j(i)} \in S_j$ for all the $i \in \{1, 2, \ldots, n(S_j)\}$, and c_j is the vector having as components the means of all the components of vectors $x_{j(i)}$. It is worth noting that different methods for clustering may use different representatives for a cluster. For instance, the k-medoids method uses as representative one of the samples in the cluster.

Let us consider the set of points shown in Figure 3.1. Even though there is no previous knowledge about the data, the figure clearly shows that two subsets of points can be defined. For simplicity, let us refer to the cluster whose points are marked with the symbol $\star$ as $C^{\star}$. Similarly, C^{+} denotes the cluster containing the points marked with the symbol $+$. Such subsets represent the inherent patterns that clustering algorithms try to discover by partitioning the data. The two points marked by a circle in the figure represent the centers of the two clusters. Let us consider computing the distances between one of the samples and the two centers. The distance between a sample and the center of its cluster is smaller than the distance between the sample and the center of another cluster. This shows that samples that are similar belong to the same cluster or that a cluster contains similar samples. The center, or the mean of the cluster, can be considered as a sample similar to all samples contained in this cluster. Since the similarity is here measured using the distance function, the smaller the value of the distance between samples, the more similar samples are.

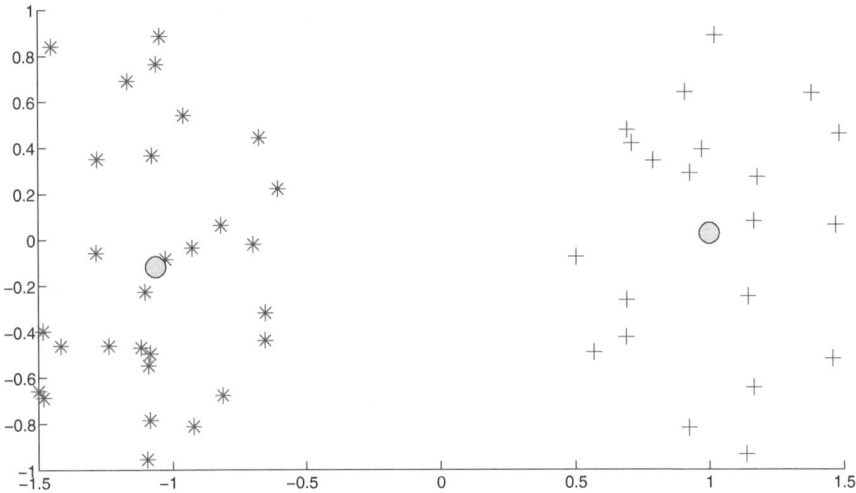

Fig. 3.1 A partition in clusters of a set of points. Points are marked by the same symbol if they belong to the same cluster. The two big circles represent the centers of the two clusters.

The example shown in Figure 3.1 is a very simplified one. Actually, clusters are usually not so easily defined, and the dimension of the set of data does not allow analyzing the samples visually. Therefore, a general formulation of the clustering problem is to find k disjoint clusters that minimize the *error function*:

$$f(S_1, S_2, \ldots, S_k) = \sum_{j=1}^{k} \sum_{i=1}^{n(S_j)} d(c_j, x_{j(i)}), \qquad (3.1)$$

where d represents a suitable distance function. In the example provided in Figure 3.1, d is the Euclidean distance. Regardless of the distance function used, the error function (3.1) consists of a sum of positive real numbers, because the distance function always has non-negative values. Therefore, minimizing the error function means minimizing all its terms. Each term represents the distance between the sample $x_{j(i)}$ and the center of its cluster. The optimal partition of the data is obtained when all the samples are closer to the representative of their own cluster.

Note that the error function (3.1) can be considered as the objective function of an optimization problem. The optimization problem can be formulated for finding a partition that minimizes the error function. Therefore, optimization methods may be used to solve this partitioning problem (see Section 1.4).

An easier way to solve this partitioning problem is to use the k-means algorithm. The basic k-means algorithm is also known in literature as the Lloyd's algorithm, and it is based on a simple idea. Let us suppose that an arbitrary partition is currently associated with a certain set of data. The aim of the algorithm is to find a partition which minimizes the objective function (3.1), or equally, a partition which minimizes all its terms. Then, all the terms of the objective function must be checked for finding out whether the current partition is optimal or not. Let us consider the general sample $x_{j(i)}$ that is currently assigned to cluster S_j. The distances between $x_{j(i)}$ and c_j, the center of S_j, should be minima, in order to minimize the error function. Therefore, all the distances between $x_{j(i)}$ and the k centers of the k clusters are computed. If the distance $d(x_{j(i)}, c_j)$ is the smallest one, then sample $x_{j(i)}$ stays in the cluster. If $x_{j(i)}$ is instead closer to the center $c_{\bar{j}}$ of another cluster $S_{\bar{j}}$, then $d(x_{j(i)}, c_j)$ should be replaced by the distance $d(x_{j(i)}, c_{\bar{j}})$. In function (3.1), the distances are computed between the sample and the center of its cluster. In order to substitute the distance, then, the sample $x_{j(i)}$ must be moved from cluster S_j to cluster $S_{\bar{j}}$, so that the new distance associated to this sample is $d(x_{\bar{j}(i)}, c_{\bar{j}})$. In general, at each iteration of the algorithm, a sample is moved from its current cluster to the cluster whose center is closer to the sample. Note that every time a sample moves, the centers of the two clusters, the one where the sample was and the one where the sample is moved to, change.

The k-means algorithm is shown in Figure 3.2. At the start, each sample is randomly assigned to one of the k clusters. The centers c_j of the clusters are then calculated. The main loop of the algorithm (**while** loop) analyzes all the samples, from the first one to the last one. Each time a sample is considered, its distances from the k centers are calculated and it is assigned to the cluster with the closest center. Even though the sample already belongs to such a cluster, it is reassigned to it another time. The algorithm can be modified to reassign samples to clusters only when a sample changes cluster. However, this approach makes the algorithm more computationally expensive, and therefore, it is not used in practice. When a sample changes cluster, the centers of two clusters, the one where the sample was and the new cluster, are recomputed. The **while** loop is executed until no samples are effectively moved from one cluster to another. The clusters are considered as *stable* when there are no movements of samples from one cluster to another during one iteration of the **while** loop. The stability of the clusters can also be checked by controlling their centers: if the centers do not change during an iteration of the **while** loop, then the clusters are stable. When the stability of the clusters is reached, then an optimal partition is obtained, and the algorithm stops. Figure 3.3 shows the result of the execution of the algorithm on a set of points defined in a two-dimensional space. Such execution has been performed using the MATLAB® function kmeans,

```
randomly assign each sample to one of the k clusters S(j), 1 ≤ j ≤ k
compute c(j) for each cluster S(j)
while (clusters are not stable)
    for each sample Sample(i)
        compute the distances between Sample(i) and all the centers c(j)
        find j* such that c(j*) is the closest to Sample(i)
        assign Sample(i) to the cluster S(j*)
        recompute the centers of the changed clusters
    end for
end while
```

Fig. 3.2 The Lloyd's or k-means algorithm.

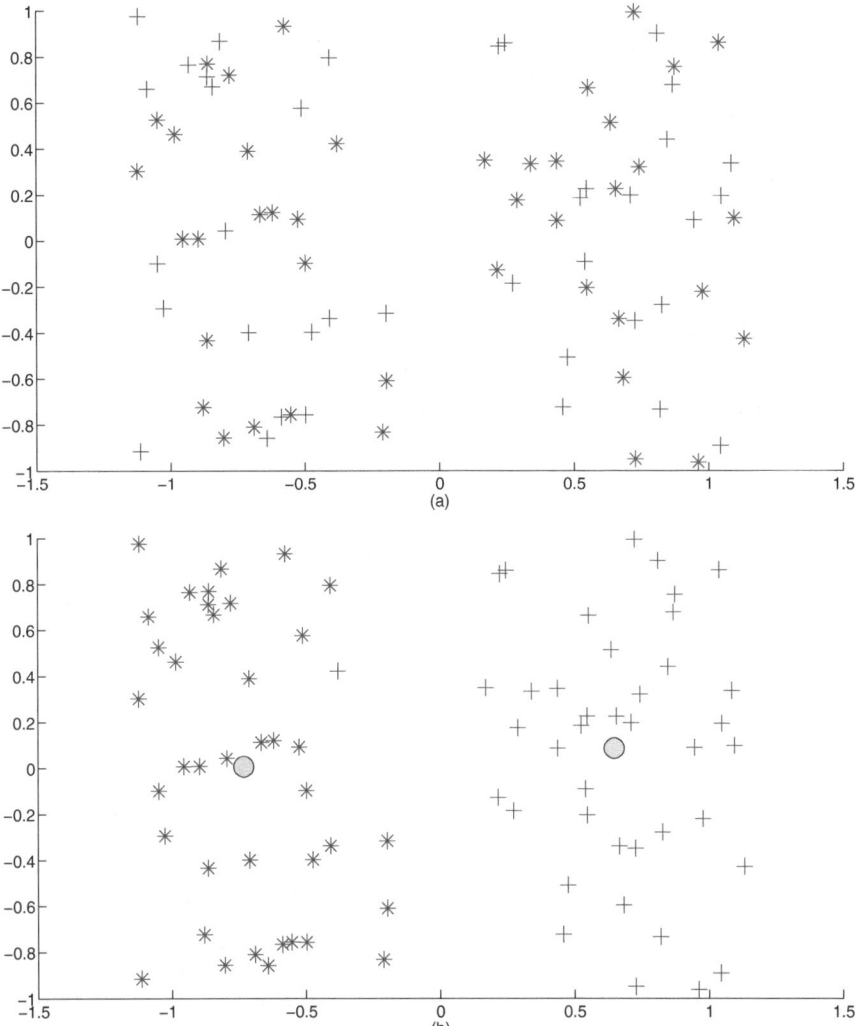

Fig. 3.3 Two possible partitions in clusters considered by the k-means algorithm. (a) The first partition is randomly generated; (b) the second partition is obtained after one iteration of the algorithm.

whose code is provided in Section 3.6. Figure 3.3(a) shows the initial distribution of the points and Figure 3.3(b) shows the distribution of the points after one iteration of the algorithm. Note that the algorithm almost converges after only one iteration. In Figure 3.3(b) there is only one point which is still contained in the wrong cluster. Let us compute its distance from the centers of the two clusters: the distance from the center of the cluster C^+ is greater than the distance from the center of the cluster $C^\star$. Such point needs then to be assigned to the cluster $C^\star$ and removed from cluster

C^+. Performing one more iteration of the algorithm, the convergence is reached and the optimal partition is obtained.

As already mentioned, the k-means algorithm solves an optimization problem having as objective function the error function (3.1). At each iteration of the algorithm, a partition with a smaller value of the error function is obtained. Indeed, if at least one of the samples $x_{j(i)}$ is moved, then the corresponding distance $d(x_{j(i)}, c_j)$ has a smaller value, and therefore the error function has a smaller value. The error function decreases at each iteration, until an optimal partition is obtained. Since the error function can never increase after one **while** loop, the algorithm defines a strictly decreasing path on the domain of the objective function. Therefore, if the function has more than one local minimum, the k-means algorithm stops at one of these, which may or may not be the global minima. The algorithm can reach one local minima or another depending on the starting random partition, which represents the root of the decreasing path followed on the domain of the function. Therefore, the k-means algorithm is actually a method for local optimization, because it provides one of the *local* optimal partitions in clusters. For this reason, often the k-means algorithm is applied more than one time using different starting partitions. The best result obtained over a certain number of trials is then considered as the global optimal solution.

The k-means algorithm can be represented in terms of a Voronoi diagram. The Voronoi diagram is a partition of a metric space in disjoint parts referred to as *cells*. The diagram is related to a given set V of points, and each point defines a cell in the diagram. A point y of the metric space lies in the cell corresponding to the point $x_p \in V$ if and only if

$$d(x_p, y) \leq d(x_q, y) \quad \forall x_q \in V.$$

As a result, different sets of points define different Voronoi diagrams. Such diagrams are able to capture information on the relative distances between the points in a metric space.

In order to build a Voronoi diagram related to a certain set of points, the boundaries between its cells need to be drawn. Let us start describing a simple case with a set containing 2 points only. For simplicity, all the figures presented in this chapter refer to Voronoi diagrams built in two-dimensional spaces. If only 2 points x_1 and x_2 are considered, then the diagram divides the Euclidean plane in two parts only: the diagram has only two cells. The border between the two cells can then be defined by all points on the plane which are equidistant from x_1 and x_2. Figure 3.4(a) shows the Voronoi diagram of two points. The infinite line that divides the two cells separates the points which are closer to x_1 and the points which are closer to x_2. If the set contains more than two points and they are aligned, the Voronoi diagram is the one in Figure 3.4(b). If a point is randomly selected in any of the cells, this point is closer to point x_i defining the cell than to any other x_j with $i \neq j$.

The Voronoi diagrams shown in Figure 3.4 are quite simple to draw. Such diagrams, however, can become more complex, when the number of points in the set increases and when they do not satisfy particular conditions. If this is the case, the

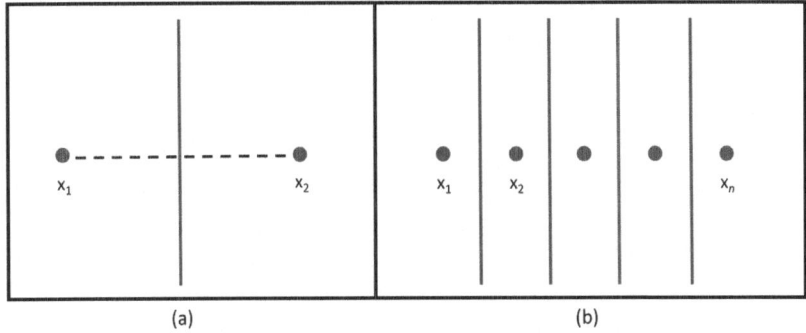

Fig. 3.4 Two Voronoi diagrams in two easy cases: (a) the set contains only 2 points; (b) the set contains aligned points.

following procedure can be used for drawing the diagram. Figure 3.5 shows the procedure in the case in which three non-aligned points are considered. For each couple of points, the perpendicular bisector between them needs to be computed (Figure 3.5(a)). As mentioned before, the borders between two cells contain points which are equidistant from the two points generating the two cells. If all the points that violate this equidistance rule are removed from the lines drawn in Figure 3.5(a), then Figure 3.5(b) is obtained. This is exactly the Voronoi diagram related to the three considered points. The same procedure can be used if more than three points are considered. Other more efficient procedures have been developed for building Voronoi diagrams, as for instance the one implemented in the MATLAB function voronoi. By using this function, the Voronoi diagram related to a random set of two-dimensional points is built and it is shown in Figure 3.6.

The k-means algorithm can be presented in terms of the Voronoi diagrams. A sketch of the algorithm is shown in Figure 3.7. Figure 3.8(a) shows the initial distribution of a set of points and Figure 3.8(b) shows the distribution of points after one iteration of the algorithm. The parameter k is set to 5. The Voronoi diagrams are built

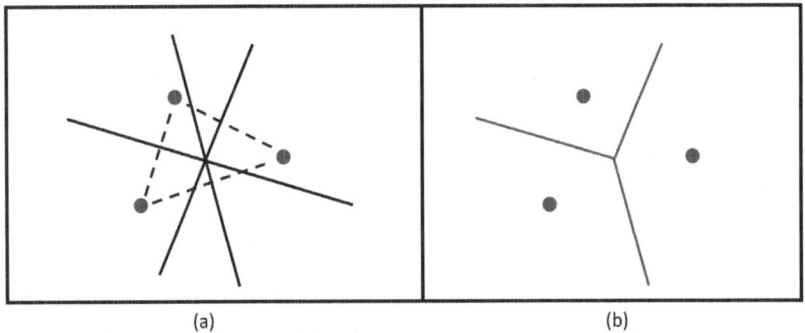

Fig. 3.5 A simple procedure for drawing a Voronoi diagram.

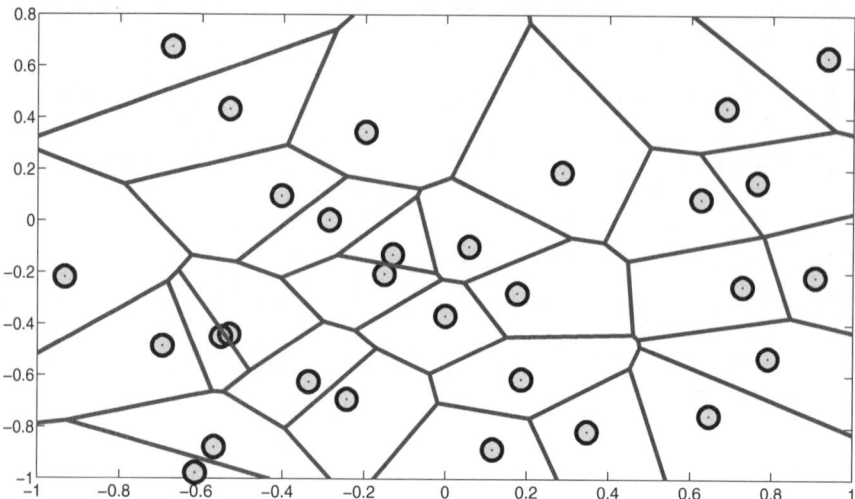

Fig. 3.6 The Voronoi diagram of a random set of points on a plane.

using the centers of the five clusters. As shown in Figure 3.8(a), the Voronoi cells do not coincide with the k clusters, because the same cell contains points belonging to different clusters. Figure 3.8(a) shows the encircled points that belong to clusters C^+ and $C^\times$ and to a cell that is different from cells containing clusters C^+ and $C^\times$. The algorithm moves these points to the cluster whose center generates the cell containing these points. It is worth noting that the algorithm only moves these points to another cluster, i.e., the algorithm only changes the symbol representing the points. After this step, the new centers are calculated, and a new Voronoi diagram is built. Figure 3.8(b) shows an optimal partition. In this case, the Voronoi cells coincide with clusters and provide the optimal partition.

```
randomly assign each sample to one of the k clusters S(j), 1 ≤ j ≤ k
compute c(j) for each cluster S(j)
while (clusters are not stable)
    build the Voronoi diagram of the set of centers c(j)
    for each sample Sample(i)
        locate the cell Sample(i) is contained in
        assign Sample(i) to the cluster whose center generates such cell
        recompute the centers of the changed clusters
    end for
end while
```

Fig. 3.7 The k-means algorithm presented in terms of Voronoi diagram.

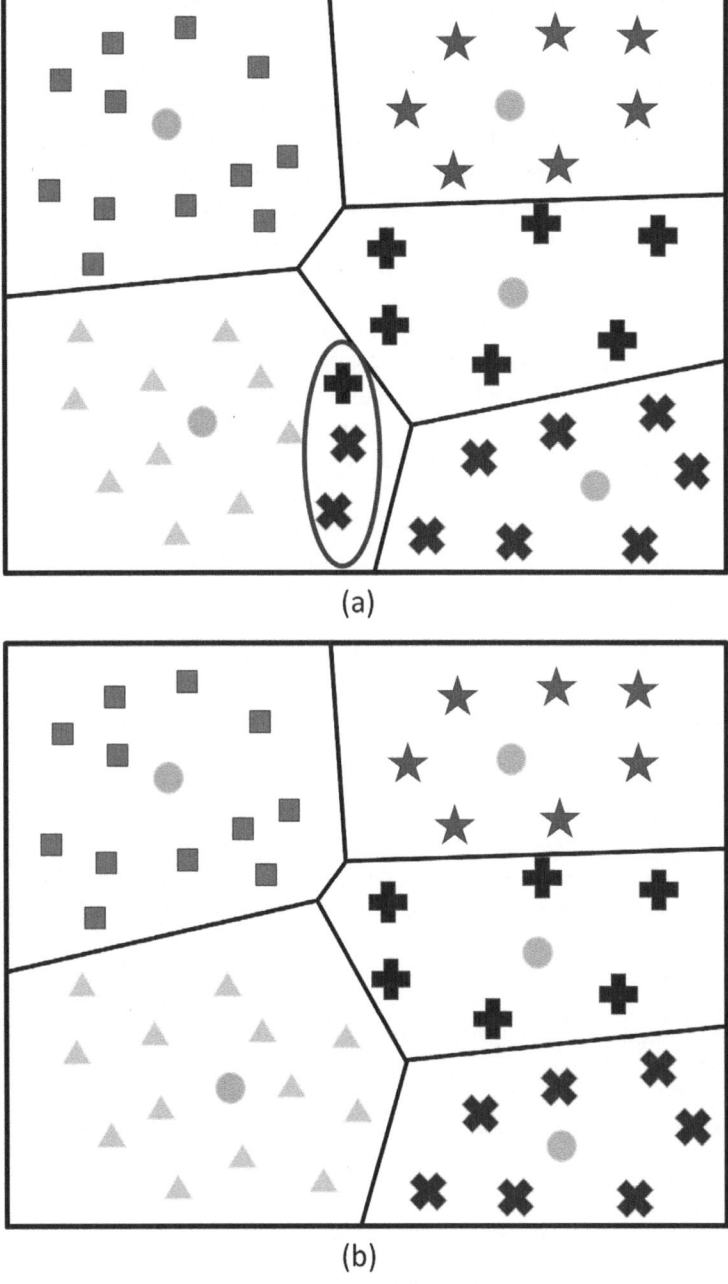

Fig. 3.8 Two partitions of a set of points in 5 clusters and Voronoi diagrams of the centers of the clusters: (a) clusters and cells differ; (b) clusters and cells provide the same partition.

3.2 Variants of the k-means algorithm

Over the years, many variants to the standard version of the k-means algorithm have been proposed as there are a few well-known issues with the standard algorithm. The standard algorithm may be slow to converge, it may reach a local optimal solution which is not the global one, and it may provide empty clusters. The convergence speed depends on the number of iterations needed to stabilize clusters. The computational cost is mainly due to the evaluations of the distances and to the computation of the centers of the clusters. Finally, in the k-means algorithm, the k clusters are not constrained to contain a predefined number of samples, and hence the algorithm may provide empty clusters. An empty cluster does not have any practical meaning, and therefore constraints need to be considered so that the algorithm avoids creating empty clusters. The following focuses on ideas and strategies developed with the aim of overcoming these problems. The k-means algorithm is often found in literature with other names representing various modifications to the main algorithm. In [237], an inventory of the 10 most known algorithms for data mining is presented, and none of them but the k-means algorithm is mentioned. This means that the basic algorithm is mostly used rather than its variants. However, some of the following ideas for overcoming some of the k-means limitations can be useful in particular practical cases.

In order to improve the performances of the algorithm, a simple variation of the Lloyd's algorithm is proposed by [110]. In literature, this variation of the algorithm is sometimes referred to as h-means algorithm, and sometimes as the k-means algorithm itself as the h-means algorithm is very similar to Lloyd's algorithm. Figure 3.9 shows the h-means algorithm. The only difference between k-means and h-means algorithms stands in the computation of the centers of the clusters. In the algorithm in Figure 3.2 the centers are computed into the **for** loop, whereas in the h-means algorithm they are computed just after the **for** loop. Therefore, even when a sample migrates from one cluster to another, the new centers are not recomputed. Centers are recomputed only after the **for** loop. In terms of Voronoi diagram, the algorithm changes as it is shown in Figure 3.10. Even though a very small change is applied to the standard algorithm, the h-means algorithm can provide different solutions. The optimal partition obtained depends on the random initial partition in clusters. Just like

```
randomly assign each sample to one of the k clusters S(j), 1 ≤ j ≤ k
compute c(j) for each cluster S(j)
while (clusters are not stable)
   for each sample Sample(i)
      compute the distances between Sample(i) and all the centers c(j)
      find j* such that c(j*) is the closest to Sample(i)
      assign Sample(i) to the cluster S(j*)
   end for
   recompute all the centers
end while
```

Fig. 3.9 The h-means algorithm.

```
randomly assign each sample to one of the k clusters S(j), 1 ≤ j ≤ k
compute c(j) for each cluster S(j)
while (clusters are not stable)
   build the Voronoi diagram of the set of centers c(j)
   for all the samples Sample(i)
      locate the cell Sample(i) is contained in
      assign Sample(i) to the cluster whose center generates such cell
   end for
   recompute all the centers
end while
```

Fig. 3.10 The *h*-means algorithm presented in terms of Voronoi diagram.

k-means, the *h*-means algorithm can be seen as a method for local optimization. The *h*-means algorithm improves the current partition iteration after iteration by reducing the value of the error function (3.1). After one iteration, the obtained partition can be either better than the previous one or exactly the same. The obtained values of the error function create a decreasing sequence of values and therefore the algorithm converges toward a local minimum of the function. Therefore, both *k*-means and *h*-means algorithms are usually carried out many times using different starting partitions. Different partitions in clusters can be randomly generated and the algorithm can be carried out as many times. In general, the algorithm can provide a different solution for each run. The greater is the number of executions of the algorithm, the greater are chances to find the global optimal partition. This procedure can be used for both *k*-means and *h*-means algorithms independently.

Moreover, the two algorithms can be used together. The *h*-means algorithm is faster than the *k*-means algorithm, but the latter has better chances to obtain optimal solutions. Therefore, the two algorithms can be used together: the *h*-means algorithm can be used to obtain a partition close to the optimal one, and then, the *k*-means algorithm can be used to locate an optimal solution. The two-phase algorithm is often referred to as *hk*-means algorithm.

Both the *k*-means and *h*-means algorithms need that a predetermined value *k* is decided before any of the algorithms is executed. The *k* value is the number of clusters in which the data are partitioned. In some applications, this number is unknown. Different *k* values can be used and the one providing a partition with the minimum error is retained. The choice of *k* plays an important role in the success of the algorithm. In some cases, indeed, the *k*-means and the *h*-means algorithms may provide a final partition with one or more empty clusters. This situation is to be avoided, since the *k* value represents the number of clusters expected in the partition. Empty clusters have no practical meanings. The *k*-means+ and the *h*-means+ algorithms use a particular strategy (described in [219]) for avoiding that the found optimal partitions include empty clusters. The strategy works as follows. Both *k*-means or *h*-means can be carried out until their halting criteria are satisfied. Then, the obtained partition can be checked for the presence of empty clusters. If *t* clusters are empty, then all samples are considered and *t* samples with the greatest distance from their respective centers are selected and each of them is moved in one of the empty clusters. In this way, the new partition has t clusters with only one sample. At this point, the k-means or the h-means algorithm can restart from this new partition and halt when the stopping criteria is satisfied. This procedure can be iterated until a partition having only non-empty clusters is obtained. Figure 3.11(a) shows that an optimal solution for $k = 4$ is obtained and the optimal solution contains an empty cluster. This figure shows three cells of the Voronoi diagram, each cell coinciding with a cluster of the optimal partition. Clusters of the optimal partition are $C^\star$, $C^\times$ and C^+. The encircled point in cluster $C^\times$ (which has the greatest distance from the center of cluster $C^\times$) is considered to move to the empty cluster and a new cell is therefore created. The newly created cluster contains only one sample. The new partition just created, as shown in Figure 3.11(b), is then used by the k-means algorithm as the initial partition and a new optimal solution without empty clusters is obtained. Figure 3.12 shows the k-means+ algorithm, while Figure 3.13 shows the h-means+ algorithm. In the algorithms, a **repeat...until** loop is iterated until an optimal partition including only non-empty clusters is obtained.

In [101] another variant of the k-means algorithm is presented, referred to as J-means algorithm. In the cases when k is large, some of the centers of the clusters may coincide with or be very close to some of the samples. When a cluster contains one sample only, then its center corresponds to the sample. Generally a cluster has more than one sample, and its center can be very close to one or more of its samples. All the samples in the same clusters are similar to their common center. Moreover, if a threshold distance or positive tolerance *tol* is set up, then samples with distance from centers smaller than *tol* can be considered as *very* similar. Only few samples around the center satisfy this rule, and, in the J-means algorithm, these samples are referred to as *occupied* samples. The basic idea behind this algorithm is to *jump* from a partition to another by selecting as center of a cluster an *unoccupied* sample. At each iteration of the algorithm, a new cluster is added to the partition whose center is an unoccupied sample. When a new cluster is added to the partition, another cluster is deleted in order to keep the k value constant. Therefore, the unoccupied sample defining the new cluster and the old cluster to delete are chosen so that the error function (3.1) decreases as much as possible. The J-means algorithm is able to reduce the error function value at each iteration. When the algorithm halts, an optimal partition is reached. Hybrid algorithms can be developed using the k-means(+), h-means(+) and J-means algorithms. For instance, the partition obtained at each step of the J-means algorithm can be improved by applying one iteration of the k-means or h-means algorithm.

As mentioned before, the k parameter needs to have a value before the k-means(+), h-means(+) or J-means algorithm can be carried out. Sometimes the k value can be easily obtained from the real-life application at hand. Some other times more than one value may be suitable for the parameter k. In these cases, the algorithms can be carried out more than once and the value providing the best partition can be selected for k.

Another variant of the k-means is the Y-means algorithm, designed for cases when no information on k is available. The k value is defined during the execution of the algorithm. k can range from 1 to the total number of samples. During the execution

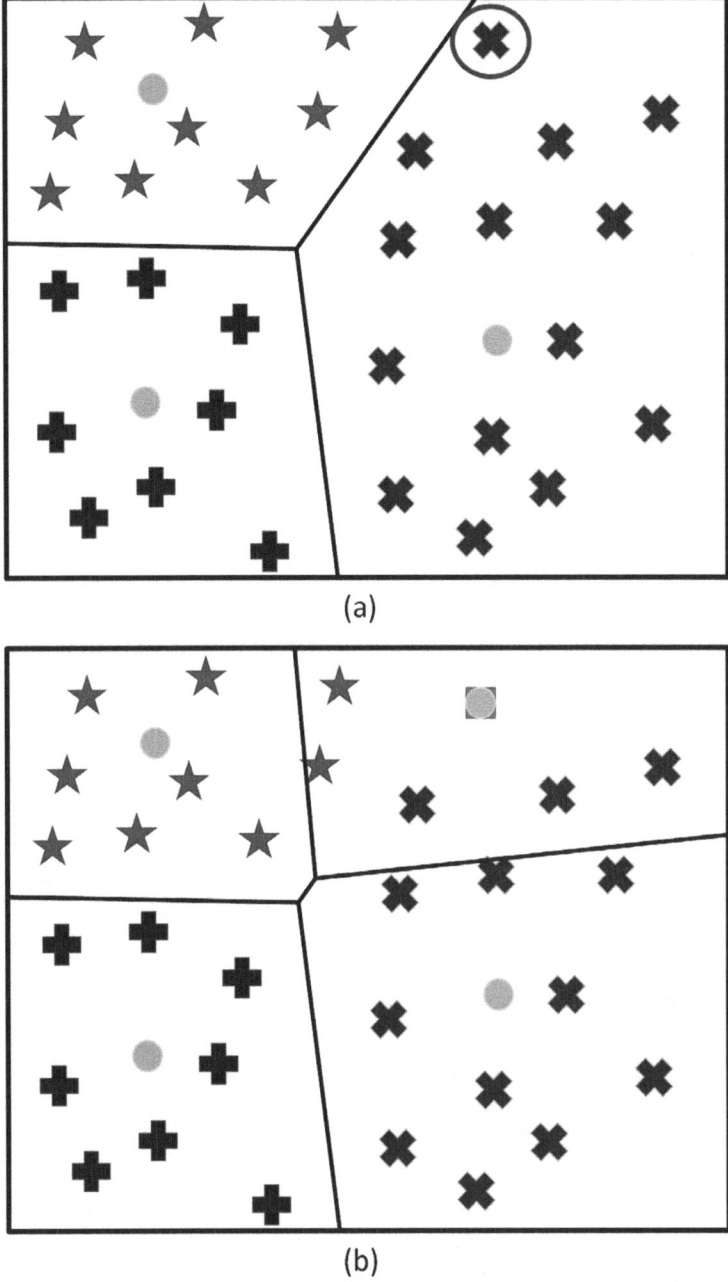

(a)

(b)

Fig. 3.11 (a) A partition in 4 clusters in which one cluster is empty (and therefore there is no cell for representing it); (b) a new cluster is generated as the algorithm in Figure 3.12 describes.

```
randomly assign each sample to one of the k clusters S(j), 1 ≤ j ≤ k
repeat
   if (some of the clusters S(j) is empty)
       compute the number t of empty clusters
       find the t samples farther from their centers
       for each of these t samples
           move the sample to an empty cluster
       end for
   end if
   compute c(j) for each cluster S(j)
   while (clusters are not stable)
       for each sample Sample(i)
           compute the distances between Sample(i) and all the centers c(j)
           find j* such that c(j*) is the closest to Sample(i)
           assign Sample(i) to the cluster S(j*)
           recompute the centers of the changed clusters
       end for
   end while
until (all the clusters are non-empty)
```

Fig. 3.12 The k-means+ algorithm.

of the algorithm, clusters are deleted and other clusters are added to the current partition, until an optimal partition is obtained. The algorithm searches, for instance, for empty clusters. If there are empty clusters, they are deleted. The algorithm also searches for outliers, i.e., for samples which are different from the majority of the samples in the same cluster. If outliers are detected, they are removed from their clusters and used for generating new clusters. This operation splits one cluster in two

```
randomly assign each sample to one of the k clusters S(j), 1 ≤ j ≤ k
repeat
   if (some of the clusters S(j) is empty)
       compute the number t of empty clusters
       find the t samples farther from their centers
       for each of these t samples
           move the sample to an empty cluster
       end for
   end if
   compute c(j) for each cluster S(j)
   while (clusters are not stable)
       for each sample Sample(i)
           compute the distances between Sample(i) and all the centers c(j)
           find j* such that c(j*) is the closest to Sample(i)
           assign Sample(i) to the cluster S(j*)
       end for
       recompute all the centers
   end while
until (all the clusters are non-empty)
```

Fig. 3.13 The h-means+ algorithm.

parts and therefore, in this case, the k value increases. The algorithm also looks for adjacent clusters that may overlap with each other. If such clusters are found, they are merged to form one unique cluster. This operation merges clusters: the k value is decreased. When the optimal partition is obtained at the same time k has its optimal value.

Krishna and Murty [134] combined the basic k-means algorithm with genetic algorithms (GAs) [88]. As explained in Section 1.4, GAs are meta-heuristic methods for global optimization that simulate the evolutionary process of living organisms according to Darwinian theory. The *genetic k-means algorithm* is a GA in which the crossover operator is substituted with one iteration of the k-means algorithm. At the start, an initial population of chromosomes is randomly generated. Each chromosome represents a partition in clusters of the data. As in GAs, the selection and mutation operators are used. Here, the mutation operator is defined such that the probability of performing a change on a sample is higher if the sample is closer to one of the centers. At each iteration of the algorithm, a partition in clusters is selected from the current population, a mutation is performed on the partition and one step of the k-means algorithm is performed on the whole set. The genetic k-means algorithm performs better than the basic k-means algorithm, because it couples the basic idea of the k-means with the heuristic evolutionary searches. Variations of this algorithm have been proposed in [157, 158].

Many other variants of the standard k-means algorithm can be found in the literature. One of these variants is the so-called *global k-means algorithm* [154]. This is a global optimization method which uses the k-means algorithm as a local search procedure. In [68], the k-means algorithm has been modified to avoid unnecessary distance calculations and to perform faster. The well-known triangle inequality is used in this algorithm. In [26], the performances of the k-means algorithm have been improved by refining the initial and randomly generated partition in clusters. Another variant of the basic algorithm is the *symmetry-based k-means algorithm* [50, 51, 223].

Finally, we mention a technique for clustering efficiently a feature-extended set of samples [207]. Precisely, it is supposed that a partition of a set of samples is known, and that a new partition is searched after that some features have been added for representing the samples. The technique in [207] has been applied to hierarchical clustering. However, as the authors pointed out, it takes the concept of center of a cluster from the k-means approach, and therefore it can be applied to partitioning clustering as well. The idea is to avoid to partition the set of data again after features are added to the samples, but rather to exploit the previous partition in clusters. The easiest strategy could brutally divide the samples in the clusters as they were in the previous partition. However, the introduction of new features for representing these samples may change the clustering, and samples can migrate from a cluster to another. In [207], a rule based on the centers of the clusters has been proposed for removing samples from the clusters where they should not belong anymore. The new samples can then be used for generating new clusters. In the agglomerative hierarchical approach, the new set of clusters are successively merged and the samples are assigned to the correct cluster. In the k-means approach, the removed samples could be assigned to the least populated cluster or equally distributed to all the clusters in a random way. The obtained partition would anyway be better than a random partition, and the k-means algorithm would reach another optimal partition faster.

In this section, many variants on the standard k-means algorithm have been presented. As discussed above, they are able to overcome some of the issues arising when the k-means approach is used. However, there are still other problems that may arise when this approach is used. First of all, the basic idea of the method is to use the centers for representing such clusters. The centers are computed as the mean among all the samples in the same cluster. Unfortunately, a mean is not a good representative of a set of samples if there are outliers. Indeed, the presence of one outlier can modify the center of a cluster, and it can become closer to a certain subgroup of samples. If this happens, and the k-means algorithm is executed, this subgroup of samples is then moved in the cluster having such center. The partition in clusters can therefore drastically change if outliers are contained in the considered set of data. For avoiding this problem, outliers have to be removed prior the application of the algorithm.

In some cases, for instance when the parameter k is not known, the quality of a partition is evaluated through the value of the error function (3.1). With fixed k value, the better partitions correspond to the smallest values of the error function. It is much more difficult to compare instead the error function values in correspondence with partitions in which the k value changes. Indeed, the error values tend to decrease when k is larger. When only one cluster is considered, the error is the sum of all the distances between the samples and the only center. Intuitively, if two clusters are considered, then the average distances are smaller in general, and many non-optimal partitions in clusters could have an error function value which is smaller than the one corresponding to the partition in one cluster. The extreme case is the one in which the number of clusters equals the number of samples. In such a case, there is only one possible partition and the value of the error function is zero. This tendency of the error value to decrease when k is increased makes it difficult to find out if a partition in a larger number of clusters is better than any other with fewer clusters. Indeed, a reduction on the error function value might be due to the increase of the k value only, and not because the quality of the partition is higher.

3.3 Vector quantization

Data compression represents an important field in informatics. Large sets of data are usually stored in single *files* on computer memories. Each file can represent a text, a sound, a movie. The memory of the computer is limited, therefore it needs to be managed efficiently. If the data files are compressed, less memory space is needed. Thus, data compression allows saving this memory space, and it also allows exchanging files over the Internet faster. Great interest is currently given to methods for compressing images, sounds and movies, which are the kind of files mostly used on the Internet [57, 201, 218, 234]. For instance, a good compression of images can be obtained by using the well-known JPEG format. Music is currently exchanged on the Internet through MP3 files, which can have only 10% of the size of the corresponding standard WAV file. Movies in MPEG format are sold on standard DVDs where up to 9 GB of data can be stored. The same movies are also exchangeable over the Internet in DVIX format. The DVIX format is very efficient and an entire movie requires less than 1 GB of space. When compressing data, some information can be lost, and their quality can decrease. For instance, a movie in DIVX format in general has a lesser quality than the same movie in MPEG format.

Vector quantization is a method for data compression [92]. It is based on the same idea as the k-means algorithm. In the k-means approach, a set of data is partitioned in k clusters where similar samples are contained. The idea behind vector quantization is that the representative of each cluster, i.e., the center of each cluster, can be a good approximation of each sample in the same cluster. In fact, all the samples in a cluster are similar and the center is similar to all the samples in the cluster. In order to compress a given set of data, representing a certain data file, all the samples belonging to the same cluster can be substituted by the center of the cluster. Therefore, only k different samples are contained in the set of data. These samples can be substituted by a numeric label referring to the k centers. For instance, 0 can refer to the center of the first cluster, 1 to the center of the second cluster, and so on. These labels can have values from 0 to $k - 1$ and they can be efficiently stored on a computer memory. Indeed, if the data can be partitioned into 2 clusters, then the label can have either value 0 or 1. In this simplified case only 2 possibilities are allowed, and only one bit is sufficient for storing this information. If the clusters were 4, the labels would be 4, and 2 bits would be needed. In general, $\log_2 k$ bits are needed for storing a label when the data are partitioned in k clusters. The bits needed for a label are less than the ones for storing a sample. For this reason, substituting a sample with the label associated to the center of the cluster containing the sample actually compresses the data.

The k-means algorithm, or the vector quantization algorithm, can hence be used as an encoding algorithm for data compression. As in many applications, the number k of clusters is usually not known a priori. Several k values could be used and the one providing the smallest value of error should be chosen. In this case, small k values should be tried before. Indeed, the aim here is to compress data, and the smaller k is, the more important the compression rate is. Very small k values can provide a very good compression rate, but the error function (3.1) value in the optimal partition may be high, meaning that a lot of information is lost during the compression. Therefore, small k values can be tried at the start, and the k value can be increased until an acceptable value for the error function is obtained. The error function (3.1) provides indeed the total error occurring when all the samples of the set are substituted by the centers of the clusters. The smaller the value of error is, the less information is lost during compression. Once the data are compressed and stored as a sequence of labels, they need to be decompressed to be used. The decoding algorithm associated to the vector quantization is simply the algorithm which associates a sample (the old centers in the encoding algorithm) with each label. The decoding process aims at restoring the original data. However, not all the information comes back as before. All samples previously contained in one cluster are all represented by the center. The error function (3.1) sums all the distances between the samples and the corresponding center, and hence it provides the encoding/decoding distortion of a set of data, or data file.

3.4 Fuzzy c-means clustering

In the k-means algorithm and in all its variants presented in Section 3.2, each sample can belong to only one cluster. In this section we will analyze another variation of the k-means algorithm in which samples can belong to more than one cluster. As mentioned before, none of the variants on the k-means algorithm is in the list of the top 10 algorithms for data mining [237]. However, in some application, the following ideas might be helpful, and therefore we decided to present them in this section.

The standard k-means algorithm and its variants discussed in Section 3.2 perform a *crisp* partition of a certain set of samples. The term "crisp" is used here to indicate that each sample can belong to one and only one cluster per time. *Fuzzy* clustering instead refers to methods and algorithms for partitioning data where a single sample can belong to more than one cluster. In this case, a sample is assigned to a certain cluster with a certain "membership." The *membership* of a sample indicates the degree to which the sample belongs to different clusters. If a sample is assigned to a cluster with a full membership, then it belongs to that cluster only. However, one sample may belong, for instance, to both clusters C_1 and C_2, simultaneously. The cluster C_1 might be more representative of the sample than the cluster C_2. This is considered by giving a different membership to the sample when it is considered as belonging to C_1 and C_2. A larger membership of a sample when referring to a certain cluster corresponds to a better representability of the sample in the cluster.

In the case of fuzzy clustering, the error function (3.1) becomes:

$$f(U, V; X) = \sum_{j=1}^{c} \sum_{i=1}^{n} (u_{ji})^m d(v_j - x_{j(i)}), \qquad (3.2)$$

where

$$U \equiv (u_{ji}), \quad \forall j = 1, \ldots, c; \quad \forall i = 1, \ldots, n$$

is the fuzzy partition matrix, and

$$V \equiv (v_j), \quad \forall j = 1, \ldots, c$$

identifies all the centers of the clusters. The parameter $m \in [1, \infty)$ controls the fuzziness of the membership values. In the formulas, n represents the number of data samples and c is the number of clusters in which the data are partitioned. As in the crisp case, the generic cluster is denoted by the symbol S_j, and $n(S_j)$ refers to the number of samples assigned to the generic cluster. In the case of fuzzy clustering, the sum

$$\sum_{j=1}^{c} n(S_j)$$

can be greater than the total number n of samples. If the sum equals n, then the partition is actually a crisp partition.

The general element u_{ji} of the fuzzy partition matrix U belongs to the interval $[0, 1]$ and represents the membership degree of the sample $x_{j(i)}$ in the cluster S_j. The matrix U is constrained such that

$$\sum_{j=1}^{c} u_{ji} = 1, \quad \forall i = 1, 2, \ldots, n,$$

and

$$0 < \sum_{i=1}^{n} u_{ji} < n, \quad \forall j = 1, 2, \ldots, c.$$

The first constraint indicates that the sum of the elements on the same column of matrix U must be equal to 1. On the same column of U there are elements pertaining to the same sample $x_{j(i)}$, in particular there is the membership of the same sample on the different clusters. The constraint forces therefore the sum of all the memberships pertaining to the same sample to be equal to 1. In this way, the memberships of the same sample can be seen in terms of percentages. The second constraint forces instead the sum of the elements on the same row of matrix U to be in the open interval $(0, n)$, where n is the total number of samples. On the same row there are elements pertaining to the same cluster S_j, and the constraint indicates that the sum of all the membership values in the same cluster cannot be less than or equal to 0, or greater or equal to n. Indeed, if it were 0, the cluster S_j would be empty, and, as previously mentioned, empty clusters need to be avoided. Moreover, if it were n, the cluster S_j would have all the samples with full membership. This must be avoided as well, since the partition would correspond to one cluster only containing the whole data. Sum values smaller than 0 or greater than n make no sense.

A fuzzy partition matrix related to a crisp clustering would be like

$$\begin{pmatrix} 1 & 1 & 1 & 0 & 0 & 1 & 1 & 1 & 0 & 0 & 0 & 0 & 0 \\ 0 & 0 & 0 & 1 & 1 & 0 & 0 & 0 & 1 & 1 & 1 & 1 & 1 \end{pmatrix}.$$

This matrix satisfies the constraints showed above, and, in particular, the sum on each column is always 1 because one of its elements is 1 while the other one is 0. This means that each sample has a full membership in the first class or the full membership in the second one. A hypothetical matrix

$$\begin{pmatrix} 1 & 1 & 1 & 1/4 & 0 & 0 & 1 & 3/4 & 1/4 & 0 & 0 & 0 & 0 \\ 0 & 0 & 0 & 3/4 & 1 & 1 & 0 & 1/4 & 3/4 & 1 & 1 & 1 & 1 \end{pmatrix}$$

corresponds to a fuzzy partition, where three points belong to two clusters, but with different degrees of membership.

The most popular and effective method for fuzzy clustering alternates between the optimization of function (3.2) over U with fixed V and over V with fixed U. At each step, the new centers are computed by

$$
v_j = \frac{\sum\limits_{i=1}^{n} \left(u_{ji}\right)^m x_{j(i)}}{\sum\limits_{i=1}^{n} \left(u_{ji}\right)^m}
$$

by using the U values obtained during the previous step. Then, the fuzzy matrix U is updated by using the V values just computed:

$$
u_{ji} = \left[\sum_{p=1}^{c} \left(\frac{||x_{j(i)} - v_j||^2}{||x_{j(i)} - v_p||^2} \right)^{\frac{2}{m-1}} \right]^{-1} .
$$

The fuzzy c-means algorithm stops when the difference in norm between two consecutive matrices V is less than a certain tolerance $\epsilon > 0$.

As for the k-means algorithms, fuzzy c-means clustering algorithm is an optimization algorithm. Its convergence depends heavily on the choice of the initial values such as the starting partition in clusters and the starting degrees of membership. As discussed before, the k-means algorithm and its variants should be carried out several times using different initial parameters. The best solution obtained after a certain number of trials can then be chosen. However, choosing good initial parameters is not an easy task, in the case of fuzzy clustering. In [212], for instance, an algorithm that can automatically and adaptively select these parameters with optimal values is proposed.

The fuzzy algorithm can be very sensitive to noise and to outliers. For overcoming these two problems, Hathaway et al. [104] tried to use L_p norms for computing distances between samples and between a sample and the centers of the clusters. After bench-marking the algorithm using different p values, they concluded that the best results can be obtained for $p = 1$ or $p = 2$, and that $p = 1$ should be chosen when the data are affected by noise and if there are outliers. The weighting exponent m also plays a crucial rule, and it has been studied in [181].

The features used for representing the data are usually expressed by numerical values. Such values can be sometimes not completely known, or, in other words, some of the data can be incomplete. The fuzzy algorithm is able to deal with partition problems where the set of data is incomplete. In [105], for instance, different strategies have been considered. If the proportion of incomplete data is small, then it may be useful to simply delete all the incomplete data and apply the fuzzy algorithm to the remaining "complete" data. This strategy allows working on the known data, but it does not provide any information on the missing ones. As an alternative strategy, the fuzzy algorithm can be applied using a distance which does not consider the missing data explicitly, but they are considered implicitly. Another way to take them into account is to consider the missing data as additional variables. These variables are optimized during the execution of the fuzzy algorithm in order to obtain the smallest possible value of the objective function (3.2). Numerical results suggest that although the simple approach of deleting incomplete data works fine for small percentages of missing data (less than 20%), the other approaches usually perform better if a larger proportion of data are incomplete. For more details, refer to [105].

3.5 Applications

The k-means algorithm and all its variants, including the ones discussed in Section 3.2 and the fuzzy approach presented in Section 3.4, have been applied to a wide variety of real-life problems. The k-means and fuzzy c-means algorithms are for instance used for analyzing and categorizing gene expression data [9, 80, 245], in order to analyze and presume the function of unknown genes. The k-means algorithm has been applied to solve the problem of segmenting images with smooth surfaces [182]; the genetic k-means algorithm has been applied for compressing images [133]; the Y-means algorithm has been developed for monitoring intrusions in computer systems [94]; the fuzzy c-means algorithm has been applied for detecting crime hot-spots or geographic areas of elevated criminal activity [93].

One of the classic applications of the k-means algorithm is text mining [61, 251]. Text mining generally refers to the process of deriving high-quality information from texts. Nowadays, there is a growing amount of text documents, and many of them are also available on the Internet. Text categorization is the text mining process aimed at the classification of a set of text documents with a certain criterion. If the criterion refers to the document topic, then text categorization techniques try to classify the documents by their topic.

Let us suppose that documents related to agriculture and computer science need to be categorized. The aim is to partition the documents in two clusters, one containing only documents related to agriculture, and the other one containing only computer science-related documents. Let us suppose that the topic of interest can be recognized by words used in the text of the document. For instance, the word *agriculture* can be used as a criterion for grouping documents in the first cluster, and two words *computer* and *science* together can be used for grouping documents in the second cluster. The k-means algorithm can be applied to this clustering problem. However, the standard Euclidean distance cannot be used in this application, because the text documents are not points in a Euclidean space. Therefore, another kind of distance needs to be used. It is defined as follows. Let T_1 and T_2 be two text documents. If the word *agriculture* is used as criterion, the similarity between T_1 and T_2 can be measured as the difference between the number of recurrences of this word in T_1 and in T_2. If the word *agriculture* occurs 5 times in T_1 and 50 times in T_2, then $d(T_1, T_2) = 45$.

A distance function defined in this way is not very meaningful. Indeed, in the previous example, the distance $d(T_1, T_2)$ is 45: it is quite far from 0 and hence T_1 and T_2 are different. This may be true, if T_1 is a computer science-related document, and there is a small part of the document that deals with agriculture-related matters, while T_2 is agriculture-related. However, this distance value does not preclude the possibility that T_1 and T_2 are both on agriculture. In such a case, the two documents are similar and their relative distance should be smaller. For instance, T_1 might be a short text: shorter texts have fewer words in general, and in particular they may contain less occurrences of the word *agriculture*. For this reason, this distance is not a good measure of text similarities.

In general in text mining, the *cosine similarity* function is used. The samples are normalized for overcoming the problem discussed above. The distance function consists of the inner product between two vectors representing two samples. Since the vectors are normalized, the inner product corresponds to the cosine of the angle between them. If the samples are similar, the angle between the vectors is small and then the cosine is close to 1. Inversely, the more different samples are, the wider is the angle, and the smaller is the cosine value. The k-means algorithm applied to text mining by using the cosine similarity function is also referred to as *spherical k*-means.

In the field of agriculture, the k-means algorithm has been applied, for instance, for

- Forecasting pollution in the atmosphere [123];
- Soil classifications using GPS-based technologies [233];
- Classification of plant, soil, and residue regions of interest by color images [165];
- Predicting wine fermentation problems [230];
- Grading apples before marketing [146];
- Monitoring water quality changes [132];
- Detecting weeds in precision agriculture [225].

In the next sections, two applications in agriculture are discussed in detail. The problem of predicting the fermentation process of wine and classifying it as good or bad is presented in Section 3.5.1. The problem of classifying apples on the basis of their grade is discussed in Section 3.5.2.

3.5.1 Prediction of wine fermentation problem

Problems occurring during the fermentation process of wine can impact the productivity of wine-related industries and also the quality of wine. The fermentation process of wine can be too slow or it can even become stagnant. Predicting how good the fermentation process is going to be may help enologists (wine specialists) who can then take suitable steps to make corrections when necessary and to ensure that the fermentation process concludes smoothly and successfully. In order to monitor the wine fermentation process, metabolites such as glucose, fructose, organic acids, glycerol and ethanol can be measured, and the data obtained during the entire fermentation process can be analyzed in order to obtain useful information [229]. Data mining techniques can help extract this information from large databases, which may be able to predict the fermentation process. In the work which is the focus of this section, a k-means algorithm has been applied for exploring data accumulated from measurements sampled regularly of 24 industrial vinifications of *cabernet sauvignon* [230]. Data measured during the first three days of fermentation has been compared to those obtained during the whole fermentation process. Information on the behavior of the fermentation during the first three days can provide important clues about the final classification.

The data come from a winery in Chile's Maipo Valley, and they are related to the 2002 harvest. Between 30 and 35 samples are taken per fermentation depending on the duration of a vinification. The levels of 29 compounds are analyzed. Among them, sugars are analyzed, such as glucose and fructose, organic acids, such as the lactic and citric acids, nitrogen sources, such as alanine, arginine, leucine, etc., and alcohols. The whole set of data consists in approximately 22,000 data points. The used compounds are actually 28, since taking glucose and fructose as a single variable (sugar) is the same as considering the two sugars as independent variables. Four sets of data are defined in order to perform the analysis. Datasets A and B just consider 8 variables, including "sugar," alcohols and organic acids, whereas datasets E and F include all 28 components. The data contained in datasets A and E are related to the first three days of fermentation, whereas datasets B and F are related to data measured during the whole fermentation process. Figure 3.14 shows a graphic representation of the considered databases.

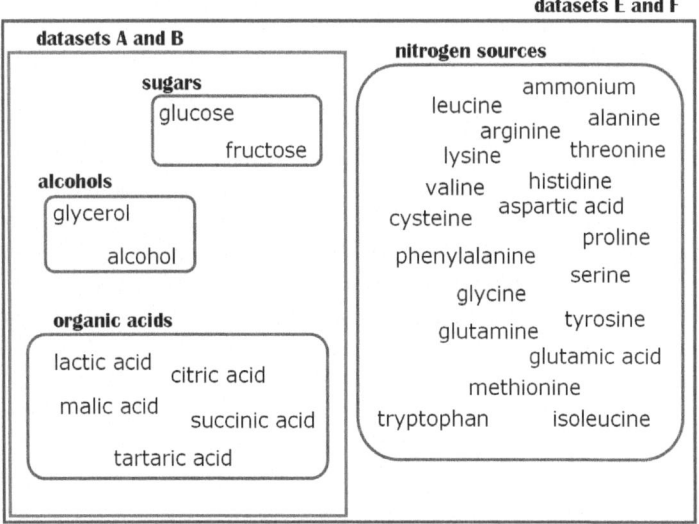

Fig. 3.14 A graphic representation of the compounds considered in datasets A, B, E and F. A and E are related to data measured within the three days that the fermentation started; B and F are related to data measured during the whole fermentation process.

These datasets have been reduced by applying a principal component analysis (PCA) before the k-means algorithm is applied [163]. PCA is able to reduce the dimension of a set of data, as discussed in Section 2.1. The k-means algorithm has then been applied with the aim of classifying fermentations using data from the first three days.

To establish if it is possible to classify fermentations early, results from applying k-means to samples from the first three days, datasets A and E, are compared with those in datasets B and F, where the whole set of data is contained. A and B showed similar cluster patterns that are essentially sugar concentration-linked. Additionally, around 80% of fermentations with datasets E and F are clustered similarly. Consequently, these classification results show that information contained in data taken during the first three days of fermentation (datasets A and E) are sufficient to classify fermentations early. For this reason, the following analysis considers datasets A and E only.

In these studies, the samples of datasets A and E are partitioned into 5 clusters, arbitrarily named as five colors: the blue (B), red (R), pink (P), brown (Br) and green (G) clusters. The k-means algorithm is applied to classify the samples, by using $k = 5$ and considering the data related to a certain time t smaller than 3 days. Due in large part to the time-variable nature of the fermentation process, the algorithm often partitions the data in different ways for different times t. Some of the fermentations are then assigned to more than one cluster for different t. Twenty-four fermentations are considered and there are 15 of them with fermentation problems, due to slow fermentation processes or to processes getting stuck. When the dataset A is used, only one fermentation process is always assigned to the same cluster, whereas all the others are assigned to two or three different clusters. When the dataset E is instead used, three fermentations are always found in the same cluster, whereas all the others are assigned to two or three clusters. The 5 clusters are then grouped, in order to put in evidence the properties of the fermentations belonging to more than one cluster.

Groups of clusters containing from one to three clusters have been obtained. On the base of the fermentation processes found in them, a percentage of problem fermentation is assigned to each of them. For instance, when using dataset A, three fermentation processes are assigned to the group containing the red (R) and pink (P) clusters. Two of them are good fermentation processes, while the third one is related to a sluggish and stuck fermentation. For this reason, each fermentation process classified in this same group has the 33% possibility to be bad, and the 67% probability to be good. Other groups just contain good or bad fermentation processes, and hence any other process found in the same group should be 100% good or 100% bad. Figure 3.15 shows more details about the classification of the fermentations in clusters and groups, and therefore it also shows how another unknown fermentation can be considered on the basis of these classifications. In these studies, the bad fermentations are the ones in which there is a high residual sugar content, which will probably not finish properly, and the ones that take more than 13 days. Among the fermentation processes used in the analysis, 9 of them are good, 10 are sluggish, because they require more than 13 days, 3 of them get stuck, because the final sugar content is too high, and finally 2 of them are both sluggish and stuck.

Grouping	Clusters			Good ferm.	Bad ferm.	% problem
1	B			0	1	100
2	R	P		2	1	33
3	B	Br		1	0	0
4	R	Br		0	2	100
5	G	Br		0	2	100
6	P	Br		0	1	100
7	B	R		0	2	100
8	B	R	P	4	2	33
9	R	G	Br	1	1	50
10	R	P	G	1	0	0
11	R	P	Br	0	1	100
12	B	P	Br	0	1	100
13	B	R	Br	0	1	100

Fig. 3.15 Classification of wine fermentations by using the k-means algorithm with $k = 5$ and by grouping the clusters in 13 groups. In this analysis the dataset A is used.

The previous results have been obtained by using the dataset A. When the dataset E is instead used, the nitrogen compounds are also considered. Nitrogen deficiency is widely reported to be an important factor in problem wine-making fermentations. The clustering process in which nitrogen compounds are also included produced 12 groupings. Five groupings contain just problem fermentations, other five groupings contain only normal fermentations, and the remaining two groups only provide a percentage for the fermentation to be good or bad. It seems that dataset E does not provide any additional information.

3.5.2 Grading method of apples

Machine vision offers a great potential to extract and identify target features, based on color, shape, etc., of fruits, soil, etc. Fresh market fruits like apples are graded into quality categories according to their size, color and shape and the presence of defects. This process can be performed by humans, but it is expensive, repetitive and therefore it cannot be considered reliable. For this reason, the interest in creating machines able to classify fruits on the basis of their grading has created interest in the research community. These machines are able to acquire images of the fruit, analyze and interpret images, and finally classify the fruit. The main issue that needs to be addressed is to find a reliable way to identify the fruit defects.

In [146], a real-time grading method for classifying apples is proposed. The first step consists in acquiring images of the surface of the apples. In order to successfully grade the fruits, two requirements must be addressed: the images must cover the whole surface of the fruit, and a high contrast must be created between the defects and the healthy tissue. There are machines able to take pictures of the fruits while they are passing through them. Usually, fruits are placed on rollers which make the apples rotate on themselves and the pictures are taken from a camera located above. In this case, the parts of the fruit close to the points where the rotation axis crosses its surface may not be observed. Hence, if some defect is there, it may not be identified, but this problem can be overcome by placing mirrors on each side of the rollers. More complex systems have also been developed, in which fruits are free to move on ropes while three cameras take pictures from different places, or where robot arms are used to manipulate the fruit. The system which uses robot arms was able to observe 80% of the fruit surface with four images, but it is quite slow, since it takes about 1 second for analyzing 4 fruits. Another important issue is the lighting system used. Commonly the images are monochrome images, but they can also be color images.

After the image (or images) has been acquired from an apple, the segmentation process must be applied. The result of an image segmentation is the division of such image in many regions, related for instance to different gray levels, that represent the background, the healthy tissue of the fruit, the calyx, the stem and possible defects. The contrast between the fruit and the background should be high to simplify the localization of the apple, even though calyx, stem ends and defects may have the same color of the image background. The hard task is how to separate the defects from the healthy tissue, the calyx and the stem. On monochrome images, the apple appears in light gray, the mean luminance of the fruit varies with its color and decreases from the center of the fruit to the boundaries [241, 243]. Defects are usually darker than the other regions, but their size and their shape can vary strongly.

Supervised or unsupervised techniques can be used to segment the obtained images. As it has been pointed out in Chapter 1, supervised techniques tend to reproduce a pre-existent classification or segmentation, whereas unsupervised techniques produce a segmentation on their own. For instance, in [177], neural networks have been used (see Chapter 5) for classifying pixels into six classes including a class representing the fruit defect. The work which is the focus of this section is instead based on a k-means approach, which is an unsupervised technique, since it is able to partition the data without having any previous knowledge about them.

This approach is different from the others because it manages several images representing the whole surface of the apple at the same time. In previous works, indeed, each image taken from the same fruit was treated separately and the fruit was classified according to the worst result of the set of representative images. The method discussed here combines instead the data extracted from the different images of a fruit moving on a machine in order to dispose information related to the whole surface of the fruit. The method is applied on Jonagold apples characterized by green (ground color) and red (blush) colors.

Images representing different regions of the fruit are analyzed and segmented as described in [144]. The regions issued from the segmentation process including the defects, over-segmentation and calyx and stem ends are called *blobs*. These regions are characterized by using color (or gray scale), position, shape and texture features. In total, 15 parameters are considered for characterizing a blob, five for the color, four for the shape, five for the texture and only one for the position. The k-means algorithm, the blob and fruit discriminant analysis are made off-line by the program, whereas the blobs and afterwards the fruits can be graded *in-line* by using the parameters of the discriminant analysis [145].

Once the clusters have been defined, apples are classified with a global correct classification rate of 73%. These results have been obtained by using a set containing 100 apples, i.e., 100 apples have been partitioned for obtaining the set of clusters successively used for classifying other unknown apples.

3.6 Experiments in MATLAB

In this section we will present some programs written in MATLAB for performing some of the algorithms we discussed in this chapter. In Appendix A there is a description of the MATLAB environment and of its potentialities. The k-means algorithm will be carried out on a set of randomly generated samples. We will also write a MATLAB function for visualizing the clusters that the k-means algorithm can locate in the random set. After the presentation of each code, we will discuss it in a very detailed way, in order to give to the reader the possibility to work and modify such codes for his personal purposes. Initially, simple examples will be introduced, but they can anyway show the difference between the theory and the practical work of a programmer. Interested readers can find exercises at the end of this chapter.

In order to apply the k-means algorithm to a set of data for partitioning it in clusters, a MATLAB function which generates such set of data is needed. Figure 3.16 shows a short code for generating points in a two-dimensional space. The function generate has two input parameters. The first one is the number of samples n. The second one is a real variable eps which can be used for separating the samples with a certain margin. In practice, the algorithm generates about 50% of the samples having a negative x value, and about 50% of the samples with a positive x value. If eps is greater than 0, then all the samples will have at least distance eps from the y axis and the double of eps from any other sample on the other side. The output parameters of this function are x and y, which will contain, respectively, the x and y coordinates of the generated samples. The function consists of a simple for loop on the number n of samples to generate, and at each step it decides whether to generate a sample on the left or the right of line $x = 0$ by using a random mechanism. The built-in function rand in MATLAB generates a uniform random real number in the interval $(0, 1)$, and hence there is exactly 50% probability that this number is in $(0, \frac{1}{2}]$ or in $[\frac{1}{2}, 1)$. The y coordinates are generated with values in the interval $(-1, 1)$. When eps is zero, the x coordinates belong to the same interval, and it increases as eps becomes larger. In Figure 3.17 a set of randomly generated points is shown. The parameters used for generating this set of data are n = 100 and eps = 0.2. The points are then simply plotted by the MATLAB function plot. Another execution of this function would generate a different set of data, because it is based on a random number generator.

```
%
% this function generates a random sets of data
% in the two-dimensional space;
%
% input:
% n    - number of random samples to be generated
% eps - predefined margin between samples separated by the line x = 0
%
% output:
% x - x coordinates of the samples
% y - y coordinates of the samples
%
% [x,y] = generate(n,eps)

function [x,y] = generate(n,eps)

   for i = 1:n,
     random = rand();
     if random < 0.50,
       x(i) = -eps - rand();
     else
       x(i) =  eps + rand();
     end
     y(i) =  2.0*rand() - 1.0;
   end

end
```

Fig. 3.16 The MATLAB function generate.

Before starting working on the k-means algorithm, let us work on one of its sub-problems. k-means is based on the distances between the samples and the centers of the clusters. One of the tasks to be carried out during the algorithm is the computation of the new centers. This task is required many times, and precisely every time a sample migrates from a cluster to another. In Figure 3.18 the function centers for

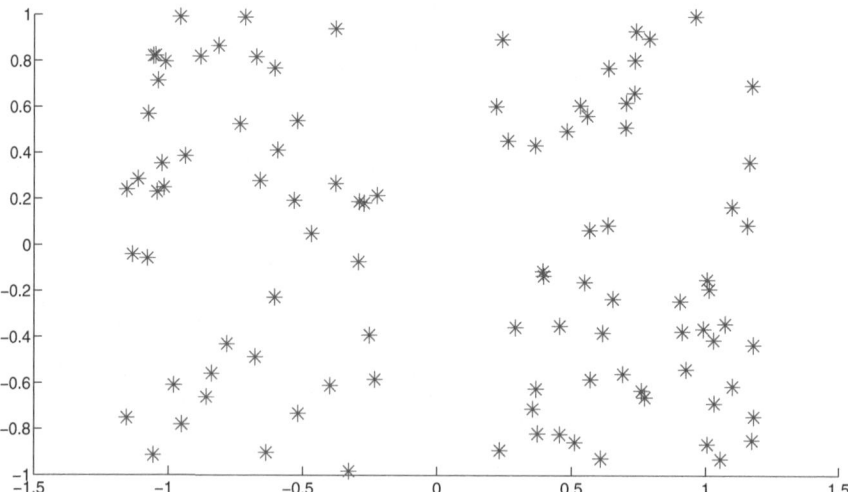

Fig. 3.17 Points generated by the MATLAB function generate.

```
%
% this function computes the centers of k classes or clusters
%
% samples (x,y) are in the two-dimensional space
%
% input:
% n      - number of samples
% x      - x coordinates of the samples
% y      - y coordinates of the samples
% k      - number of classes
% class - classes to which each sample belongs
%
% output:
% cx - x coordinates of the k centers
% cy - y coordinates of the k centers
%
% [cx,cy] = centers(n,x,y,k,class)

function [cx,cy] = centers(n,x,y,k,class)

    % initializations

    for j = 1:k,
      cn(j) = 0;
      cx(j) = 0.0;  cy(j) = 0.0;
    end

    % summing the coordinates of the points in the same class

    for i = 1:n,
      cn(class(i)) = cn(class(i)) + 1;
      cx(class(i)) = cx(class(i)) + x(i);
      cy(class(i)) = cy(class(i)) + y(i);
    end

    % computing the centers

    for j = 1:k,
      if cn(j) ~= 0,
        cx(j) = cx(j)/cn(j);  cy(j) = cy(j)/cn(j);
      else
        cx(j) = 0.0;  cy(j) = 0.0;
      end
    end

end
```

Fig. 3.18 The MATLAB function `centers`.

the computation of the centers of the clusters is presented. This function has 5 input parameters: n is the number of points contained in the set of data to be partitioned; x is the vector containing all the *x* coordinates of such points in the two-dimensional space; y contains the *y* coordinates; k is the number of clusters in which the data must be partitioned; class is a vector containing the cluster or class code of the corresponding point in x and y. If k is 2, the first class is simply coded by 1 and the second one by 2. The output of the function consists of two vectors cx and cy containing, respectively, the *x* and *y* coordinates of the centers.

First of all, the algorithm initializes the needed variables, including the ones in which the centers will be stored. cn is a vector with the same length as cx and cy in which the number of samples belonging to one class or another is counted. This is

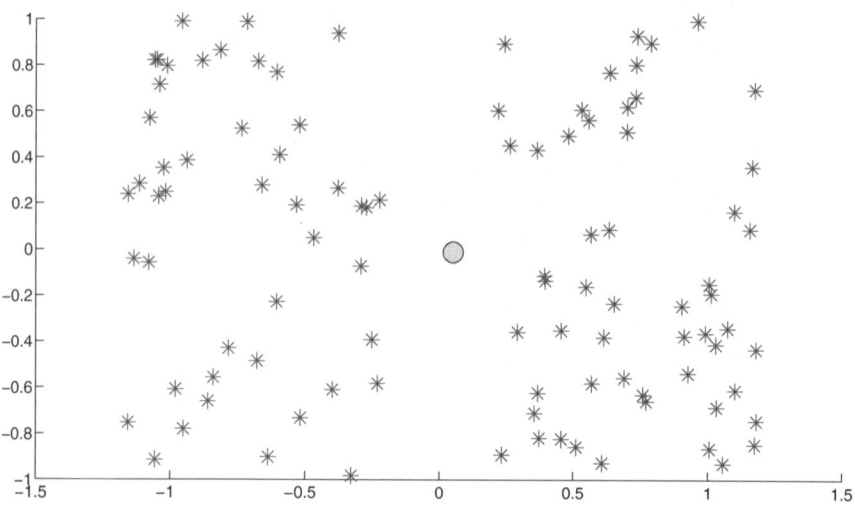

Fig. 3.19 The center (marked by a circle) of the set of points generated by `generate` and computed by `centers`.

needed because in the second `for` loop, sums of x and y components are accumulated cluster by cluster and they need to be divided by the corresponding `cn` for obtaining the average value. It might happen that some cluster does not have any sample, and in this case the corresponding values in `cx` and `cy` would be both 0, as well as the `cn` value. This situation must be treated as a particular case, because there is a division by `cn`, and so `cn` cannot be zero. The function `centers` simply returns $(0, 0)$ as center of the cluster when it is empty. This does not cause any problems on the convergence of the k-means algorithm. In Figure 3.19 the center of the whole set of data previously generated is shown. By using the MATLAB `plot` function, it is possible to change the color and the symbols used for marking points. The function `centers` has been used with `k` set to 1 and the vector `class` containing all the elements equal to 1.

The function `kmeans` is an implementation in MATLAB of the k-means algorithm (see Figure 3.20). Its input parameters are the number `n` of points in the set, the `x` and `y` coordinates of the samples and the number `k` of clusters in which these data have to be partitioned. The output parameter is a vector containing the code of the cluster for each point. These codes are numbers from 1 to `k`. At the start, points are randomly assigned to one of the clusters, using random generated numbers. As mentioned earlier, the MATLAB function `rand` generates a random real number in the interval $(0, 1)$. If this number is multiplied k times by itself then it becomes a real number in $(0, k)$, and its integer part is one of the natural numbers between 0 and $k - 1$. This number increased by 1 is therefore a random integer number between 1 and k, and it can be used to randomly assign a point to one of the k clusters. Note that the function `int16` is used for extracting the integer part of a real number.

As shown in Section 3.1, the k-means algorithm consists of a main `while` loop which terminates when the centers cannot be rearranged any longer. In the function, this is controlled by the variable `stable`, which assumes value 1 when the centers

```
%
% this function performs a k-means algorithm
% on a two-dimensional set of data
%
% input:
% n - number of samples
% x - x coordinates of the samples
% y - y coordinates of the samples
% k - number of classes
%
% output:
% class - classes to which each sample belongs
%
% [class] = kmeans(n,x,y,k)

function [class] = kmeans(n,x,y,k)

  % initializing the clusters

  for i = 1:n,
    class(i) = int16(k*rand());
    if class(i) == 0,
      class(i) = k;
    end
  end

  % computing the cluster centers

  [cx,cy] = centers(n,x,y,k,class);
  for j = 1:k,
    cxnew(j) = cx(j);  cynew(j) = cy(j);
  end

  stable = 1;  % unstable

  while stable == 1,

      % computing the distances between samples (x,y) and centers (cx,cy)
      for i = 1:n,
        mindist = 10.e+100;
        minindex = 0;
        for j = 1:k,
          dist = (x(i) - cxnew(j))^2 + (y(i) - cynew(j))^2;
          dist = sqrt(dist);
          if dist < mindist,
            mindist = dist;
            minindex = j;
          end
        end
        % changing cluster
        class(i) = minindex;
        [cxnew,cynew] = centers(n,x,y,k,class);
      end

      % checking the algorithm convergence
      stable = 0;
      for j = 1:k,
        if abs(cxnew(j) - cx(j)) > 1.e-6 | abs(cynew(j) - cy(j)) > 1.e-6,
          stable = 1;
        end
      end

      % preparing for the next iteration
      for j = 1:k,
        cx(j) = cxnew(j);  cy(j) = cynew(j);
      end

  end % while

end
```

Fig. 3.20 The MATLAB function kmeans.

are not stable and 0 otherwise. The `for` loop on `i` is performed for each sample. Once a sample has been fixed, its distance from every center is computed and at the same time the smallest distance and the corresponding center are located. In the algorithm, `mindist` contains the value of the minimum distance between the sample and the centers. It is initialized with a huge number that is soon substituted with the first computed distance. `minindex` contains instead the code of the cluster whose center has distance `mindist` from the prefixed sample. In the `for` loop on `j`, a distance is computed and `mindist` and `minindex` are updated if the new distance is smaller than the previous one already computed. After the `for` on `j`, the variable `minindex` contains the code of the center which is closer to the predetermined sample. Then, `class` is updated with this code. After that, all the clusters are recomputed by the function `centers` and the algorithm starts working on another sample.

When all the samples have been processed, the current centers need to be compared to the previous ones. If this is the first iteration of the algorithm, the new centers will be compared to the centers of the clusters randomly generated at the start of the algorithm. The convergence of the centers is checked through the variable `stable`. In the algorithm, it is set to 0 (which means that the centers did not change) and then it is eventually reset to 1 if at least one condition in the `if` construct is verified. Such conditions are verified when the difference between the centers on their x or y coordinates is greater than 10^{-6}. Before the algorithm starts another iteration, the variables `cx` and `cy` are updated with the new centers of the clusters.

This function is able to partition the set of points previously generated in two clusters, where one cluster contains all the points on the left of the y axis and the other cluster contains all the points on the right of the y axis. In order to view these results on a figure, let us consider the MATLAB function in Figure 3.21. The function `plotp` displays the points in the set by using different symbols and different colors for each cluster. It receives as input parameters the number of points to display, the x and y coordinates of such points, and the code of the cluster to which they belong. For our purposes, a function distinguishing among no more than 6 clusters or classes is sufficient. The function can be easily improved by adding other colors and/or other symbols for other classes. In Figure 3.22 the partition found by function `kmeans` is shown.

When the set of points is generated, the eps variable in the function `generate` is set to 0.2. This means that points at the left of the y axis and points on the right of the axis have a relative distance equal to or greater than two times eps. This helped the k-means algorithm in re-finding this pattern with which the data have been generated. However, if eps gets smaller, then it may be more difficult for the algorithm to find a set of clusters that partition the data in the same way. In Figure 3.23 and Figure 3.24 more executions of the k-means algorithm have been performed using different sets of points. Such sets have been generated by using the same function `generate` but with decreasing eps values. The algorithm is able to correctly find the partition in points generated by the function `generate` when eps = 0.10 and eps = 0.05 (Figure 3.23). Finally, when eps is set to 0.02 or 0, the algorithm cannot identify any pattern and randomly divides the set in two balanced parts. These last two examples are shown in Figure 3.24.

```
%
% this function plots the n samples in (x,y) by using different colors
% for visualizing their belonging to different classes
%
% note that no more than 6 colors are used
%
% input:
% n     - number of samples
% x     - x coordinates of the samples
% y     - y coordinates of the samples
% class - classes to which each sample belongs
%
% plotp(n,x,y,class)

function plotp(n,x,y,class)

   hold on

   for i = 1:n,
     if class(i) == 1,  col = 'r*';        % red/star
       elseif class(i) == 2,  col = 'b+';  % blue/plus
       elseif class(i) == 3,  col = 'kx';  % black/x-mark
       elseif class(i) == 4,  col = 'ms';  % magenta/square
       elseif class(i) == 5,  col = 'gp';  % green/pentagran
       else col = 'yd';                    % yellow/diamond
     end

     plot(x(i),y(i),col,'MarkerSize',16)

   end

end
```

Fig. 3.21 The MATLAB function `plotp`.

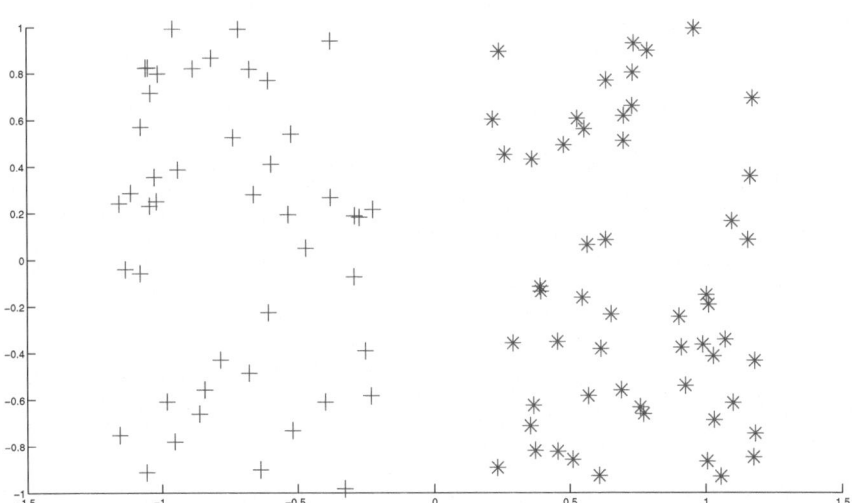

Fig. 3.22 The partition in clusters obtained by the function `kmeans` and displayed by the function `plotp`.

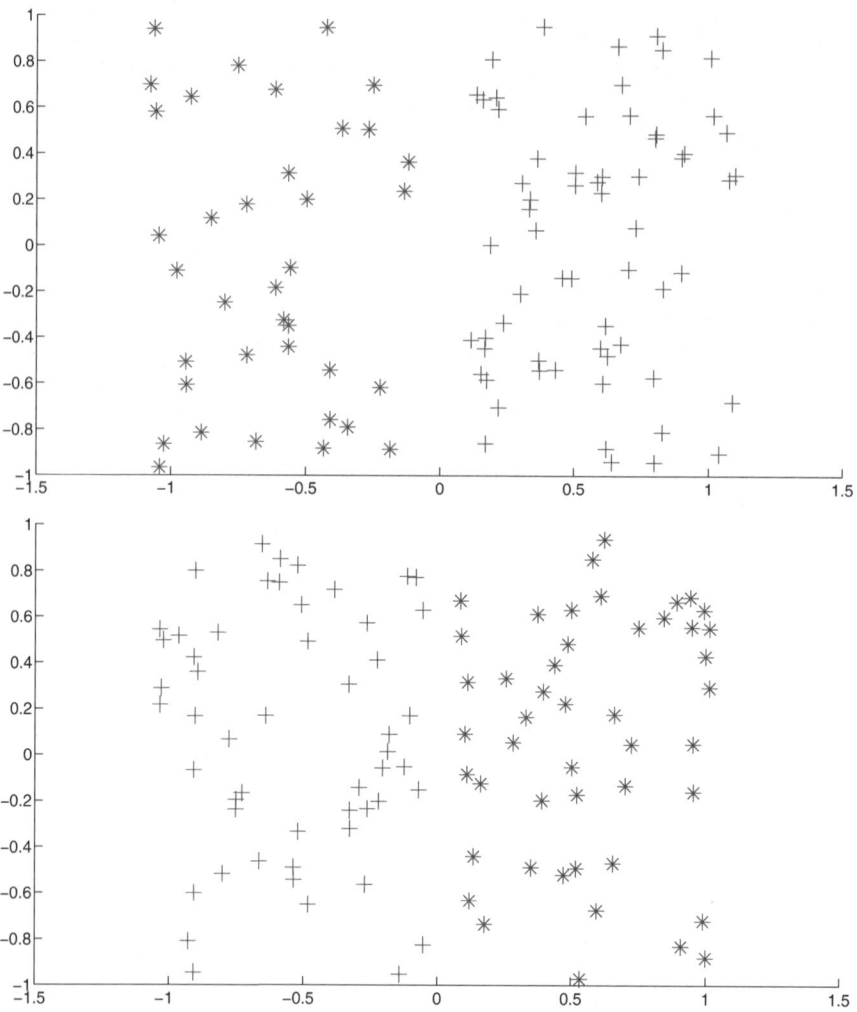

Fig. 3.23 Different partitions in clusters obtained by the function `kmeans`. The set of points is generated with different eps values. (a) eps = 0.10, (b) eps = 0.05.

3.7 Exercises

In this section some exercises regarding the data mining technique discussed in this chapter are presented. Some of the exercises require programming in MATLAB. All the solutions are reported in Chapter 10.

1. Consider 6 samples in a two-dimensional space:

$$(-1, -1), (-1, 1), (1, -1), (1, 1), (7, 8), (8, 7).$$

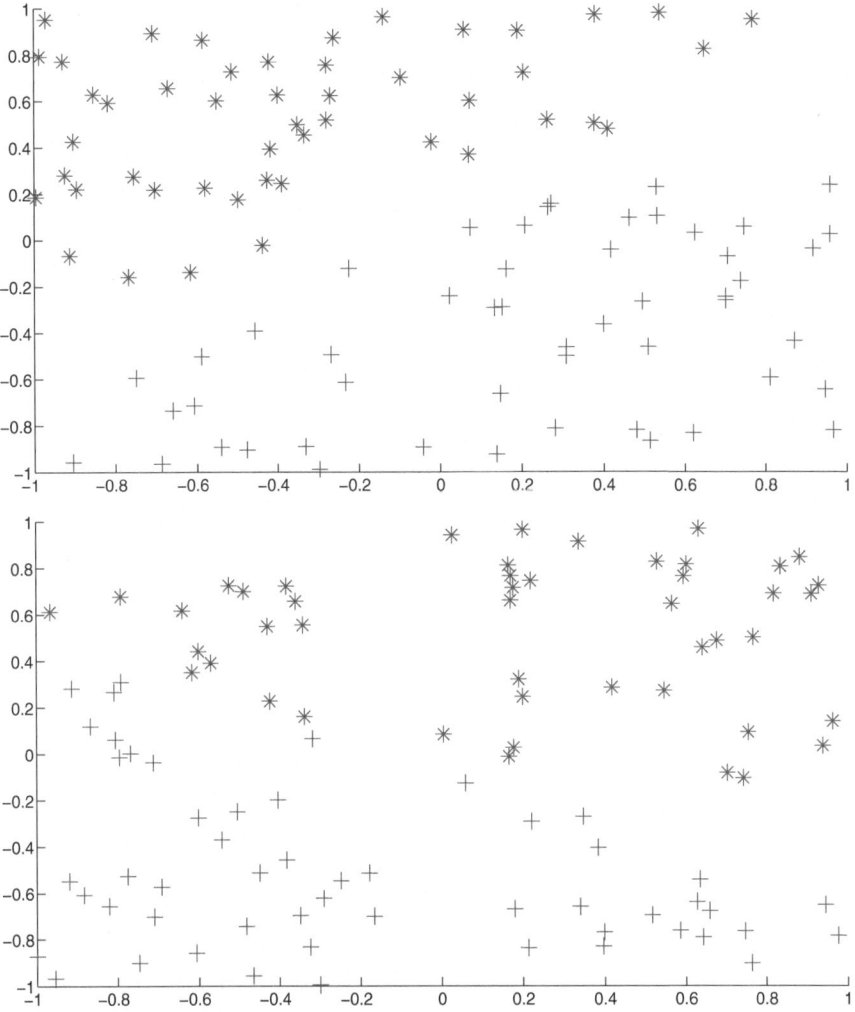

Fig. 3.24 Different partitions in clusters obtained by the function kmeans. The set of points is generated with different eps values. (a) eps = 0.02, (b) eps = 0.

Assuming that the 1st, 3rd and 5th samples are initially assigned to cluster 1, and that the 2nd, 4th and 6th samples are assigned to cluster 2, run the Lloyd's algorithm or the k-means algorithm.

2. Consider 7 samples in a two-dimensional space:

$$(1, 0), (1, 2), (2, 0), (0, 1), (1, -3), (2, 3), (3, 3).$$

Assuming that the 1st, 3rd, 5th and 7th samples are initially assigned to cluster 1, and that the 2nd, 4th and 6th samples are assigned to cluster 2, run the k-means algorithm.

3. Run the h-means algorithm on the set of samples described in Exercise 1. Observe the obtained results.

4. Provide an example in which the k-means algorithm can find 4 different partitions in clusters corresponding to the same error function value (3.1).

5. Give an example of 8 points on a Cartesian plane that can be partitioned by k-means in 2 different ways that correspond to the same error function value (3.1).

6. Consider the 7 samples described in Exercise 1. Suppose that the samples have to be partitioned into 3 clusters. Assume that the samples are currently assigned to cluster 1 and 2 as described in Exercise 1, while cluster 3 is empty. Apply the k-means+ algorithm.

7. Using the same set of samples and the same initial conditions of Exercise 6, apply the h-means+ algorithm and compare the solution to the one obtained in the previous exercise.

8. By using the MATLAB function `plotp`, build a figure in which the points described in Exercise 1 are drawn with two different symbols showing how they are partitioned. Use the partition in clusters found in Exercise 1.

9. Starting from the MATLAB function `kmeans` presented in Section 3.6 (Figure 3.20), write the function `hmeans` which implements the h-means algorithm.

10. Prove that the sum of squares of distances from the samples of a class to its center is equal to the sum of squares of all pairwise distances between the samples in the class divided by the number of samples in the class:

$$\sum_{j \in S_i} ||x_j - c_j||^2 = \frac{1}{|S_i|} \sum_{j_1 \in S_i} \sum_{j_2 \in S_i, j_2 > j_1} ||x_{j_1} - x_{j_2}||^2.$$

Chapter 4
k-Nearest Neighbor Classification

4.1 A simple classification rule

The k-nearest neighbor (k-NN) method is one of the data mining techniques considered to be among the top 10 techniques for data mining [237]. The k-NN method
uses the well-known principle of Cicero *pares cum paribus facillime congregantur*
(*birds of a feather flock together* or literally *equals with equals easily associate*). It
tries to classify an unknown sample based on the known classification of its neighbors. Let us suppose that a set of samples with known classification is available, the
so-called *training set*. Intuitively, each sample should be classified similarly to its
surrounding samples. Therefore, if the classification of a sample is unknown, then it
could be predicted by considering the classification of its nearest neighbor samples.
Given an unknown sample and a training set, all the distances between the unknown
sample and all the samples in the training set can be computed. The distance with the
smallest value corresponds to the sample in the training set closest to the unknown
sample. Therefore, the unknown sample may be classified based on the classification
of this nearest neighbor.

However, in general, this classification rule can be weak, because it is based on
one known sample only. It can be accurate if the unknown sample is surrounded by
several known samples having the same classification. Instead, if the surrounding
samples have different classifications, as for example when the unknown sample is
located amongst samples belonging to two different classes (and hence with different classifications), then the accuracy of the classification may decrease. In order to
increase the level of accuracy, then, all the surrounding samples should be considered and the unknown sample should then be classified accordingly. In general, the
classification rule based on this idea simply assigns to any unclassified sample the
class containing most of its k nearest neighbors [42]. This is the reason why this
data mining technique is referred to as the k-NN (k-nearest neighbors). If only one
sample in the training set is used for the classification, then the 1-NN rule is applied.

Figure 4.1 shows the k-NN decision rule for $k = 1$ and $k = 4$ for a set of samples
divided into 2 classes. In Figure 4.1(a), an unknown sample is classified by using

A. Mucherino et al., *Data Mining in Agriculture*, Springer Optimization and Its Applications 34,   83
DOI: 10.1007/978-0-387-88615-2_4,

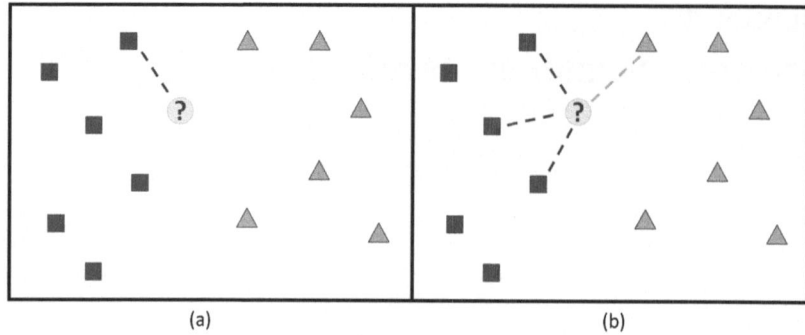

Fig. 4.1 (a) The 1-NN decision rule: the point **?** is assigned to the class on the left; (b) the k-NN decision rule, with $k = 4$: the point **?** is assigned to the class on the left as well.

only one known sample; in Figure 4.1(b) more than one known sample is used. In the last case, the parameter k is set to 4, so that the closest four samples are considered for classifying the unknown one. Three of them belong to the same class, whereas only one belongs to the other class. In both cases, the unknown sample is classified as belonging to the class on the left. Figure 4.2 provides a sketch of the k-NN algorithm.

The distance function plays a crucial role in the success of the classification, as is the case in many data mining techniques. Indeed, the most desirable distance function is the one for which a smaller distance among samples implies a greater likelihood for samples to belong to the same class. The choice of this function may not be trivial. Another important factor is the choice of the value for the parameter k. This is the main parameter of the method, since it represents the number of nearest neighbors considered for classifying an unknown sample. Usually it is fixed beforehand, but selecting an appropriate value for k may not be trivial. If k is too large, classes with a great number of classified samples can overwhelm small ones and the results will be biased. On the other hand, if k is too small, the advantage of using many samples in the training set is not exploited. Usually, the k value is optimized by trials on the training and validation sets (see Chapter 8). Moreover, assigning a classification on the basis of the majority of the k "votes" of the nearest neighbors could not be accurate in some particular cases. For example, if the nearest neighbors vary widely in their distance, then an unknown sample may be classified considering samples

```
for all the unknown samples UnSample(i)
   for all the known samples Sample(j)
      compute the distance between UnSamples(i) and Sample(j)
   end for
   find the k smallest distances
   locate the corresponding samples Sample(j1),..,Sample(jk)
   assign UnSample(i) to the class which appears more frequently
end for
```

Fig. 4.2 The k-NN algorithm.

that are located far from it. Therefore, a more sophisticated approach could be to weight the vote of each sample by its distance, so that the closest samples have more importance during the classification. The two applications discussed in Section 4.4.1 and 4.4.2 use this approach.

The k-NN method is said to be a *lazy* classifier [237], because it actually does not generate a classifier from the data in a training set, but it rather exploits the training set every time a classification needs to be performed. This makes the method easier, but computationally expensive. The main computational cost is due to the computation of the distances between known and unknown samples. This task can be expensive if the training set or the number of unknown samples is large. Therefore, the next section introduces several strategies for reducing the size of the training set while keeping the accuracy of the classification as high as possible.

4.2 Reducing the training set

As described in the previous section, the k-NN algorithm searches the k-nearest neighbors of an unknown sample computing all the distances between the unclassified sample and the samples in the training set. Therefore, if the training set is large, or the number of samples to classify is large, then the computational complexity may impact the performance of the algorithm.

In [102], a *condensed* nearest neighbor (CNN) rule has been introduced with the goal of reducing the computational effort needed for carrying the algorithm out. Let T_{NN} be the available training set. Instead of using T_{NN}, one of its subsets, T_{CNN}, may be used. If T_{CNN} is able to correctly classify every sample in the set $T_{NN} - T_{CNN}$, then it is referred to as a *consistent* subset of T_{NN}. A sample is correctly classified when k-NN is able to reassign to it the correct classification using its nearest neighbors. When this is verified, the correctly classified sample is discarded. If k-NN provides a wrong classification when it tries to classify a sample, then the sample will be incorrectly classified. In this case, the sample cannot be discarded and it remains in the training set. A minimal consistent subset is a consistent subset containing the minimum possible number of samples.

The algorithm for obtaining T_{CNN} can be summarized as follows. Two sets of samples X and Y, initially empty, are defined. Then a random sample from T_{NN} is placed in X. For each sample in $T_{NN} - X$, the k-NN rule is applied for classifying it and using X as the training set. All samples which are correctly classified are placed in Y, whereas the ones that are incorrectly classified are placed in X. The set X can change at each iteration, and in particular it becomes bigger every time a sample is incorrectly classified. After this first phase in which the starting set T_{NN} is used, the algorithm starts considering samples stored in Y. Iteratively, each sample in Y is classified using X as the training set and the current sample is moved to X if the classification is incorrect. The algorithm can stop only when Y is empty or when no samples are moved from Y to X after an entire loop on Y. The final samples contained in X are used as reference samples: $T_{CNN} = X$. If Y is empty at the end

```
X = ∅,  Y = ∅
copy the first sample from T_NN to X
for all the samples Sample(i) in T_NN − X
    classify Sample(i) by using X as training set
    if (Sample(i) is correctly classified)
       copy Sample(i) in Y
    else
       copy Sample(i) in X
    end if
end for
repeat
    nmoves = 0
    for all the samples Sample(i) in Y
        classify Sample(i) by using X as training set
        if (Sample(i) is not correctly classified)
           nmoves = nmoves + 1
           move Sample(i) to X
        end if
    end for
until (Y = ∅ or nmoves = 0)
T_CNN = X
```

Fig. 4.3 An algorithm for finding a consistent subset T_{CNN} of T_{NN}.

of the algorithm, then X is equal to the whole set T_{NN}. A sketch of this algorithm is shown in Figure 4.3.

Figure 4.4 shows a training set in which samples are classified in four different classes. Samples belonging to different classes are marked with different symbols. The encircled samples are classified by using the k-NN rule, with $k = 3$. According to the previous algorithm for finding a condensed training set, the samples that are correctly classified can be discarded. For instance, the sample A is correctly classified: its three neighbors have the same classification. Sample B is classified in the wrong way by k-NN. Two neighbors have classification □ and only one has classification ⋆. Therefore B is classified as □, whereas its original classification was ⋆. Sample C is classified incorrectly as well. In this case, C is closest to a sample of its own class, but the other two neighbors have classification +.

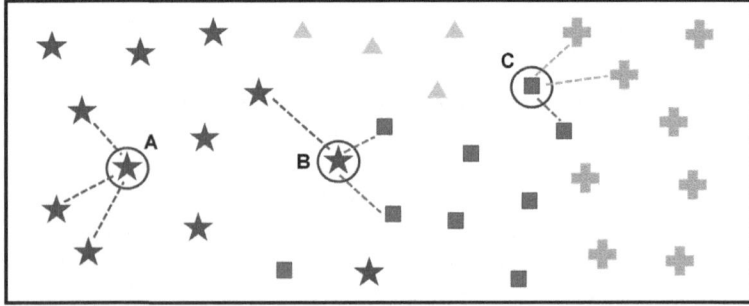

Fig. 4.4 Examples of correct and incorrect classification.

```
X = T_NN ,   Y = ∅
for all the samples Sample(i) in X
    move Sample(i) from X to Y
    classify all the samples in Y by using X as training set
    if (at least one sample is not correctly classified)
        move Sample(i) from Y to X
    end if
end for
T_RNN = X
```

Fig. 4.5 An algorithm for finding a reduced subset T_{RNN} of T_{NN}.

In [81], a *reduced* nearest neighbor (RNN) rule is proposed. As the author points out, this rule is able to reduce the training set T_{NN} to a set T_{RNN} which is smaller than T_{CNN}, which is the subset that can be obtained using the algorithm in Figure 4.3. Since T_{RNN} is smaller than T_{CNN}, it provides a more efficient way of reducing the initial training set. The algorithm for obtaining T_{RNN} is presented as follows. X is set equal to T_{NN}, whereas Y is empty. At each step of the algorithm, a sample migrates from X to Y and all the samples in Y are classified using X as the training set. If one of these samples is not correctly classified, then the sample just moved from X to Y is reassigned back to X. The final set X will be considered as the reduced training set T_{RNN}. A sketch is given in Figure 4.5.

Over the years, many other variations on the algorithms presented above have been proposed with the aim of finding the most efficient consistent subset of T_{NN}. In the following, we will present the main ideas behind proposed methods and algorithms without providing many details. Interested readers may refer to the quoted references. In [195], for example, the condensed rule is used coupled with other requirements, in order to improve the quality of the obtained subset. In [90], the concept of mutual nearest neighborhood and mutual neighborhood value are introduced and used for selecting samples in a more effective way. Moreover, in [44], the training set is iteratively reduced by merging the closest two samples. The closest pair of samples is located in the training set at each step of the algorithm. They are then replaced by another sample, which may simply be the average of the two deleted samples or their weighted average. It is required that merged samples have the same classification. The procedure stops when there are samples in T_{NN} that are not correctly classified using the obtained subset as training set.

More recently, a *modified condensed* nearest neighbor (MCNN) rule has been proposed in [59]. At the start, the set X is empty and samples are added to it until it becomes consistent. Actually, such samples can be either samples from the original set T_{NN} or samples computed from the ones in T_{NN}. Therefore, the samples iteratively added to X are called *prototypes*. At the first step of the algorithm, the set X has one prototype per each class. The set Y contains all other samples not included in X. At each step of the algorithm, all samples in Y are classified using X as training set. Then, all incorrectly classified samples are considered and a representative prototype for each class is determined and added to the set X. More than one method for finding the representative prototypes can be used, and it may depend on the selected data representation. The easiest method computes the representative as the mean of all the incorrectly classified samples. Once these representatives have been added to X and Y is updated, the classification algorithm is carried out again on all the samples in Y and using the enriched set X. Other representative prototypes are then generated if incorrectly classified samples are found. The algorithm is repeated until all the samples in the training set are classified correctly. This algorithm converges in a finite time and the generated prototypes give 100% accuracy on the training set. Another example of a recently proposed method for reducing the training set is the *fast condensed* nearest neighbor (FCNN) method [6].

As pointed out in [236], strategies for decreasing the computational complexity of the k-NN algorithms may impact the accuracy of the algorithm. In [236], two strategies have been proposed that may be able to speed the algorithm up while the accuracy does not decrease. The first strategy reduces the training set T_{NN} as in the previous cases, but using a different approach. The basic idea is that, if a large number of classified samples are close to each other, then the number of classified samples in the neighborhood of an unknown sample is usually greater than k. Therefore, some of the classified samples can be discarded as they are not relevant to the classification of the unknown sample. The other strategy dynamically reduces T_{NN} by performing a preprocessing phase on the training set. The L_1 or L_2 norm of a vector representing a sample can be considered as a particular characteristic of that sample. The L_1 norm is the sum of all its components in absolute values:

$$||x||_1 = \sum_{i=1}^{n} |x_i|.$$

The L_2 norm is the square root of the sum of the squares of the components of the vector x:

$$||x||_2 = \sqrt{\sum_{i=1}^{n} (x_i)^2}.$$

If x is an unknown sample and $x^i \in T_{NN}$, then x can be considered to be a distorted variant of x^i if

$$\text{abs}(||x|| - ||x^i||) < \delta$$

where δ is a certain positive threshold and $||\cdot||$ is either the L_1 norm or L_2 norm. The larger is δ, the more samples are considered similar to each other and precluded from participating in the matching process. The choice of δ is important in such a strategy, because a too small value may reject samples that are very close to an unknown sample x, whereas large values of δ may make the preprocessing phase inefficient.

4.3 Speeding k-NN up

In the previous section we discussed different proposed strategies for reducing the training sets which are used in the k-NN algorithm. Another way to speed the k-NN algorithm up is to accelerate the matching algorithm. Since the computational cost is due to the computation of distances, a quick method which is able to locate samples close or far from each other would be very useful. The KD-tree method is one of the well-known methods for accelerating k-NN [19].

This method works with the individual components of the vectors representing the samples. If the samples closer to an unknown sample need to be identified, then vectors having a set of components similar to the one the unknown sample has are searched. This pre-process, applied before the distance function is used, can help increase the speed of the algorithm, because only a subset of distances may be chosen for the computation. However, as pointed out in [28], this method was more efficient on problems of low dimension and with simple distance functions.

The template trees method [28] is more general than the KD-tree. The main difference is that it directly works with distance functions rather than with the vector components representing the samples. The template trees method is able to construct large template trees that correctly identify all the samples in a training set. Since it increases the speed of the k-NN algorithm, larger training sets can be used and therefore this method is helpful even for increasing the classification accuracy.

A recent review of strategies for locating the nearest neighbors of a given sample can be found in [4, 52]. Most of these strategies are based on suitable approximations of such neighbors. When the 1-NN rule is applied, the aim is to find one sample in a training set which is the closest to a given unknown sample. The triangle inequality can be used for approximating the distances without computing them explicitly. As it is well known, the triangle inequality allows one to define bounds on distance values. Indeed, the sum of lengths of any two sides of a triangle is always greater than the third side. Therefore, if d_{12} is the distance between x_1 and x_2 and d_{23} is the distance between x_2 and x_3, then the distance d_{13} between x_1 and x_3 must be smaller or equal to the sum $d_{12} + d_{23}$. Moreover, an approximation of the known sample can be for instance a sample in the training set whose distance from the unknown one is at most a prefixed value R. There are many algorithms following these two ideas for speeding the k-NN classification up, and many of them are reviewed in the above quoted papers. In [52], moreover, general considerations on the metric space in which the classification method is applied are provided.

4.4 Applications

The k-NN algorithm is one of the most popular algorithms for text categorization or text mining. Some of the most recent works on this topic are for instance [14, 85, 95, 214, 227]. When working on a particular problem, and in this case in the field of text mining, the standard algorithm for data mining can be tailored to the particular problem to be solved. Just to quote an example, in [14], the k-NN algorithm has been modified for solving text mining problems. Different numbers of nearest neighbors are used for different classes in this approach, rather than a fixed number across all classes. In this way, the only parameter that needs to be chosen by the user when using k-NN, the k value, becomes less sensible and hence it does not need to be carefully chosen as in the standard algorithm. Indeed, the probability that an unknown sample belongs to a class is computed by using only some top k_n nearest neighbors for that class. The k_n value is derived from k according to the size of the corresponding class in the training set. This modified k-NN was efficient and less sensible to the k values when applied to text mining problems. The k-NN algorithm has been also applied for analyzing micro-array gene expression data [149], where the k-NN algorithm has been coupled with genetic algorithms, which are used as a search tool. Other applications include the prediction of solvent accessibility in protein molecules [216], the detection of intrusions in computer systems [150], and the management of databases of moving objects such as computer with wireless connections [16].

In general, k-NN is applied less than other data mining techniques in agriculture-related fields. It has been applied, for instance, for simulating daily precipitations and other weather variables [192]. Another interesting application is the evaluation of forest inventories and for estimating forest variables [15, 108]. In these applications, satellite imagery is used, with the aim of mapping the land cover and land use with few discrete classes. In [97], the studied area includes Lake, Carlton, Cook, Koochiching, Lake, and St. Louis counties in Northeast Minnesota. Figure 4.6 shows this study area. The dots represent the samples that are taken in consideration. The white parts represent clouds, where data have not been obtained.

The next sections present details of the use of the k-NN method in climate forecasting (Section 4.4.1) and for estimating soil water parameters (Section 4.4.2).

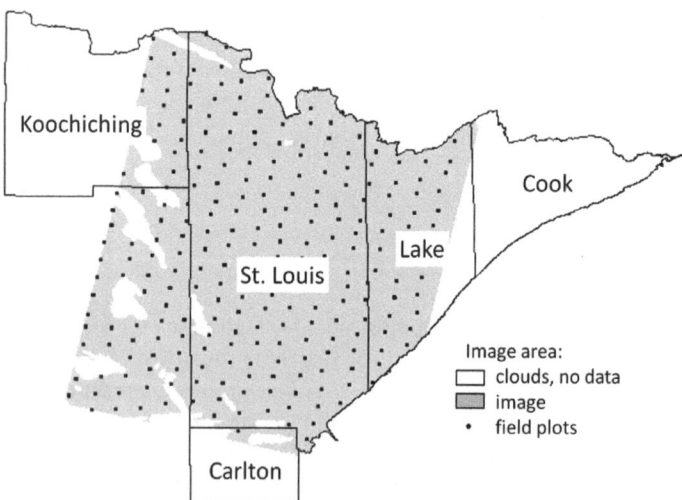

Fig. 4.6 The study area of the application of k-NN presented in [97]. The image is taken from the quoted paper.

4.4.1 Climate forecasting

Knowing the weather a day or a week in advance is very important especially in agriculture. Weather forecast can influence decisions, in order to avoid unwanted situations or to take advantage of favorite weather conditions. The variability of the climate is indeed one of the most important factors that seriously impacts agricultural production. While TV channels or journals are able to provide quite accurate forecasts of the weather in the next few days, it is still a big challenge forecasting the weather conditions 3 to 6 months ahead of time. These are the kinds of time intervals to deal with when working in agriculture. The uncertainty about the weather can be devastating in agriculture, because farmers may not be prepared to face the weather conditions that might occur. It can cause also poor productivity, because of the use of conservative strategies that sacrifice productivity to reduce the risk of losses. If the future weather conditions were known, this could be exploited for decreasing unwanted impacts and for taking advantage of expected favorable conditions.

Most of the current climate forecasts are based on analysis on the *El Niño-Southern Oscillation* (ENSO). This phenomenon is characterized by three phases: warm (*El Niño*), neutral and cool (*La Niña*) phases. Even though the ENSO phenomenon occurs within the tropical Pacific, it affects inter-annual weather variability across much of the globe, and, in particular, it affects the climate of the southeastern USA. In this region, lower winter temperatures with higher precipitations occurs during the *El Niño* events, whereas *La Niña* events show the reverse of the climate anomalies associated with the *El Niño* [87, 121].

In the studies presented in [117], a *k*-NN algorithm is used for the recalibration of the precipitation outputs from the FSU-GSM (Florida State University *Global* Spectral Model) and FSU-RSM (Florida State University *Regional* Spectral Model) climate models. These climate models may not produce sufficiently accurate daily weather variable outputs to use in crop models. For details on the FSU-GSM and FSU-RSM, please refer to [39, 117].

The studies presented in [117] are related to 10 sites chosen in Florida and Georgia (see Figure 4.7). They have been selected to represent increasing distances from the Atlantic Ocean and the Gulf of Mexico. A set of monthly forecasts, from March to August, related to the years from 1987 to 2003, with the exception of 2002, has been used. The forecasts come from both FSU-GSM and FSU-RSM models. As pointed out by the authors, all climatology models of the FSU-GSM are very accurate. For instance, they can predict higher precipitation and excessive wet days. Even though FSU-RSM has a higher resolution, it behaves almost the same, simulating only a better average of the rainfall in March.

In order to recalibrate monthly rainfall forecasts, 10 combinations of the data from FSU-GSM and FSU-RSM are used. In this way, results from both models are taken into consideration. In the following, R_{ij} refers to the forecast output obtained by the FSU-RSM model and related to the ith month of the jth year. Similarly, G_{ij} refers to the forecasts obtained by the FSU-GSM model related to the ith month of the jth year. Different combinations of R_{ij} and G_{ij} have been selected, as Figure 4.8 shows. Combinations number 1 and 8 just consider the output from the regional

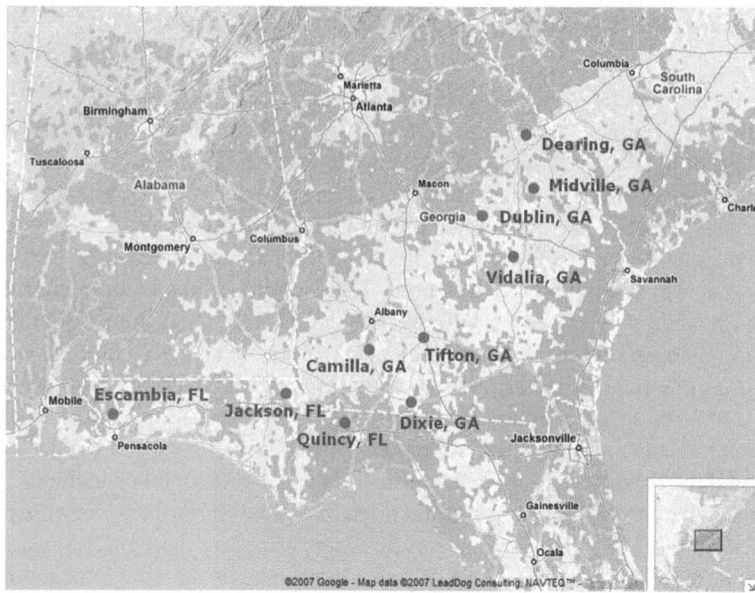

Fig. 4.7 The 10 validation sites in Florida and Georgia used to develop the raw climate model forecasts using statistical correction methods.

and global model, respectively. Some combinations consider the outputs related to all the months taken into account, and other combinations consider the current and the neighboring months (as for instance combination number 6). In the following, P_{ijq} refers to the qth forecast output related to the ith month and j^{th} year, for a prefixed target combination. In general, $q \in \{1, 2, \ldots, Q\}$, where Q refers to the number of considered outputs in a given combination. For instance, $Q = 3$ when the combination number 6 is used, because R_{i-1j}, R_{ij} and R_{i+1j} are used in this case.

The objective is to find k neighboring years which have the forecasts closest to those of a target year n. It is therefore assumed that the climate during a target year is

Counter	P_{ij} values
1	R_{ij}
2	R_{ij}, G_{1j}
3	$R_{ij}, \forall j$
4	$G_{1j}, R_{ij}, \forall j$
5	$R_{ij}, G_{ij}, \forall j$
6	$R_{i-1j}, R_{ij}, R_{i+1j}$
7	$R_{i-1j}, R_{ij}, R_{i+1j}, G_{1j}$
8	G_{ij}
9	$G_{ij}, \forall j$
10	$G_{i-1j}, G_{ij}, G_{i+1j}$

Fig. 4.8 The 10 target combinations of the outputs of FSU-GSM and FSU-RSM climate models.

a replication of the weather recorded in the past. Once a combination of the forecast outputs has been chosen, the distances between target year and all the others can be computed on the basis of the variables P_{ijq} and P_{inq}, the first ones being related to the jth year, and the second ones related to the target year n. The distance function is defined as:

$$d_{ij} = \sqrt{\sum_{q=1}^{Q} \left(P_{ijq} - P_{inq}\right)^2}, \quad \forall j \neq n.$$

The k-NN algorithm has been applied for classifying the ith month of the target year n on the base of the k closest d_{ij} distances. Then, the k neighboring years have been sorted in ascending order, and the function $j(r)$ has been defined for providing the years in the correct order, and the weights of the k years have been defined as:

$$w_r = \frac{\frac{1}{r}}{\sum_{i=1}^{k} \frac{1}{i}}, \tag{4.1}$$

where $r \in \{1, 2, \ldots, k\}$. The corrected precipitation for month i in a target year n has then been estimated as a weighted average of measured precipitations from the same k analog years. If $O_{ij(r)}$ represents the precipitation during the ith month of the $j(r)^{th}$ year, then the corrected precipitation is computed by the following formula:

$$F_{in} = \sum_{r=1}^{k} w_r O_{ij(r)}.$$

A considerable variability has been observed in the FSU-RSM predictions from year to year, and this appears to be independent of variations in seasonal rainfall. Negative correlations indicate that forecast rainfall is high when observed rainfall is actually low and vice versa. The k-NN method was able to improve the accuracy of the monthly precipitation forecasts across all sites. The best results in March and April are obtained by FSU-GSM outputs, while FSU-RSM gives better results for later months. This suggests that, in the predominantly flat topography of the area under study, the corrected FSU-GSM output is able to closely mimic observed rainfall in early season simulations.

4.4.2 Estimating soil water parameters

In recent years, several crop simulation systems have been developed. Examples are DSSAT [122], CROPSYST [221], and GLEAMS [147], to name a few. Such systems include components that are able to simulate soil dynamics, when certain soil parameters are specified. Among these parameters, the ones usually denoted by the symbols LL, DUL, and PEWS are mostly used. LL is the lower limit of plant water availability; DUL is the drained upper limit; PESW is the plant extractable soil water. Unfortunately, these parameters are usually unknown. The available information about the soils usually concerns their texture, indicating the percentage of clay, silt, sand and organic carbon in the soil. If there is a relationship between the texture information and the parameters needed for the simulation models, then this relationship can be used for obtaining the needed parameters.

As explained in [118], regression models may be used for finding these kinds of relationships (Section 2.2). However, this approach may not be easy and may not provide satisfactory results. In fact, when dealing with regression models, a function needs to be defined able to fit the data. Examples of such functions are the linear and quadratic functions, just to name two of the ones mentioned in Section 2.2. The function that better fits the data is not known a priori, and usually it is chosen by trying different functions and choosing the one that better fits the data.

The *k*-NN method can be considered a reasonable alternative to address this category of problems [118]. The application discussed in the following and the one discussed in the previous section have some authors in common. This shows how the same methodology can be applied to different problems.

Experimental observations show that soils having similar textures also have similar values for the LL, DUL and PESW parameters. Let us suppose then that a database is available where soil data are collected by their textures and LL, DUL and PESW parameters. Let us consider now another soil, whose LL, DUL and PESW parameters are unavailable. In order to find an approximation of the needed parameters, the texture of the new soil can be compared to the textures of the soils in the database. The soil under study most likely has LL, DUL and PESW parameters similar to those of the nearest soils in the database. The distances between soils are based in this case on percentages of clay, silt, sand and organic carbon in the soils. This strategy is nothing else but the *k*-NN method.

In the quoted paper, an explicative picture has been used for showing the basic idea of the approach. We present this picture in Figure 4.9. It describes an example in which a target soil is considered such that its texture can be represented by a pair (20, 60). The pair specifies that 20% of the soil is clay and that the 60% is sand. In the database there are no pairs with the same values, but pairs having "similar values." In Figure 4.9, the four nearest pairs are shown. They correspond to the four soils having more similarities with the target soil. If the 1-NN rule is applied, then only the nearest soil in the database is considered. If the *k*-NN rule is applied and $k > 1$, more soils are used, and the mean of their LL, DUL and PESW paramenters can be considered as the best fit of the target soil.

In general, more parameters need to be used for representing a soil. If m is the number of parameters, then

$$d_i = \sqrt{\sum_{j=1}^{m} s_i (v_{ij} - v_{tj})^2}$$

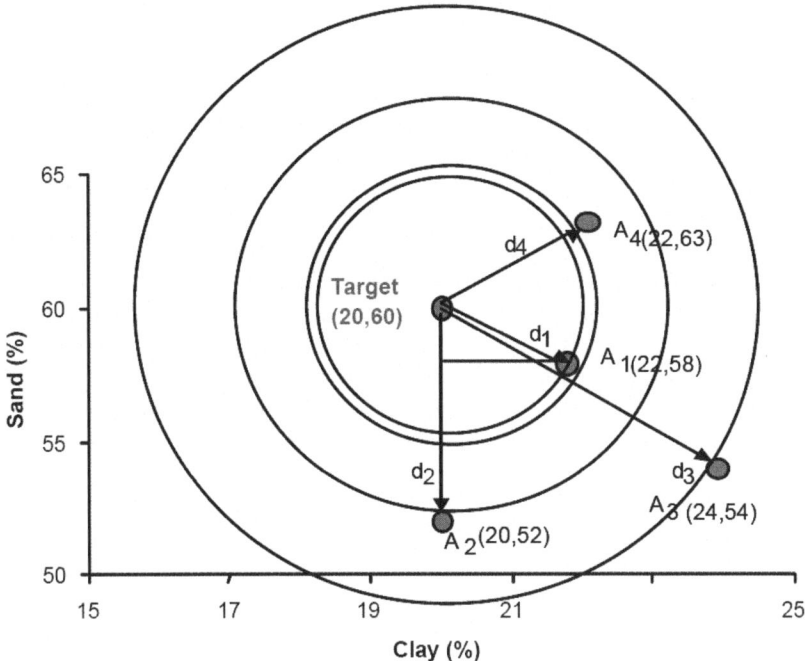

Fig. 4.9 Graphical representation of k-NN for finding the "best" match for a target soil. Image from [118].

is the distance between the target soil t and the ith soil in the database. The parameter s_i represents the scaling factor of the ith soil. Scaling the variables can be helpful if they have different ranges of variability. Scaling can prevent having variables that predominate on the others in the computation of the soils nearest to the target soil. Let y be a vector containing the values of one of the searched parameters (LL, DUL and PESW) for all the k nearest neighbors of the target soil. Then, the value of such parameter for the target soil can be estimated by applying the following formula:

$$\hat{y} = \sum_{i=1}^{k} w_i y_i,$$

where w_i is the weight associated to the ith nearest neighbor of the target soil. Weights can be associated to the nearest neighbors on the basis of their distances to the target. If the neighbors are sorted in ascending order, then the formula (4.1) can be used, as in the application discussed in the previous section. By using this approach, soil water retention parameters can be efficiently estimated with a high degree of accuracy using a database containing data pertaining to percentages of clay, sand and organic matter of soils.

4.5 Experiments in MATLAB®

This section presents some experiments in the MATLAB environment. The *k*-NN will be implemented in the simple case in which the samples are points in a two-dimensional space.

In Figure 4.10 the MATLAB function knn is shown. It has 8 input parameters and only one output parameter, which is the vector class containing the classification of the samples obtained by the *k*-NN algorithm. As inputs, the function needs: the number n of unknown samples to classify; the *x* and *y* coordinates of such samples, stored in x and y, respectively; the number k of nearest neighbors that will be used for classifying the unknown samples; the number ntrain of known samples used as training set for classifying the unknown ones; the *x* and *y* coordinates of such known samples are stored in xtrain and ytrain; finally, the classes to which each known sample belongs are stored in ctrain. Note that this MATLAB function has more parameters than the function in Figure 3.20 performing the *k*-means algorithm discussed in Chapter 3. Indeed, *k*-NN is a classification method whereas *k*-means is a clustering method, and then *k*-NN needs information about a training set for classifying the unknown samples.

The main loop in the algorithm is a for loop on i, which counts all the unknown samples. For each of them, three main operations must be performed: all the distances between this unknown sample and all the ones in the training set need to be computed; then the smallest *k* computed distances need to be checked and the corresponding known samples need to be located; finally the unknown sample is classified according to the known classification of these *k* known samples. The Euclidean distances between the current unknown sample (x(i),y(i)) and the samples in the training set (xtrain(j),ytrain(j)) are collected in a vector dist. The vector ind collects the index of the known samples used to compute the distances. It is needed for the identification of the *k* smallest distances. This task is performed by partially sorting in ascending order the vector dist by using one of the well-known methodologies for sorting data.

The methodology used for sorting the vector dist works as follows. It starts considering the last element in dist, it compares this current element to its neighbor in the vector and it exchanges their positions if they are not sorted in an ascending order. Step by step, the current element moves one step toward the first element in the vector. When it reaches the first element, and it eventually exchanges the last two elements, the first element of dist refers the minimum distance contained in the whole vector. At this point, the procedure can restart another time from the last element of the vector and repeat the same instructions. This time it is not needed to reach the first vector element, because it already contains the global minimum. It can stop at the second element. If this procedure is repeated a number of times equal to the vector size minus 1, then the vector will be completely sorted. In this case, instead, only the smallest *k* distances are searched, and therefore the procedure can be iterated only *k* times. Not only the distance values are important, but even the indices of the points having these distances from the unknown sample. Therefore,

```
%
% this function performs a k-NN algorithm
% on a two-dimensional set of data
%
% input:
% n      - number of samples to classify
% x      - x coordinates of the samples to classify
% y      - y coordinates of the samples to classify
% k      - kNN parameter
% ntrain - number of training samples
% xtrain - x coordinates of the training samples
% ytrain - y coordinates of the training samples
% ctrain - classes to which each training sample belongs
%
% output:
% class - classes to which each unknown sample belongs
%
% [class] = knn(n,x,y,k,ntrain,xtrain,ytrain,ctrain)

function [class] = knn(n,x,y,k,ntrain,xtrain,ytrain,ctrain)

  for i = 1:n,

    % computing the distance between (x(i),y(i)) and all the
    % training samples

    for j = 1:ntrain,
      dist(j) = (x(i) - xtrain(j))^2 + (y(i) - ytrain(j))^2;
      dist(j) = sqrt(dist(j));
      ind(j) = j;
    end

    % checking the k smallest distances obtained

    for kk = 1:k,
      for jj = ntrain-1:-1:kk,
        if dist(jj) > dist(jj+1),
          aus = dist(jj);  dist(jj) = dist(jj+1);  dist(jj+1) = aus;
          aus = ind(jj);   ind(jj) = ind(jj+1);   ind(jj+1) = aus;
        end
      end
    end

    % classifying the unknown sample on the base of the k-nearest
    % training samples

    for j = 1:k,
      score(j) = 0;
    end
    for kk = 1:min(k,ntrain),
      score(ctrain(ind(kk))) = score(ctrain(ind(kk))) + 1;
    end
    maxscore = 1;  val = score(1);
    for j = 2:k,
      if score(j) > val,
        val = score(j);
        maxscore = j;
      end
    end
    class(i) = maxscore;

  end

end
```

Fig. 4.10 The MATLAB function knn.

during the sorting process, the elements of the vector ind are exchanged according
to the changes applied to dist.

For classifying the current unknown sample (x(i),y(i)), a "score" is assigned
to each class and the one having the maximum score is considered to be the class to
which the unknown sample belongs. At the start all the scores are set to 0, then each
of them is updated according to the classes in which the *k* closest known samples are.
The expression score(ctrain(ind(kk))) refers to the score related to the class
located by ctrain when it refers to the known sample having index ind(kk). After
all the scores are updated, the maximum score and related class are identified and
the unknown sample is assigned to the class coded by maxscore.

A training set of 50 points has been generated using the MATLAB function
generate (Figure 3.16) and setting eps to 0.1. The value of eps allows one to
separate the points in 2 groups with a certain margin. The points so obtained are not
assigned yet to a class or another. The function kmeans (Figure 3.20) has been used
for clustering these data and therefore for assigning them a classification. Therefore,
the generated set of points can now be considered as a training set for the *k*-NN algorithm (see Figure 4.11). Another set of 200 points is then generated by the function
generate and by setting eps = 0, so that these points contain no inherent patterns.
Figure 4.12 shows the points marked in accordance with the classification obtained
by the function knn using the training set in Figure 4.11. The boundary between the
two classes is not precisely located, but most of the points can be considered as well
classified.

As previously shown, the *k*-NN algorithm can be computationally expensive if
the used training set contains more information than needed. This can happen when
the number of samples it contains is too large. In these cases, subsets of the original training set can be identified for obtaining the same classification in a shorter

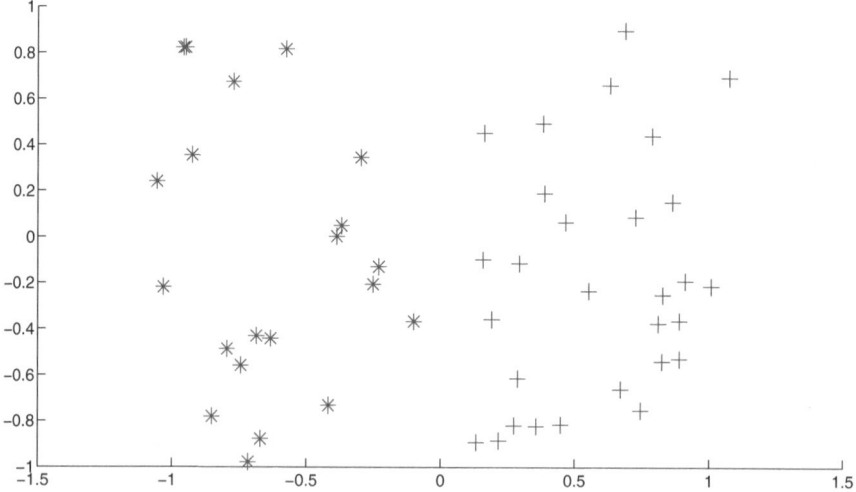

Fig. 4.11 The training set used with the function knn.

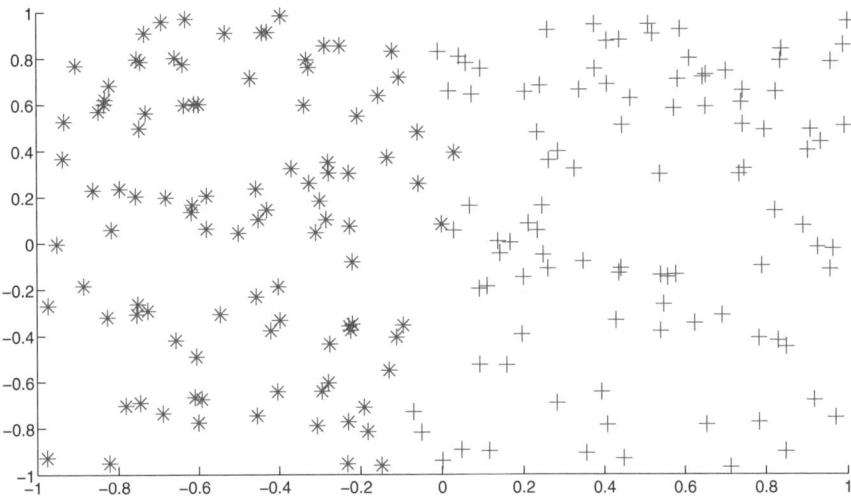

Fig. 4.12 The classification of unknown samples performed by the function knn.

time, but with a good accuracy. Figures 4.13 and 4.14 show a MATLAB function implementing the algorithm for finding a consistent subset of a starting training set T_{NN}. The function has as input parameters: the number ntnn of points contained into the original training set T_{NN}; the x coordinates xtnn of these points and the corresponding y coordinates ytnn; the numerical code indicating the class the point belongs to; finally, the k value related to the k-NN algorithm, i.e., the number of classes. The function output parameters are: the number ntcnn of points contained in the condensed subset T_{CNN}; the x and y coordinates of these points, xtcnn and ytcnn, respectively; the codes ctcnn of classes these points belong to. This MAT-LAB function uses the function knn as sub-procedure, because classifications are performed in the algorithm for finding T_{CNN}.

This function is the translation of the algorithm in Figure 4.3 in the MATLAB language. The roles played by sets X and Y are played by wellclass and badclass in the function. These two sets contain the points that are "well" classified and "bad" classified during the algorithm. Each of them is represented in MATLAB by an integer number counting its size and three vectors. The set of bad classifications is for instance considered through the variables: nbadclass counting the number of points in the set; xbadclass containing the x coordinates of its points; ybadclass containing the y coordinates of such points; cbadclass containing the corresponding class codes. At the start, badclass is initialized by copying the first sample of T_{NN} into it. Recursively, then, all the other samples are classified by the knn function using badclass as training set (first for loop). Even though badclass contains one point only at the start, it gets bigger every time a sample is not classified correctly by knn. More precisely, every time knn runs for classifying one sample, such sample is moved to wellclass if it is classified well and it is moved to badclass if the classification is not correct. After this starting phase, a while loop starts. This loop

```
%
% this function computes a condensed subset T_CNN of
% a given training set T_NN
%
% input:
% ntnn - number of points in T_NN
% xtnn - x coordinates of points in T_NN
% ytnn - y coordinates of points in T_NN
% ctnn - classes each point in T_NN belongs to
% k    - kNN parameter
%
% output:
% ntcnn - number of points in T_CNN
% xtcnn - x coordinates of points in T_CNN
% ytcnn - y coordinates of points in T_CNN
% ctcnn - classes each point in T_CNN belongs to
%
% [ntcnn,xtcnn,ytcnn,ctcnn] = condense(ntnn,xtnn,ytnn,ctnn,k)

function [ntcnn,xtcnn,ytcnn,ctcnn] = condense(ntnn,xtnn,ytnn,ctnn,k)

  % the first point is added to class "badclass"

  nbadclass = 1;
  xbadclass(nbadclass) = xtnn(1);
  ybadclass(nbadclass) = ytnn(1);
  cbadclass(nbadclass) = ctnn(1);
  nwellclass = 0;

  % checking the classification

  for i = 2:ntnn,

    % classifying points in (1,xtnn(i),ytnn(i))
    % by using (nbadclass,xbadclass,ybadclass,cbadclass) as training set
    class = knn(1,xtnn(i),ytnn(i),k,nbadclass,xbadclass,ybadclass,cbadclass);

    if class == ctnn(i),
      nwellclass = nwellclass + 1;
      xwellclass(nwellclass) = xtnn(i);
      ywellclass(nwellclass) = ytnn(i);
      cwellclass(nwellclass) = ctnn(i);
    else
      nbadclass = nbadclass + 1;
      xbadclass(nbadclass) = xtnn(i);
      ybadclass(nbadclass) = ytnn(i);
      cbadclass(nbadclass) = ctnn(i);
    end
  end
```

Fig. 4.13 The MATLAB function condense: first part.

stops when wellclass does not have any point anymore or the variable nmoves is zero when an iteration of the while loop is over. At each iteration of this while loop, each point in wellclass is classified by using badclass as training set. If the point is classified incorrectly, then it is moved from wellclass to badclass. At the end of the procedure, the points in badclass are able to classify correctly those in wellclass. The final condensed set is therefore given by the current points in badclass.

As before, a training set can be generated by using the function generate and by applying the kmeans algorithm for assigning a class to each point of the set. In

```
% checking the points in "wellclass"

while nwellclass > 0,

    nmoves = 0;
    i = 0;

    while i < nwellclass,

      i = i + 1;

      % classifying points in (1,xwellclass(i),ywellclass(i),cwellclass(i))
      % by using (nbadclass,xbadclass,ybadclass,cbadclass) as training set

      class = knn(1,xwellclass(i),ywellclass(i),k,nbadclass,xbadclass,
                  ybadclass,cbadclass);

      % if the point is not well-classified
      % it is moved from wellclass to badclass
      if class ~= cwellclass(i),
        nmoves = nmoves + 1;
        del(nmoves) = i;
        nbadclass = nbadclass + 1;
        xbadclass(nbadclass) = xwellclass(i);
        ybadclass(nbadclass) = ywellclass(i);
        cbadclass(nbadclass) = cwellclass(i);
        nwellclass = nwellclass - 1;
        xwellclass(i) = [];
        ywellclass(i) = [];
        cwellclass(i) = [];
        i = i - 1;
      end

    end

    if nmoves == 0,
      nwellclass = 0;
    end

end

% the class "badclass" is the condensed subset

ntcnn = nbadclass;
for i = 1:ntcnn,
  xtcnn(i) = xbadclass(i);
  ytcnn(i) = ybadclass(i);
  ctcnn(i) = cbadclass(i);
end

end
```

Fig. 4.14 The MATLAB function condense: second part.

this case, 200 points have been generated by setting eps = 0.1. Then, the function kmeans is used with k = 4. The generated training set is presented in Figure 4.15(a). In Figure 4.15(b) there is the corresponding condensed set T_{CNN}. In Figure 4.16 the performances of the knn function using the obtained reduced set are shown. After the reduction of the training set, the quality of the classification remains the same. Just few points close to the borders among different classes are misclassified. This

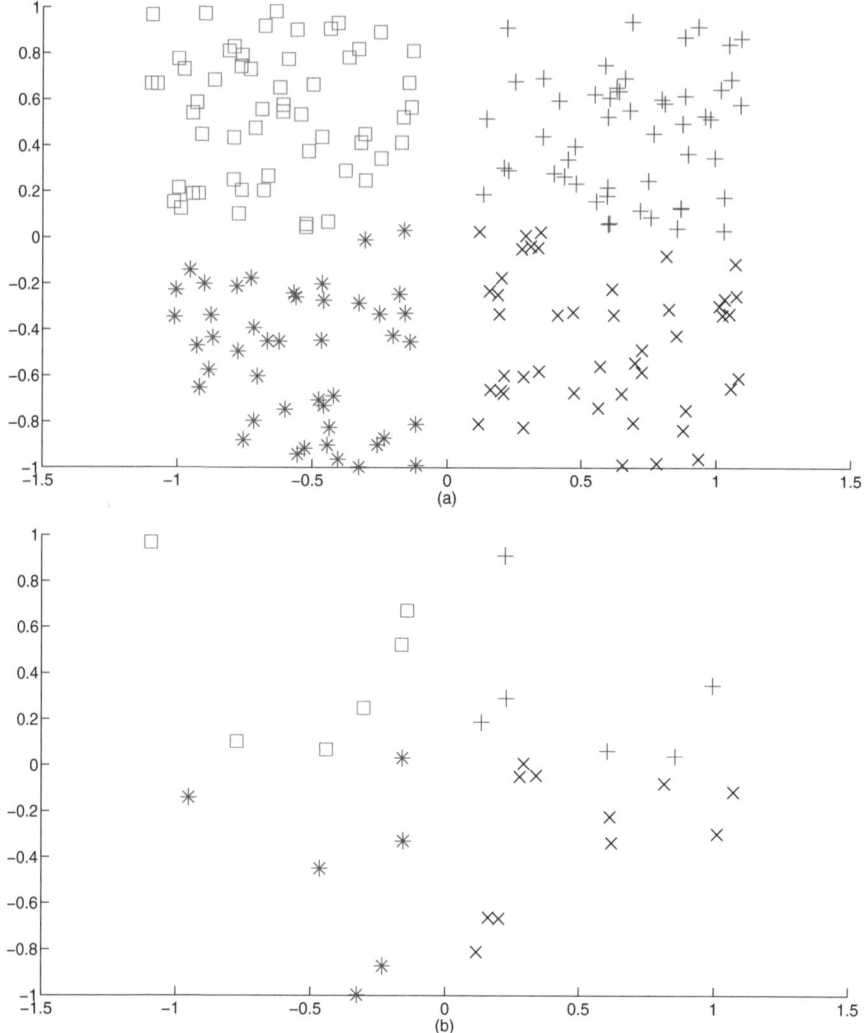

Fig. 4.15 (a) The original training set; (b) the corresponding condensed subset T_{CNN} obtained by the function condense.

can be avoided if the margin among the classes is larger. The set of points to classify has been generated by the function generate with n = 50 and eps = 0.

Figure 4.17 shows the MATLAB function implementing the algorithm in Figure 4.5 for finding a reduced subset T_{RNN} of a training set. The input and output parameters of this function are similar to the ones of the function condense. The integer number ntnn and the three vectors xtnn, ytnn and ctnn represent the original training set T_{NN}. The parameter k always refers in this context to the number of nearest neighbors used during the classification algorithm. As outputs, the integer

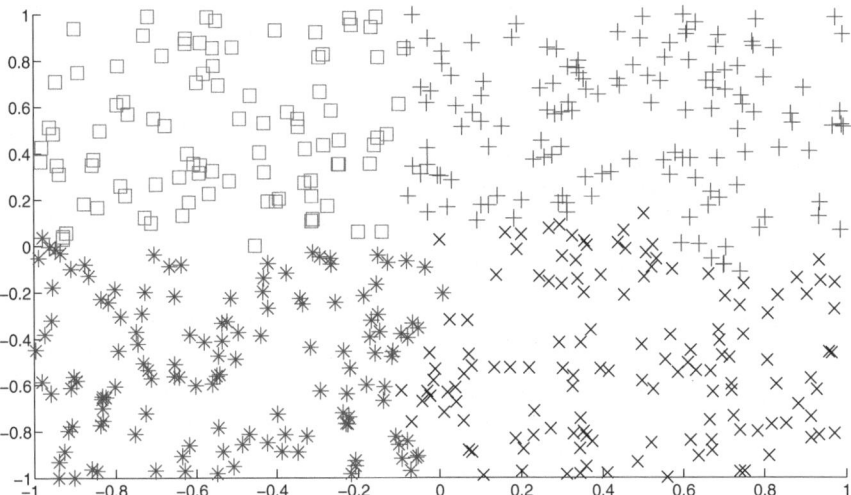

Fig. 4.16 The classification of a random set of points performed by knn. The training set which is actually used is the one in Figure 4.15(b).

ntrnn and the vectors xtrnn, ytrnn and ctrnn represent the reduced set this function provides.

At the beginning, the entire original set is copied into the variables that will contain the reduced set at the end. At each iteration of the for loop, a sample has a chance to be removed from it, allowing the reduced set to get smaller. The integer count counts the points of the original training set. Each of them is moved from the reduced set to an auxiliary set. If the points currently in the auxiliary set cannot be correctly classified using the current reduced set as training set, then the point is moved back to the reduced set. In the other case, however, the point remains in the auxiliary set, and therefore the reduced set is actually reduced. Figure 4.18(a) shows the subset obtained by function reduce from the set in Figure 4.15(a). Figure 4.18(b) shows the classification provided by knn using the reduced training set on the same set of 500 random points. As before, the classification accuracy remains the same after the reduction of the training set, while the computational cost of the classification decreases.

4.6 Exercises

Exercises related to the k-NN algorithm follow.

1. Let us suppose it is necessary to distinguish between points on a Cartesian system having positive x value and negative x value. Let us call these two classes as C^+ and C^-. By using the training set

```
%
% this function computes a reduced subset T_RNN of
% a given training set T_NN
%
% input:
% ntnn  - number of points in T_NN
% xtnn  - x coordinates of points in T_NN
% ytnn  - y coordinates of points in T_NN
% ctnn  - classes each point in T_NN belongs to
% k     - kNN parameter
%
% output:
% ntrnn - number of points in T_RNN
% xtrnn - x coordinates of points in T_RNN
% ytrnn - y coordinates of points in T_RNN
% ctrnn - classes each point in T_RNN belongs to
%
% [ntrnn,xtrnn,ytrnn,ctrnn] = reduce(ntnn,xtnn,ytnn,ctnn,k)

function [ntrnn,xtrnn,ytrnn,ctrnn] = reduce(ntnn,xtnn,ytnn,ctnn,k)

   % copying the original training set

   ntrnn = ntnn;
   for i = 1:ntrnn,
     xtrnn(i) = xtnn(i);
     ytrnn(i) = ytnn(i);
     ctrnn(i) = ctnn(i);
   end

   % an auxiliary set (n,xtrain,ytrain,ctrain) is needed
   n = 0;

   % performing the reduction algorithm

   for count = 1:ntnn,

     % moving one point from T_RNN to the auxiliary set
     n = n + 1;
     xtrain(n) = xtrnn(1);
     ytrain(n) = ytrnn(1);
     ctrain(n) = ctrnn(1);

     ntrnn = ntrnn - 1;
     xtrnn(1) = [];
     ytrnn(1) = [];
     ctrnn(1) = [];

     % classifying points in the auxiliary set by T_RNN

     aux_class = knn(n,xtrain,ytrain,k,ntrnn,xtrnn,ytrnn,ctrnn);

     % counting the number of misclassifications

     nbadclass = 0;
     for i = 1:n,
       if ctrain(i) ~= aux_class(i),
         nbadclass = nbadclass + 1;
       end
     end

     % if there is one misclassification at least
     % the point is moved back

     if nbadclass > 0,
       ntrnn = ntrnn + 1;
       xtrnn(ntrnn) = xtrain(n);
       ytrnn(ntrnn) = ytrain(n);
       ctrnn(ntrnn) = ctrain(n);
       n = n - 1;
       xtrain(n) = [];
       ytrain(n) = [];
       ctrain(n) = [];
     end

   end

end
```

Fig. 4.17 The MATLAB function reduce.

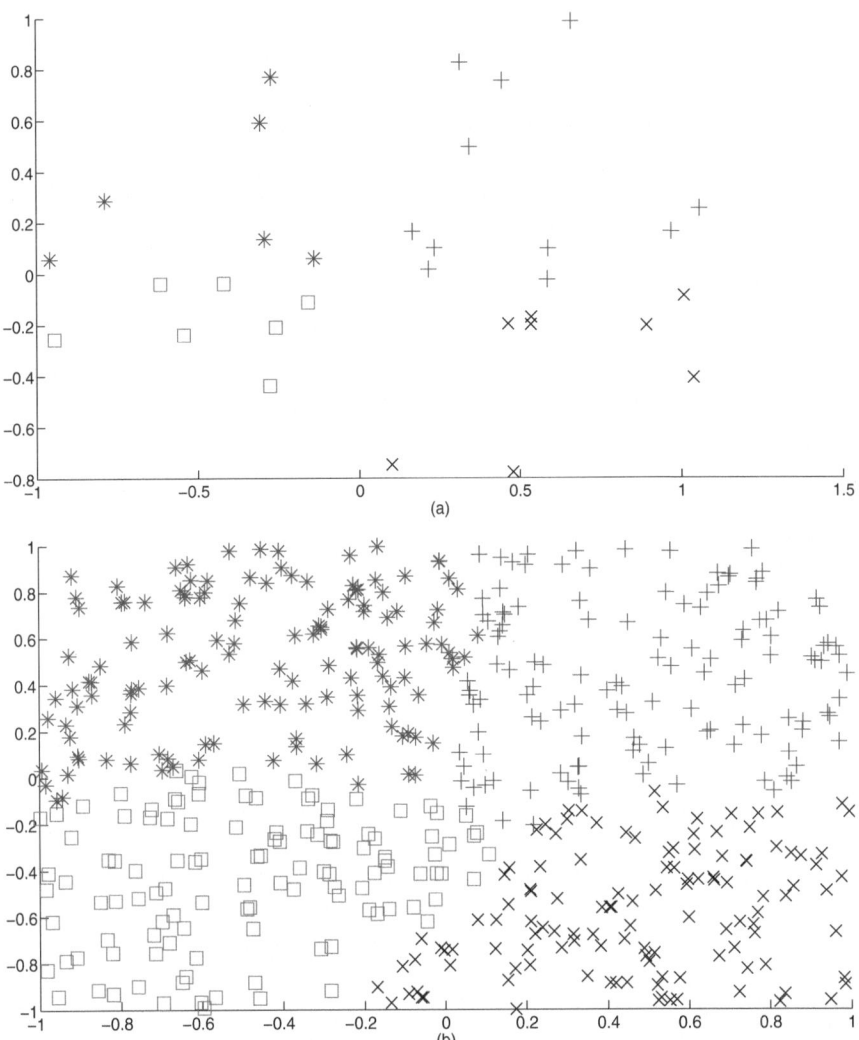

Fig. 4.18 (a) The reduced subset T_{RNN} obtained by the function `reduce`; (b) the classification of points performed by knn using the reduced subset T_{RNN} obtained by the function `reduce`.

$$\left\{\{(-1, -1), C^-\}, \{(-1, 1), C^-\}, \{(1, -1), C^+\}, \{(1, 1), C^+\}\right\},$$

classify the points (2,1), (-3,1) and (1,4) with the k-NN algorithm and $k = 1$.

2. Given the training set:

$$\{\{(0, 1), C_A\}, \{(-1, -1), C_A\}, \{(1, 1), C_A\}, \{(-2, -2), C_B\}, \{(2, 2), C_B\}\},$$

classify the points

$$(7, 8), (0, 0), (0, 2), (4, -2)$$

using the 1-NN rule.

3. By using the training set in Exercise 2, classify the points $(5, 1)$ and $(-1, 4)$ applying the 3-NN decision rule.

4. Provide an example of a training set such that the same unknown sample can be classified in different ways if k is set to 1 or 3.

5. Plot the training set and the unknown sample that satisfies the requirements of Exercise 4 by using the MATLAB function `plotp`.

6. Solve the classification problem proposed in Exercise 1 in the MATLAB environment and with the function `knn`.

7. In MATLAB, create a training set and find the corresponding condensed and reduced set with the functions `condense` and `reduce`.

8. In MATLAB, build figures showing the original training set, and the condensed and the reduced subsets obtained in the previous exercise.

9. In MATLAB, use the function `knn` for classifying a set of unknown points. Use the training set originally generated in Exercise 7. Create a figure showing the obtained classification.

10. In MATLAB, use the function `knn` for classifying a set of unknown points. Use the condensed and reduced subsets obtained in Exercise 7. Create a figure showing the obtained classifications.

Chapter 5
Artificial Neural Networks

5.1 Multilayer perceptron

In the early days of artificial intelligence (AI), artificial neural networks (ANNs) were considered a promising approach to find good learning algorithms to solve practical application problems [189]. Perhaps, a certain unjustified hype was associated to their use, since, nowadays, ANNs seem to have less appeal for researchers. In fact, they are not considered to be among the top 10 data mining techniques [237]. Moreover, publications using ANNs are found not to be backed by a sound statistical analysis [75] and that statistical evaluation of ANNs experiments is a necessity [74]. There are, however, applications in which ANNs have been successfully used. Among such applications, there are the applications in the agricultural-related areas which are discussed in Section 5.4 of this chapter. Therefore, even though they may not be so appealing for some researchers anymore, we decided to dedicate this chapter to ANNs.

ANNs can be used as data mining techniques for classification. They are inspired by biological systems, and particularly by research on the human brain. ANNs are developed and organized in such a way that they are able to learn and generalize from data and experience [99]. Despite their origin related to brain studies, the networks discussed in this chapter have little to do with biology.

In general, ANNs are used for modeling functions having an unknown mathematical expression. In Chapter 2 we showed that, given a set of independent variables (inputs) and corresponding dependent variables (outputs), interpolation and regression techniques can be used for modeling such data. As already discussed, when interpolating polynomials are used, one problem is that their degree grows with the dimension of the set of data. This problem is avoided when splines or regression approaches are used. However, there are reasons that brought researchers to use ANNs instead of interpolation and regression models. First of all, ANNs do not become more complex if the set of data used is larger. Moreover, ANNs can model very complex functions without the need of finding their (complex) mathematical expressions.

A. Mucherino et al., *Data Mining in Agriculture*, Springer Optimization and Its Applications 34, 107
DOI: 10.1007/978-0-387-88615-2_5,
© Springer Science + Business Media, LLC 2009

According to [180], ANNs consist in a number of independent and simple processors: the neurons. The network is formed by neurons, which are connected and usually organized in layers. The human brain contains tens of billions of neurons and tens of trillions of such connections. Each neuron is characterized by an activity level and by its input and output connections. The activity level represents the state of polarization of a neuron, the input connections feed the neuron with signals, whereas the output connections broadcast the neuron signal to others. All these neuron properties are represented mathematically by real numbers. Each link or connection between neurons has an associated weight, which determines the effect of the incoming input on the activation level of the neuron. The weights can be positive or negative. If a connection has a positive weight, its effect on the signal passing through is excitement, whereas effect is inhibitory if the weight is negative. In other words, if the weight sign is positive, it raises the activation; if the sign is negative, it lowers the activation. ANNs differ from each other by the way in which the neurons are connected, by the way each neuron processes its input, and by the learning method used. Usually, the network structure is defined a priori, and must be tailored to the process that must be modeled. During the learning phase, only the connection weights are optimized in a way that the network can respond with the given outputs when it has certain inputs.

The multilayer perceptron is the kind of ANNs that are the focus of this chapter. The multilayer perceptron has the neurons organized in layers, one input layer, one or multiple hidden layers and one output layer. In some applications there are only one or just two hidden layers, but it is more convenient to have more than two layers in some other applications. Figure 5.1 shows an example of a multilayer perceptron. The input data are provided to the network through the input layer, which sends this information to the hidden layers. The data are processed by the hidden layers and the output layer. Each neuron receives output signals from the neurons in the previous layer and sends its output to the neurons in the successive layer. The last layer, the output one, receives the inputs from the neurons in the last hidden layer, and its neurons provide the output values. The neurons of the input layer do not perform any computation, since they are just allowed to receive the data that they send to the first hidden layer. Layer by layer, then, the neurons communicate among them and process the data they receive. The network is able to provide the output values after the inputs have propagated from the input layer to the output layer through the entire network.

As already mentioned, initial research on ANNs presented them as a very promising approach for learning from data. For instance, many benefits in using ANNs have been discussed in [99]. In this paper, besides presenting neural networks as a good alternative to polynomial interpolation or regression, other advantages are discussed. ANNs are for instance said to be able to handle imperfect and incomplete data. Therefore they may be useful when working with data from the real world, which are noisy and imprecise. Moreover, data from the real world are often complex, and since multilayer perceptron is nonlinear, it can capture complex iterations among the input variables of the system. Finally, ANNs are highly parallel, so that

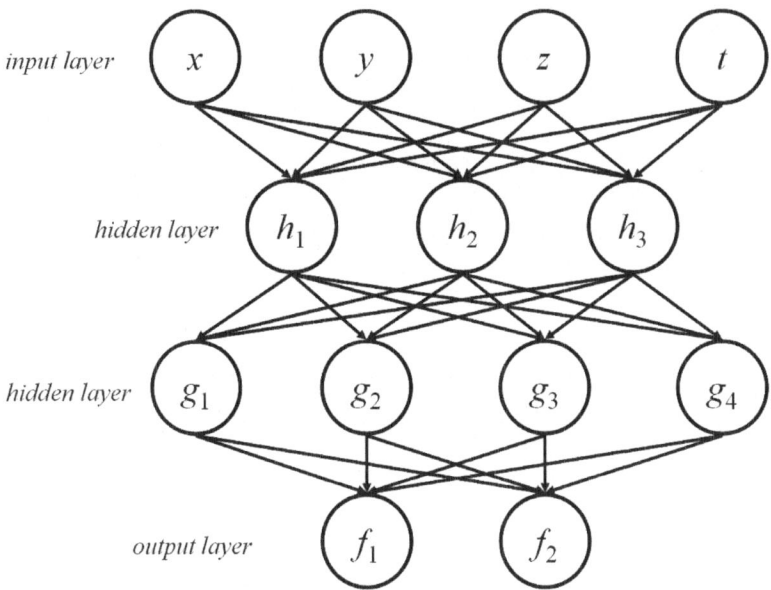

Fig. 5.1 Multilayer perceptron general scheme.

they can naturally be developed in a parallel environment. In fact, an implementation for parallel computing of ANNs is provided in Section 9.3.3.

As already pointed out, ANNs can be used for mathematically modeling a certain unknown process. A network having n neurons in the input layer and m neurons in the output layer can be used for describing a function having n independent variables and m dependent variables. Using a mathematical language, ANNs can model functions defined in $\Re^n$ and having values in $\Re^m$ (where $\Re$ is the set of real numbers), if the network has n input neurons and m output neurons.

Each neuron receives as input the outputs from the neurons in the previous layer. Passing through the connections, these outputs are lowered or raised, depending on the connection weights. All these values are assumed to sum linearly yielding an activation value for the current neuron. If j is one of the neurons in the current layer, and L is the number of neurons in the previous layer connected to j, then the function

$$\text{net}_j = \sum_i^L w_{ij} o_i$$

computes the activation value for j, where w_{ij} is the weight associated to the link between the neuron i of the previous layer and the neuron j, and where o_i is the output provided by neuron i. The obtained value is then processed by neuron j in the current layer, by computing its output

$$o_j = O_j(\text{net}_j).$$

The function O_j is fixed for each neuron and it is normally a nonlinear function of its activation value. Usually it is chosen to be a smooth function, and the default choice is the standard sigmoid function:

$$O_j = \text{sigmoid}(x) = \frac{1}{1 + e^{-x}}.$$

Other functions are also used, as for instance the one used in the application described below in Section 5.4.2, which is the logistic function:

$$O_j = \text{logistic}(x) = \frac{1}{1 + e^{\frac{-x}{T}}}. \tag{5.1}$$

In the formula, the parameter T of the logistic function yields functions of different slopes. Section 5.4.1, instead, is focused on an application in which the logistic function is used only for the neurons on the output layer, while the function O_j corresponds to

$$O_j = \tanh(x) = \frac{e^x - e^{-x}}{e^x + e^{-x}}, \tag{5.2}$$

if the neurons on the hidden layers are considered. Function (5.2) is called hyperbolic tangent. Functions O_j, in general, are usually predefined a priori and they are not modified during the learning process.

The simplest neural network whose neurons are organized in layers is the one having one input layer, one output layer, but no hidden layers. This kind of network is actually called *single* perceptron and was presented in [198] in 1958. In this case, the input variables are processed only by the neurons in the output layer: the output variables are computed by functions O_j which have as input a linear combination of the input variables. This kind of network may be useful for its simplicity, but it cannot solve some types of problems, in particular when the function to model is not linearly separable. In other words, if the function to model cannot be written as a linear combination of its inputs, then the single perceptron cannot model it, just because net$_j$ is a linear combination. The hidden layers have been introduced for overcoming this problem: a multilayer perceptron having just one hidden layer can model nonlinear functions.

ANNs are commonly used as classification techniques. They can be used for supervised learning, since the network parameters (the neuron weights) are computed by computational procedures based on a certain training set of data. The hope is that the network so designed is able to generalize, i.e., to correctly classify data that are not present in the training set. As explained in [215], generalization is usually affected by three factors. The first one consists of the size and efficiency of the training set, since small sets of data cannot contain information enough for generalization, and, even when they are larger, they may not be efficient. The training set may, for instance, contain data which are representative for some classes and not for some others, providing in this way incomplete information. Another important factor is the complexity of the network. The number of hidden layers can impact the accuracy of the system: a system with a large number of hidden layers has better chances to provide better accuracy. However, if the complexity grows, the training and the normal use of the neural network may become too computationally demanding, and therefore a good trade-off must be found. Finally, a crucial factor is the complexity of the process which needs to be modeled. After the network size is determined, including the number of hidden layers and the number of neurons for each of them, the network must be trained. An important issue is to select a good algorithm for this purpose. In Section 5.2 we will overview some of the commonly used training algorithms for ANNs. After the learning phase, the network is able to use what it did learn from the data. Evaluating and using these trained networks can be computationally expensive, and some redundant links and useless neurons may be removed to make the network more efficient. This phase is called *pruning* of the network, and these issues are discussed in Section 5.3.

5.2 Training a neural network

The problem of learning in neural networks is the problem of finding a set of connection weights which allows the network to carry out the desired computations. During the learning process, the neural network must learn how to model the data. The most used method is the back-propagation method.

The basic idea of the back-propagation method is as follows. It is supposed that a set of input data and a set of corresponding output data are available. It is required that the network is able to provide the correct output when a certain input is provided. In other words, the network has to deliver certain output results $\{o_1, o_2, \ldots, o_m\}$ when it receives certain input variables $\{i_1, i_2, \ldots, i_n\}$. The back-propagation method works on the weights associated to each link between neurons. Predefined weights can be used at the start of the algorithm. In the case predefined weights are not available they can be randomly generated. The method starts feeding the network with inputs i_k and allows these signals to propagate through the network layer by layer. Every time a neuron receives inputs from the neurons in the previous layer, it computes a weighted sum of them and sends its output to the neurons in the successive layer. When the signal arrives to the output layer, its neurons compute the outputs. Let us denote the generic output obtained with the symbols co_k, meaning "current output." At this point, these current co_k outputs and the outputs o_k the network should learn to provide can be compared. The difference between o_k and co_k can be defined as the current error e_k which is present in the output neuron k. The error values are then passed back to the last hidden layer using the same weights. This backward propagation gives the name to the algorithm. By computing the weighted sums of the received errors, each neuron is able to compute its contribution to the output error, and adjust its weight for reducing the output error. The back-propagation method is iterative and it stops when the network can process the input with sufficient accuracy. The final weights represent what the network has learned.

The difficult task in this back-propagation process is to find out the connections between the neurons that are not performing correctly or that are performing worse than the others. This is a nontrivial problem, especially if the network has hidden layers. A possible way for facing the problem is to avoid finding a single connection or a set of connections to blame for the network error and considering a measure of the overall performance of the system. The performance of the network can be defined as follows:

$$E = \sum_{\xi,k} \left(co_k^\xi - o_k^\xi \right)^2 , \qquad (5.3)$$

where o_k represents the kth expected output, and where co_k is the current output provided by the network. The superscripts ξ correspond to the sets of input/output to be learned. E represents the total error of the network. Therefore the task of learning from a given training set can be seen as an optimization problem, where E must be minimized. The problem is unconstrained, and it may be solved by using one of the optimization methods discussed in Section 1.4.

Many approaches have been proposed for the learning phase of a neural network. In [129], for instance, genetic algorithms (GAs) are used [88]. GAs are meta-heuristic methods for global optimization and they use simple operators in order to simulate evolution according to Darwinian theory. GAs are among the meta-heuristic methods for global optimization listed in Section 1.4. In this case, GAs work with a population of networks, which are randomly generated when the algorithm starts. The main operator in the search is the crossover, which generates new network children starting from network parents. In these studies, a network (or individual or chromosome) is represented as a square matrix such that each single element in row i and column j has value $\eta_{ij} = 0$ if there are no connections between neurons i and j, and value $\eta_{ij} \neq 0$ if there is a connection. This matrix can contain all the information regarding a neural network, such as connectivity and weights. Two special crossover operators are used, which are tailored to the matrix representation of the network. The row-wise crossover is performed generating two children by exchanging two random rows between two parents. In the same way, the column-wise crossover is performed by exchanging two random columns between two parents. GAs have also been used in other studies with the aim of training a network as fast and efficiently as possible. In [113], for instance, GAs have been used coupled with a BFGS (Broyden-Fletcher-Goldfarb-Shano) method [204] for improving the training performance.

One problem that may occur during the learning process is overfitting. At some point, in later stages of the learning process, the network may start to fit the data in the training set very well. In the meantime, though, it may start to lose generalization. In other words, the network begins to be very good at reproducing the data on which it is trained, whereas it may be completely wrong on any other kind of data. For avoiding overfitting, the generalization ability of the network during training can be checked and the learning process can be stopped when this ability begins to decrease. The simplest method is to divide the data into a training set and a validation set. The training set can then be used during the learning process, whereas the validation set can be used to estimate the generalization ability. The learning process must therefore be stopped when the error on the validation set begins to increase. This technique can work very well for avoiding overfitting, but it may not be practical when only a small amount of data is available, since the validation data cannot be used for training.

After a network has been trained, it is expected to be able to classify samples using the parameters established during the training phase. It is desirable that the classification is as fast as possible. In order to improve the performance of a neural network, the network can be pruned. During the pruning process, all the redundant and useless connections that affect the performance of the network can be removed. In Section 5.3 we will discuss pruning strategies for neural networks.

5.3 The pruning process

As discussed in Section 5.2, ANNs can generalize well from the training set if the network does not overfit during the training process. A way for avoiding this phenomenon could be to use the smallest network able to model a certain problem [193]. However, it is not easy to determine the optimal network size for a particular problem. One possible approach is to train successively smaller networks until the smallest one is found that is able to learn from the data. This process can work but it can be time consuming. Therefore, other strategies have been proposed over time in order to improve the ability of the neural network to generalize.

Training many networks having a decreasing number of neurons and choosing the smallest one able to generalize from the data can be computationally demanding. The alternative is to try training a network with a number of neurons unnecessarily large. Training a large network can be expensive, but not as expensive as training many networks. The problem is that this large network can also be very expensive to use, and for this reason it needs to be pruned after the training process. The initial large network size allows learning reasonably quickly and the network can then work efficiently when the unnecessary neurons and connections have been removed.

A brute-force pruning method is as follows. After the network has been trained, all its weights can be considered one per time, and set to zero. The total error provided by the network can then be checked on the training set. If the error increases too much, it means that the link corresponding to the weight set to zero is indispensable and cannot be removed. Otherwise, if the total error is acceptable, the link can be eliminated from the network. If all the connections related to one neuron are removed, the neuron itself can be eliminated from the network. The brute-force method can be quite expensive. If W is the number of weights contained in the network and M the number of input/output couples from the training set, the computational cost is about MW^2, because, every time the method tries to delete one of the W weights, it has to check M errors over $W - 1$ connection.

Other methods, even more sophisticated, have been proposed over the time for pruning a neural network, and they can be divided into two main groups. One group contains methods that estimate the sensitivity of the error function to the removal of a neuron, and the ones with the least effect are then removed. The second group contains methods that add terms to the objective function that reward the network for choosing solutions in which the weights are smaller. For instance, a term proportional to the sum of all weight magnitudes favors solutions with small weights. The ones that are nearly zero are not likely to influence the output much and hence they can be eliminated. There is some overlap in these two groups, because the term added to the objective function can include sensitivity terms.

Many pruning tests have been proposed in the literature. In [37], for instance, the pruning problem is formulated in terms of solving a system of linear equations. The basic idea is to iteratively eliminate neurons and adjust the remaining weights in such a way that the network performance does not worsen over the entire training set. In [78], instead of pruning the network as a whole, it is pruned layer by layer with the use of a pruning decision based on local parameters. Other recent works on pruning algorithms can be found in [179, 238, 249].

5.4 Applications

ANNs have been used experimentally for decades in practical applications. An interesting work is for instance the one presented in [200] for detecting frontal views of faces in gray-scale images. In this approach, more than one neural network is used and each of them is trained to output the presence or the absence of a face in an image. This is a very difficult detection task. Unlike face recognition, in which the classes to be discriminated represent different kind of faces, the two classes to be discriminated in face detection are "images containing faces" and "images not containing faces." Obtaining a representative sample representing images without faces is the most difficult task. Experiments presented in [200] showed that neural network can handle this kind of problem, and one of the experiments is presented in Figure 5.2. Other general applications of neural networks include the classification of recorded musical instrument sounds [62], the development of decision making tools in the field of cancer [155], and the classification of events during high-energy physics experiments at the Super Proton Synchrotron at CERN in Geneva, Switzerland [224].

Neural networks have been successfully applied in agriculture and related fields. For instance, a neural network approach has been proposed in [135] for evaluating sugar and acid contents of a variety of oranges by a machine vision system. Machine vision can replace human visual judgment by providing a more consistent and reliable system. The measurement of the sugar or acid contents of an orange fruit is, however, a difficult task, because its skin is thick and usually light cannot penetrate the skin effectively. In the approach proposed in [135], images of the oranges have been taken and the sugar and acid contents have been measured by the standard equipment. The neural network has been used for finding the relationships between the orange aspect and the acid and sugar contents. The used three-layer network has been able to predict that reddish, low height, medium size and glossy orange fruits are relatively sweet.

Fig. 5.2 The face and the smile of Mona Lisa recognized by a neural network system. Image from [200].

However, the network could not provide a clear indication of the level of sugar content, but the feasibility to evaluate inside quality of fruits by neural networks and machine vision has been anyway demonstrated.

Other applications of neural networks in the field of agriculture are for instance:

- Classification of fertile and infertile eggs by machine vision [53];
- Prediction of flowering and maturity dates of soybean [67];
- Detection of cracks in eggs using computer vision [185];
- Forecasting water resources variables [160];
- Detection of pig coughs in farms by recorded sounds [45];
- Detection of watercores in apples by X-ray images [210];
- Wine classifications by taste sensors made from ultra-thin films [196];
- Modeling of sediment transport [22].

In the following we will focus on the problem of detecting pig coughs with the aim of identifying diseases in farms (Section 5.4.1) and on the problem of detecting watercore inside apples for a good selection of fruits for the market (Section 5.4.2).

5.4.1 Pig cough recognition

Coughing, in human and animals, is associated with the sudden expulsion of air. This is a defense mechanism of the body, against the possible entry of materials into the respiratory system. Coughing is typically accompanied by a sound, whose changes may reflect the presence of diseases affecting the airways or the lungs or of early symptoms of diseases. If someone is coughing, it is easy to say if he or she has a bad or normal cough from the sound produced. In the same way, the sound provided by pig coughing can be used for monitoring possible health problems. An expert could say if the cough of a pig signals the presence of a potential disease, and eventually check the health of the pig. Nowadays, however, human attention is not so present anymore, because big farms have a large quantity of animals and, moreover, the environment can be very harmful for the presence of contagious diseases [3].

Systems for the automatic control of the pig houses are useful. Their use can prevent the transmission of diseases from pigs to humans, and at the same time guarantee a constant control on pig health conditions. Therefore, considerable efforts have been undertaken for the development and application of sensors and sensing techniques for diagnosis in pig farms. Besides the advantages farmers can have, such as improving the health of the pigs and avoiding contaminations, the final consumer also can benefit from these techniques. The early detection of an animal disease can bring to the consumer's table better meat, by reducing, for instance, the residuals of antibiotics. The different techniques developed for cough detection have the common characteristic of being based on supervised learning methods. As a consequence, the failure or success of a technique depends highly on the quality of the training set of data. The training set is obtained by experimental observations, where the sounds produced by pigs are recorded and where each record is labeled by an expert in different ways. An expert farmer is indeed able to distinguish among coughs and other sounds pigs can issue.

We will focus in this section on the studies presented in [45, 170, 171] where neural networks have been used as a supervised learning technique. There is also a similar example in the literature that uses a fuzzy c-means algorithm [231]. In the neural network approach, a metal chanmber has been built in order to perform the experiments (see Figure 5.3). It is covered with transparent plastic for controlling the environment around the animal, and its dimensions are 2 m long, 0.80 m wide and 0.95 m high. The pigs are invited to enter in the metal chamber and sound measurements by a microphone are recorded. During this process, the environment inside the chamber is controlled by checking the temperature, the dust and the NH_3 concentration, and other variables. A full description of the experiment set up is presented in [169]. We just point out that the microphone is placed in the chamber from 0.4 m to 1.0 m from the pig, and it is positioned through an aperture into the plastic cover. The sample rate chosen is 22,050 Hz, because the frequencies of a typical cough are below 10,000 Hz. After the pig is invited to enter the chamber, normal pig sounds are recorded, such as grunting and other sounds due to respiration. Other sounds from the surrounding environment are also recorded. Animal movements can cause metal clanging, because the construction used in the experiment is metallic. Moreover, the

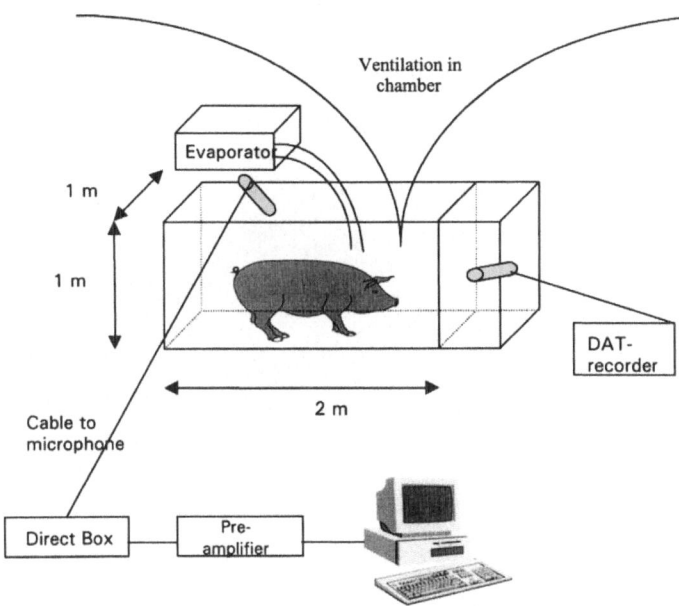

Fig. 5.3 A schematic representation of the test procedure for recording the sounds issued by pigs. Image from [45].

controlled environment that needs ventilation and the presence of researchers may cause other noises. When all these sounds are recorded, pigs are finally induced to cough for recording cough sounds.

A neural network is trained using the sounds obtained during these experiments. The training set contained 354 sounds: 212 samples are records of coughs from different pigs, 50 samples represent metal clanging, 23 samples grunts, and 69 samples background noise. Each sound is analyzed by a human expert to determine whether it is a cough or not. All these samples are then divided into two sets, the training and the testing set. The sounds have been equally distributed between the two sets, except for the sounds of coughs that are used more in the testing set, because it is important to check if the recognition of the coughs is correct. Figure 5.4 shows the time signal of a pig cough. The amplitude for the cough in all the samples recorded is 0.5 ± 0.09. The grunts have a larger duration and variability; among all the samples the duration is 1.2 ± 0.15. The time signal of these sounds is analyzed mathematically and transformed in a vector formed by 64 real numbers. For further details about this process, the reader may refer to [45] and the citations therein. This transformation is very useful, because it allows one to work on vectors and not on signals. In the following, then, two sounds are compared by comparing the components of two real vectors. They are normalized before use, because their components can vary significantly even when comparing two vectors from the same class. These variations are mainly due to the distance and direction between the pigs and the microphone.

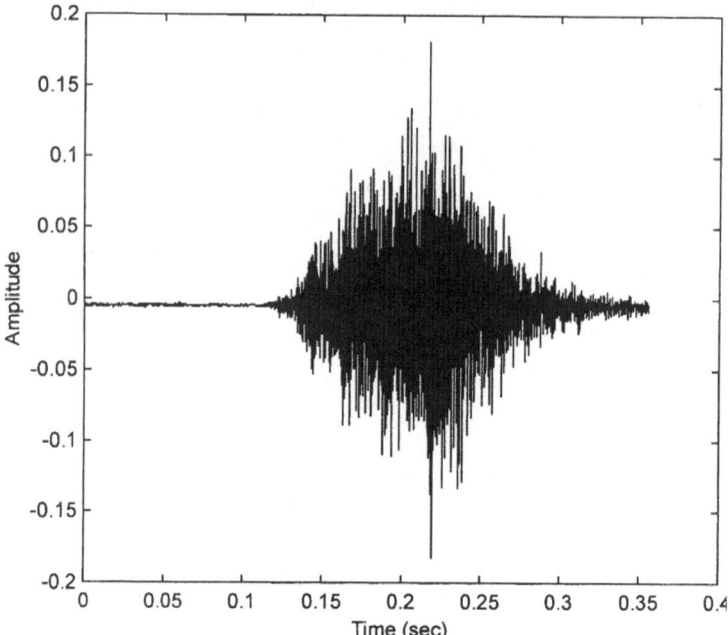

Fig. 5.4 The time signal of a pig cough. Image from [45].

Such variations do not negatively affect the quality of the sound because of the low environmental noise.

The network is trained using a BFGS optimization procedure [204]. The network is a multilayer perceptron with one or two hidden layers of hyperbolic tangent neurons (see equation (5.2)), while the output layer consists in logistic neurons (see equation (5.1)). The multilayer perceptrons with two hidden layers did not provide any improvement on the correct classification percentage. Once the network is trained to discriminate between coughs and metal clanging, it is able to reach percentages of correct recognition greater than 90%. This is a very difficult task, because these two sounds have a similar frequency range. Then the network is trained to distinguish among four sounds: coughs, metal clanging, grunting and background noise. The confusion matrix shown in Figure 5.5 describes how many of the sounds, whose correct class appears in the first column, are misclassified. The recognition accuracy remains high, as the figure shows.

5.4.2 Sorting apples by watercore

Grading fruits before marketing is a very important process that can increase the profits, since quality defects decrease the marketability of the fruit. In this section, we

Sound	Coughs	Metal Clanging	Grunting	Noise
Coughs	69.5	21.7	8.7	0.0
Metal clanging	4.3	82.6	0.0	0.0
Grunting	0.0	0.0	91.3	8.7
Noise	0.0	0.0	8.7	91.3

Fig. 5.5 The confusion matrix for a 4-class multilayer perceptron trained for recognizing pig sounds.

will focus on grading procedures of apples. Some defects, such as discoloration, poor shape, external damage, and bruising in light colored apples are visible externally and apples containing such defects are commonly removed at sorting tables. A recognition system based on the k-means algorithm and based on the external appearance of the fruit is discussed in Section 3.5.2. Unfortunately, other defects are internal. Such defects are particularly harmful to consumer acceptance since they are typically recognized after purchase. Internal defects include internal browning, internal small black regions of unknown origin, core and other rot, watercore and insect damage. Bruises are generally referred to as external defects. Codling moth problems in exported apples can be expected to increase with the phase-out of methyl bromide fumigation, resulting in more sustained insect damage.

We will focus in this section on watercore. Watercore is an internal apple disorder, found in most apple varieties, that adversely affects the longevity of the fruit. Apples with slight or mild watercore are sweeter, and this may be considered a good feature of the apple. Unfortunately, apples with moderate to severe degree of watercore cannot be stored for any length of time. Moreover, internal tissue breakdown of a few fruits during storage may damage the whole batch. For this reason, apples with a sufficient percentage of watercore need to be detected and separated from the batch. Non-destructive methods such as X-ray imaging have shown promising results for detecting internal quality defects in various horticultural products. X-ray is a radioactive method which can penetrate into the apple without serious surface reflection. In particular, radiographic imaging, which is sensitive to density differences, is a good candidate for detecting the internal defects so far neglected as well as for detecting watercore and bruises. The fact that this technology is also quite inexpensive makes the X-ray method the best choice for detecting internal disorders in apples. The major challenge in this field is thus to develop adequate image analysis and classification schemes that can successfully classify products using X-ray image data.

A normal fruit has 20–35% of the total tissue volume occupied by the intercellular air space, whereas in apples with watercore this large air space is filled with a liquid. These changes in density and water content of fruit can be exploited for watercore detection by non-destructive techniques based on X-ray. In [203] watercores in apples have been detected with an accuracy of more than 90% by using still X-ray images. In this approach, apples have been scanned by X-ray and successively sliced and photographed (see Figure 5.6). The obtained images, both normal and X-ray images, have then been used to characterize them as defective or not. In this phase, both kinds of images are inspected and evaluated by human experts. In order to create an automatic classifier, computational procedures are needed for performing some

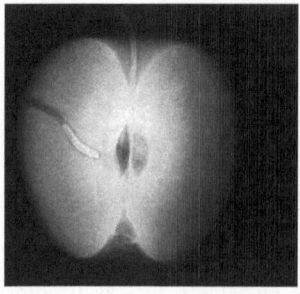

Fig. 5.6 X-ray and classic view of an apple. X-ray can be useful for detecting internal defects without slicing the fruit.

of these tasks on a computer. The inspection of the X-ray images can be carried out by a computer, which needs though to learn how to inspect such images before. Therefore, classifiers such as neural networks can be useful in these studies. In fact a method based on ANNs has been proposed in [210] for detecting watercores in apples by X-ray.

In this work, line scan images of 240 Red Delicious apples with varying degree of watercore have been acquired and three features of the images, considered good indicators of watercore, have been extracted from the images. Details about this process can be found in [209]. After scanning, the apples are cut and opened in order to check the presence of watercores from a human expert. Each fruit is scored on a scale from 0 to 2 based on watercore severity. Apples labeled with 0 do not have a watercore or they have a mild watercore, whereas apples labeled with 1 have a moderate watercore and the ones labeled with 2 a severe watercore. The final set of data obtained includes the three X-ray features and the corresponding scores. The aim is to teach a neural network to predict the score when it is fed by the three features of the images.

The set of data is randomly divided into two subsets. The first one includes 150 samples (55 having score 0, 46 having the score 1 and 49 having the score 2) and is used as training set. The second one includes 90 samples (58 having the score 0, 14 having the score 1 and 18 having the score 2) and is used as testing set. The employed network is a multilayer perceptron having three layers in total, and hence only one hidden layer. The output function O_j is the logistic function (see equation (5.1)).

Usually the number of neurons in the input layer equals the dimension of the input vector, and therefore the considered network has three neurons on the input layer, each one related to one of the features extracted from the X-ray images. There are actually two choices for selecting the number of neurons in the output layer, depending on the nature of the problem at hand. In classification problems where the network is trained to recognize well-defined classes, the number of output nodes usually equals the number of classes. A sample is recognized as belonging to a certain class when the output corresponding to this class is higher in value. However, there may be problems in using this strategy. When the classes are not well-defined by the network, the network may give two similar outputs and there may be uncertainty in assigning a sample to a class or to another. For this reason, a single output neuron is used in these studies with a continuous value coupled with two threshold levels. If the output is lower than the first threshold, the sample is considered to have mild watercore. Instead, it is considered to have severe watercore if the output value is larger than the second threshold. Samples are considered to have moderate watercore when the output value is between the two thresholds.

The number of neurons in the hidden layer is often determined either by trial and error or by ad hoc schemes. Several networks with different numbers of hidden neurons (from 2 to 10) have been evaluated to determine an optimal structure for achieving a good generalization. The network with the maximum classification accuracy is considered the optimal classifier for sorting apples. The optimal network found in this application has 4 hidden neurons. The method used for training the network is the standard back-propagation method described in Section 5.2. The 150 samples contained in the training set are divided in two other subsets. The first one, containing 105 samples, is used for normal training, while the second one having just 45 samples is used for validating the network during the training process. This strategy is used for avoiding the overfitting of the network.

The best classification accuracy has been obtained using a neural network having four neurons on the hidden layer and by using as thresholds 0.35 and 0.60. The neural classifier achieved an overall accuracy of 88% with the losses and false positives as low as 5%. The overall accuracy approached the target of 90% whereas the losses and false positives were well below the target limits of 10%. In [210] the classification of the apples has also been carried out using a fuzzy c-means method. The experiments showed that the neural network classifier performs better.

5.5 Software for neural networks

Instead of presenting experiments in MATLAB® with the technique discussed in this chapter, we just provide here a list of available software for neural networks. The main reason is that the training and use of the simplest neural network would require the need of developing relatively long codes in MATLAB. Since there is various software available for training and using neural networks, we decided it was not worthwhile to devote a section to possible implementations of the data mining technique in MATLAB. The list below includes the most popular software currently on the Internet. The reader can extend the list with a simple search with Google.

- **NeuroSolutions**, http://www.nd.com/
  It is advertised as the most powerful and flexible neural network modeling software currently available. NeuroSolutions is also available for MATLAB and Excel. It has an icon-based network design interface with an implementation of advanced learning procedures, such as conjugate gradients and backpropagation through time.
- **EasyNN**, http://www.easynn.com/
  Quoting the Web site, complex data analysis with EasyNN is fast and simple.

Prediction, forecasting, classification and time series projection is easy. Moreover, EasyNN allows one to train, validate and query ANNs with just a few button pushes.

- **MATLAB toolbox**, http://www.mathworks.com/products/neuralnet/
  Neural Network Toolbox extends the MATLAB environment with tools for designing, implementing, visualizing, and simulating neural networks. This software provides comprehensive support for many proved network paradigms, as well as graphical user interfaces (GUIs) that enable one to design and manage the networks.

5.6 Exercises

Exercises related to ANNs follow.

1. Consider a multilayer perceptron having one input neuron, two hidden neurons on only one hidden layer and one output neuron. The function O_j related to all the active neurons is just the identity function. Train the network so that it is able to model the equation:
$$y = 2x.$$

2. Prove that the network used in the previous exercise cannot model exactly the equation:
$$y = 2x + 1.$$

3. Train a multilayer perceptron having one hidden layer with 2 neurons for the AND classification problem. The network has 2 input neurons, 2 hidden neurons and only one output neuron. Suppose that the function O_j is not preassigned and choose it so that the network can perform the AND operator.

4. Consider a network with the same structure of the one in the previous exercise and with the sigmoid function (O_j) associated to the only output neuron. Suppose that all the weights have unitary value. Feed the network with the points $(6, 1)$ and $(-1, -1)$.

5. Keep working on the same network as the one in the previous exercise, but have all the weights equal to 2 and the logistic function (with $T = 2$) associated to the output neuron, and feed the network with the points $(1, 1)$ and $(0, 2)$.

6. Consider the two networks used in Exercises 4 and 5. State which of them can have the hidden layer deleted without changing the output of the network.

7. Consider the network with 2 input neurons, 3 hidden neurons on only one hidden layer and one output neuron. Suppose that the weights are equal to 0.1 if they are related to links with the input layer, and that they are equal to 0.3 if related to links with the output neuron. Suppose that the identity function is associated to all the neurons. Remove a link that caused the inactivation of one neuron.

8. Design a network having the same organization in layer and the same number of neurons as in the previous exercise, but having all the neurons from a layer connected to the following layer.

Chapter 6
Support Vector Machines

6.1 Linear classifiers

Support vector machines (SVMs) are supervised learning methods used for classification [30, 41, 232]. This is one of the techniques among the top 10 for data mining [237]. In their basic form, SVMs are used for classifying sets of samples into two disjoint classes, which are separated by a hyperplane defined in a suitable space. Note that, as consequence, a single SVM can only discriminate between two different classifications. However, as we will discuss later, there are strategies that allow one to extend SVMs for classification problems with more than two classes [232, 220]. The hyperplane used for separating the two classes can be defined on the basis of the information contained in a training set.

In this section, the basic idea behind the SVMs is introduced through examples. For this aim, let us consider the image in Figure 6.1, showing apples with a long stem and apples with a short stem (for the version of the book in color, note that green apples have a short stem and red apples have a long stem). Let us suppose that a general rule for classifying these apples is needed, i.e., a classifier is wanted that is able to decide if a given apple has a short or a long stem. In the example in Figure 6.1, areas of the Cartesian system can be easily located in which only apples with a short stem, or only apples with a long stem, can be found. Therefore, a classifier could simply follow the rule: the apple has a short stem if it is in an area defined by the apples having a short stem, and it has instead a long stem if it is in the area defined by the apples having a long stem. Apples with a known classification can be used for defining the two areas of the Cartesian system related to these two different types of apples. Such apples define the training set, which can be used for learning how to classify apples whose length of the stem is unknown. In other words, they can be used for locating the two areas of the Cartesian system in which only one type of apple is contained.

How can we define these two areas of the Cartesian system with the aim of classifying the apples? As Figure 6.2 shows, many straight lines can be used for dividing the Cartesian system into two disjoint areas such that one contains only

A. Mucherino et al., *Data Mining in Agriculture*, Springer Optimization and Its Applications 34,   123
DOI: 10.1007/978-0-387-88615-2_6,
© Springer Science + Business Media, LLC 2009

Fig. 6.1 Apples with a short or long stem on a Cartesian system.

apples with a short stem and the other one contains only apples with a long stem. Once one of these lines has been defined, the classifier can work as follows. If an unknown apple is found to be in the area defined by the apples having a short stem, then it is considered to have a short stem, otherwise it has a long stem. Note that each line drawn in Figure 6.2(a) classifies the apples of the training set correctly. However, a unique classifier is usually needed and, among all the possible choices, the best one is desirable. Intuitively, the linear classifier that provides the largest possible margin between the two classes is the best choice, because small perturbations in the data, or an operation such as adding or removing data, are least likely to cause misclassifications. Let us suppose for instance that the classifier is the dashed line in Figure 6.2(a). Such line is very close to one of the apples having a long stem. Since an apple with a short stem is found in this position, other apples having the same

Fig. 6.2 (a) Examples of linear classifiers for the apples; (b) the classifier obtained by applying a SVM.

feature are expected to be found around this position. However, this particular apple is close to the border defined by the dashed line, and hence apples close to this one may be on the separation line or in the other area. In the first case, the apples could not be classified, and in the second one the apples are classified in the wrong way. Therefore, it is important that the distance between the border and the samples close to it is as large as possible. In other words, not only a classifier able to classify the data must be searched, but also a classifier having the maximum distance from the nearest samples of each class. The larger is the margin, the higher the generalization ability of the classifier should be. Samples that the margin pushes up against are referred to as *support vectors*, and this is why this method is referred to as *support vector machines*. The image in Figure 6.2(b) shows the best linear classifier for the apples. It can be determined by computing two parallel supporting lines, one for each of the two classes, and maximizing the distance between them. In Figure 6.2(b), the two parallel supporting lines are represented by the dashed lines. In general, these supporting lines can be defined as any line such that all the points of a class are on one and only one side of that line.

Let us leave the example of the apples now and let us deal in general with samples that can be defined as points in an n-dimensional space. The data need to be classified in two disjoint classes. Let us suppose that the classes are linearly separable, and hence a hyperplane (or a line in the two-dimensional case) can be considered as a good classifier. The general equation of a hyperplane is

$$w^T x + b = 0.$$

The parameters w and b can be normalized so that $w^T x + b = +1$ is the hyperplane that goes through the support vectors of the first class, and $w^T x + b = -1$ is the hyperplane that goes through the support vectors of the other class. The first hyperplane is also called *plus-plane* and it refers to the *plus* class C^+, whereas the second one is called *minus-plane* and it refers to the *minus* class C^-. In this way, all the unknown samples x satisfying equation $w^T x + b \geq 1$ are classified as belonging to the first class, and all the samples x satisfying equation $w^T x + b \leq -1$ are classified as belonging to the second one. All the x satisfying the equation $-1 < w^T x + b < 1$ cannot be classified.

As shown in Chapter 5, the learning process of a neural network can be formulated as a global optimization problem. The training process of an SVM is formulated as an optimization problem as well. Let x^+ be a sample on the plus-plane C^+, and let x^- be the sample closest to x^+ on the minus-plane C^-. The margin width M can be expressed as the distance between x^+ and x^-:

$$M = |x^+ - x^-|.$$

However, it can be proved that M can also be expressed in terms of w:

$$M = \frac{2}{\sqrt{w^T w}}. \tag{6.1}$$

A hyperplane having a margin M as large as possible is searched as classifier for the classes C^+ and C^-. From formula (6.1), maximizing the margin M is equivalent to minimizing the quantity $\sqrt{w^T w}$. Therefore, the problem of finding the best classifier can be formulated as an optimization problem, where the objective function to be minimized is the margin M:

$$\min_{w,b} \frac{1}{2} w^T w \tag{6.2}$$

subject to separation constraints:

$$\begin{cases} w^T x^i + b \geq +1 & \forall i \in C^+ \\ w^T x^i + b \leq -1 & \forall i \in C^- \end{cases} \tag{6.3}$$

This is a convex quadratic optimization problem that can be efficiently solved. It can also be transformed into an equivalent problem by its dual formulation. We will not give details on how to compute the dual formulation of an optimization problem, but we will consider the dual reformulation of the problem (6.2)–(6.3) in the following discussion. The dual formulation is

$$\max_{\alpha} \sum_i \alpha_i - \frac{1}{2} \sum_{i,j} c_i c_j (x^i)^T x^j \alpha_i \alpha_j \tag{6.4}$$

subject to

$$\begin{cases} \sum_i c_i \alpha_i = 0 \\ \alpha \geq 0 \end{cases} \tag{6.5}$$

where α is the vector containing the dual variables, and c is a vector whose components c_i is equal to 1 if the corresponding x^i belongs to the plus class C^+, and it is equal to -1 if x^i belongs to the minus class C^-. The dual variables are also called *Lagrange multipliers*, and they are non-negative real numbers. These are the variables which SVMs use to learn from the data. They are in fact the analogue of the weights associated to the neurons of the neural networks. Once an SVM has been trained by solving this optimization problem, the optimal hyperplane is found and a number of support vectors is located. It is interesting to note that the same hyperplane can now be identified using the small training set, containing all the support vectors. In other words, all other samples can be removed from the training set and recomputing the hyperplane would produce exactly the same answer. Therefore, SVMs can also be used for summarizing the information contained in a certain set of data.

6.2 Nonlinear classifiers

The problem of classifying samples into two classes that can be separated by a hyperplane has been discussed in the previous section. However, the hypothesis of the linear separability is not always satisfied, as Figure 6.3 shows. In this case, the

Fig. 6.3 An example in which samples cannot be classified by a linear classifier.

apples cannot be separated by a line on the Cartesian system, but a more complex classifier should be used. In fact, in most real-world applications, there is no reason for expecting that the classes can be separated by hyperplanes. Therefore, SVMs need to be extended to address more general cases. Not only hyperplanes should be used, but also nonlinear surfaces may be considered. Working on nonlinear surfaces can be much more complex than working on hyperplanes. Therefore, a very smart method has been developed for taking into account the case in which the considered classes are not linearly separable.

Let us consider the one-dimensional points in Figure 6.4. The problem is to find a classifier for the samples on the one-dimensional space defined by the x axis of the Cartesian system. As one can easily note, the samples are not linearly separable, because one class (class 2) has samples ranging from 0 to 1, whereas the other class (class 1) has samples before 0 and after 1. One will never find a hyperplane (actually a point in this example) which separates the two classes. Let us project now this data in a two-dimensional space, as Figure 6.4 shows. The points belonging to class 1 and class 2 are now linearly separable, and then a linear classifier can be used. It is obvious that this data transformation may substantially increase the dimension of the problem. In this example, the dimension of the new problem is twice the original.

The function which transforms the data is called a *mapping* function and it is usually denoted by Φ. A sample x^i in the original space can be represented by $\Phi(x^i)$ in the newly transformed space. It follows that the general equation of the hyperplane in this case is:

$$w^T \Phi(x) + b = 0$$

and that the dual formulation of the optimization problem (6.4) is

$$\max_{\alpha} \sum_i \alpha_i - \frac{1}{2} \sum_{i,j} c_i c_j \Phi(x^i)^T \Phi(x^j) \alpha_i \alpha_j$$

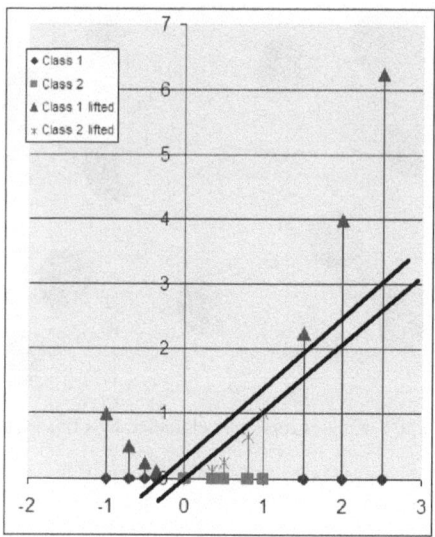

Fig. 6.4 Example of a set of data which is not linearly classifiable in its original space. It becomes such in a two-dimensional space.

subject to the constraints (6.5). For using this formulation of the optimization problem, a function Φ able to transform the set of data into another one that can be separated by hyperplanes needs to be defined. However, this is actually not required, because the function Φ participates only in the inner product $\Phi(x^i)^T \Phi(x^j)$ in the objective function of the optimization problem. Then, there is no need to compute it explicitly. This inner product can be replaced by a suitable function

$$K(x^i, x^j) = \Phi(x^i)^T \Phi(x^j),$$

which is called SVM kernel function. In this way, an SVM can be trained by solving the optimization problem (6.4), where the inner product $(x^i)^T x^j$ is substituted by the kernel function $K(x^i, x^j)$ for taking into consideration the cases where the data are not linearly separable. The corresponding optimization problem is then

$$\max_{\alpha} \sum_i \alpha_i - \frac{1}{2} \sum_{i,j} c_i c_j K(x^i, x^j) \alpha_i \alpha_j \qquad (6.6)$$

subject to the constraints (6.5). Kernel functions $K(x^i, x^j)$ can be obtained by the specific mapping needed for transforming the data. This approach, however, requires the definition of a suitable mapping. Therefore, a more common approach is to avoid defining explicitly a mapping Φ and to find a function which can work as a kernel function. In practice, pre-defined kernel functions are usually used, and different SVMs can be trained using different kernels in order to select the one that performs better for the given problem. The choice of a kernel in SVMs can be considered as analogous to the problem of choosing a suitable architecture in neural networks. The most used kernels include the polynomial kernel

$$K(x^i, x^j) = \left((x^i)^T (x^j) + 1\right)^d,$$

which is able to lift the feature space by including all monomials of the original features up to degree d. Examples of kernels also include the Gaussian kernel

$$K(x^i, x^j) = \exp\left(-\frac{(x^i - x^j)^2}{2\sigma^2}\right)$$

and the Neural-net-style kernel

$$K(x^i, x^j) = \tanh\left(\kappa (x^i)^T (x^j) - \delta\right).$$

The Gaussian kernel represents the most reasonable choice because of its simplicity and the ability to model data of arbitrary complexity. It is provided with a tuning parameter σ that adjusts the kernel's width.

SVMs coupled with these kernel functions are able to classify sets of data with a good accuracy, as is the case of the applications discussed in Section 6.5. As pointed out in [24], however, there is probably no theoretical explanation of why SVMs perform so well in practice. Following the quoted paper, we can say that, even though it is commonly accepted that maximizing the margin M is good for generalization, there is no way to prove it. Moreover, SVMs are based on ideas guided by geometric intuitions, as is the case of the example of apples on a Cartesian system. However, this intuition cannot be applied in the spaces of higher dimension obtained using the kernel functions. In the case of the Gaussian kernel, the obtained space has an infinite dimension. From here comes though the explanation of why the Gaussian kernel works well. Any two disjoint sets of data in the original space, indeed, can be separated by a hyperplane in the infinite dimensional space.

6.3 Noise and outliers

SVMs can be trained for classifying data in two classes which may and may not be linearly separable. In both cases, if the data are affected by noise and outliers, the used hyperplanes could not be able to generalize well. Indeed, a perfect separating hyperplane may be unsuccessful because of samples that badly represent their class. In these cases, it is desirable to have only the "majority" of the samples correctly classified and avoid noise data and outliers. Therefore, some violations during the training process are usually allowed. A term in the objective function of the optimization problem is added for taking these violations into account. This is done by including non-negative slack variables ξ. The optimization problem (6.2) then becomes

$$\min_{w,b} \frac{1}{2} w^T w + C \sum_i \xi_i$$

subject to

$$\begin{cases} w^T \Phi(x^i) + b \geq 1 - \xi_i \ \ \forall i \in C^+ \\ w^T \Phi(x^i) + b \leq \xi_i - 1 \ \ \forall i \in C^- \end{cases},$$

where C is the trade-off parameter between the margin M and the classification error. This problem formulation is for SVM having *soft* margins, whereas the previous formulation was for SVM with *hard* margins. The dual formulation in the soft case is very similar to the one in the hard case, since there is only a constraint more on the vector α. Finally, the optimization problem usually solved for training an SVM is

$$\max_{\alpha} \sum_i \alpha_i - \frac{1}{2} \sum_{i,j} c_i c_j K(x^i, x^j) \alpha_i \alpha_j$$

subject to

$$\begin{cases} \sum_i c_i \alpha_i = 0 \\ 0 \leq \alpha \leq C \end{cases}.$$

6.4 Training SVMs

In order to train an SVM, a suitable kernel function needs to be selected and the kernel parameters and the trade-off parameter C need to be chosen. The quality of the classifications can be greatly affected by C, since it determines how severely classification errors must be penalized. A large C value may lead to overfitting problems, thus reducing the ability of SVM to generalize. The kernel and all these parameters are usually defined by cross-validation techniques (see Chapter 8). However, strategies for finding better ways to estimate optimal parameter values have been proposed. In [202], for example, C is computed based on a slightly lower value than the largest α coefficient obtained from training with C as infinity.

The discussed SVM approach addresses only binary classifications and it was originally developed for this purpose. However, SVMs can also be used for classifying samples in n classes, where $n > 2$. Various approaches have been developed for dealing with multi-class classification problems [205]. One of these is the *one-against-rest* approach, where the data are classified in n classes using n different SVMs. The SVM $l \in \{1, 2, \ldots, n\}$ is trained so that it is able to recognize if an unknown sample belongs to class l or to one of the others. During the classification, the SVM having the maximal output defines the estimated class. In other words, an unknown sample can be classified according to the result of the SVM that recognizes it with higher confidence. Another possible approach is the *one-against-one* approach. In this case, for each pair of classes, a single SVM is considered. Therefore $\frac{n(n-1)}{2}$ SVMs are needed for considering all the possible pairs, and each of them is trained by using different training sets. In fact, each SVM just needs to select between two classes, and hence only a subset of the initial training set is needed, where only samples from the two considered classes are used. All the SVMs during the classification are combined through a majority voting scheme to estimate the final estimation. Finally, another approach is to consider decision trees of binary SVM classifiers. At the root of the tree, a classifier can select between two disjoint sets of classes. These sets may include more than 2 classes, and hence other SVMs need to be used for separating these classes into smaller subclasses. Branch by branch, at some point, the SVMs at the top of the tree can discriminate between the last two classes and provide the classification. Many tree structures can be used, but each SVM can receive only one input from its incoming edge. Hierarchies of SVMs can also be used instead of tree structures.

As mentioned above, assuming that classes are linearly separable, a simple quadratic programming problem with linear constraints needs to be solved for training a SVM (see equation (6.2)). The function to be optimized is convex. When the classes are not linearly separable, SVMs can be trained using a suitable kernel function and by optimizing the objective function (6.6). Kernels allow non-linearization of the learning algorithm while preserving the convexity of the associated optimization problem. However, due to its size, the quadratic programming problems for training SVMs are not usually solved by the standard quadratic programming techniques. The matrix of the quadratic function, indeed, has a number of elements equal to the square of the size of the training set. Different methods for solving these quadratic programming problems have been proposed [127, 188]. Many of them are based on the idea of breaking the original quadratic programming problem down into a series of smaller subproblems. In some approaches the size of the subproblems is kept constant, by adding and removing the same number of samples from the objective function. In other approaches the subproblems can have different sizes and the smallest quadratic programming problem is chosen at each step. Some of these approaches solve each subproblem by standard methods for quadratic programming, and others solve them analytically. In the last case, the manipulation of large matrices is avoided and then the algorithm is less susceptible to numerical precision problems.

6.5 Applications

There are several applications of SVMs in the literature. In [63], for instance, SVMs are used for building a handwritten Chinese character recognition system. This is a very difficult problem, since handwritten Chinese characters have complex structures and large shape variations. Moreover, there are many characters that are similar to one another. In Figure 6.5 some selected Chinese characters are displayed. There are mainly two problems to be faced when dealing with character recognition, and in particular when these characters are Chinese. First of all, the Chinese language is not similar to the English language where an alphabet of 26 letters is sufficient to create all the words written in this book. In the case of Chinese written language,

Fig. 6.5 Chinese characters recognized by SVMs. Symbols from [63].

instead, there are many existing characters that should be taken into account. As previously explained, many SVMs can be trained when a multi-class classification is needed, and the number of SVMs needed depends on the number of classes. In the easiest case, there must be at least one SVM for each class, when the one-against-rest approach is used. Therefore, the greater is the number of characters considered, the greater is the number of SVMs needed. Another problem is the representation of these characters as black and white images. The smoothness of the symbols can be lost due to the pixel representation, and it can decrease the quality of the image to the point to affect the accuracy of the classification. One of the easiest methods for avoiding this problem is to store the characters in images of higher quality.

SVMs are also used for speaker and language recognition [35]. In biology, protein function classifications [34] and cancer diseases classifications by gene selections [96] have been performed via SVMs. In medicine, SVMs have been used for analyzing signals from the macaque monkey brain during a visual discrimination task [208]. In these studies, SVMs are generalized and the selective classification concept is introduced.

In agriculture, SVMs have been applied for predicting soil moisture [86]. In fact, weather forecasts in agriculture are very important, as already pointed out in Section 4.4.1. It is difficult for a farmer to know when to irrigate the soil, especially if there is uncertainty about the weather in the following days. The irrigation schedule is a key factor in the management of a farm, and advanced knowledge or accurate forecasts can help to design an efficient irrigation scheduling and water quality monitoring. Soil moisture measurements are helpful in predicting and understanding various hydrologic processes, including weather changes, energy and moisture fluxes, and irrigation scheduling. There are different physically based approaches that can be difficult to use, and for this reason researchers are working on data driven forecasting tools. In the approach used by [86], soil moisture and meteorological data are used to generate SVM predictions for four and seven days ahead.

Other applications of SVMs in the field of agriculture include:

- Classification of crops [36];
- Classification of milk by means of an electronic nose [29];
- Detection of meat and bone meal in compound feeds [187];

- Classification of pizza sauce spread [65];
- Detection of weed and nitrogen stress in corn [124];
- Analysis on the climate change scenarios [226];
- Recognition of bird species [71].

The following presents discussions related to the classification of bird species by their sounds in Section 6.5.1 and to the verification of the presence of meat and bone meal in feedstuffs for animals in Section 6.5.2.

6.5.1 Recognition of bird species

For many people the sound of birds is the sign for the start of spring. Usually, people are able to recognize at least a few common species by their sound and experts can recognize hundreds of species only by their sound. The automatic recognition of birds by their vocalization has also use in some practical applications. For instance, collisions between aircraft and birds can cause bird death and also damages to the aircraft. In order to avoid collision with birds during the flight, different devices have been implemented on aircraft, such as radars, infrared and microphones. Radars are able to recognize objects in movement from long distances, but they cannot distinguish between harmful and non-harmful objects. Infrared cameras perform very badly when the weather is not good. Therefore, the most promising method for recognizing birds in movement is using microphones able to monitor bird sounds. This technology can also be applied to wildlife monitoring, speech enhancement in communication centers, conference rooms, aircraft cockpits, cars, buses, and so forth. It can be used for security monitoring in airport terminals and bus and train stations.

The focus of this section is a method for the automatic recognition of bird species by their sounds [71]. The final objective is to develop a fully automatic system that is able to recognize bird species from their sounds made in field conditions. Figure 6.6

Fig. 6.6 The hooked crow (lat. ab.: cornix) can be recognized by an SVM based on the sounds of birds.

shows one of the bird species considered in these studies. The name of the bird is "hooded crow," whereas its Latin name is "cornix." Bird sounds are typically divided into two categories: *songs* and *calls*. These two sounds are different because they have different functions. Generally, songs are longer, more complex than calls, occur more spontaneously, and they are mainly related to the breeding process. Many bird species sing only during the breeding season and this is generally limited to males only. Call sounds instead are typically short vocalizations that carry a function out, for example an alarm, flight, or feeding. Bird sounds can also be divided into hierarchical levels of *phrases*, *syllables*, and *elements*. A phrase is a series of mainly similar syllables that occurs in a particular pattern. Syllables are constructed from elements: there are simple syllables formed by one element only and more complex ones that can be constructed using several elements. For simplicity, the syllable can be regarded as the smallest unit of bird vocalization.

In the studies presented in [71], the sound signals have been represented by the so-called mel-cepstrum model and by the descriptive parameters model. Details about these two models can be found in [54]. The set of data obtained by recording the bird sounds has then been used for training an SVM classifier with a Gaussian kernel. This is a multi-class classification, since the aim is not to distinguish between two bird species only. A decision tree of SVMs is therefore used, where each node of the tree contains a binary SVM classifier which considers only two classes ignoring all the others. The decision tree is organized in a way that at each layer one class is rejected. The last remaining class at the bottom of the tree is considered as the winning class. The structure of the used decision tree is presented in Figure 6.7. Circles carry the Latin abbreviation of the species names on which each SVM works.

The training of the SVMs on the decision tree is performed in two steps. During the first one the optimal model parameters are searched, which are the constant C in the optimization problem to be solved and the width of the Gaussian kernel σ. Since each of the SVMs is independent from one another, different optimal C and σ values can be found for each SVM. During the second phase, the actual training process of the SVMs is performed.

n-fold cross validation method is used to find the optimal values for the model parameters (see Chapter 8). For all pairs of classes in the decision tree, the data points are divided into training and test subsets such that the test subset contains all data from one individual. The training subset is used to construct an SVM classifier and its performance is evaluated with a test subset. The classification error is the average of the test errors of the subsets. The validation procedure is repeated for a grid of parameter values C and σ. Parameters that produce the lowest classification error are selected as the final model parameters.

The actual training process is performed using the sequential minimal optimization (SMO) algorithm. The MATLAB® support vector machine toolbox implementation of the SMO algorithm is used to train individual SVM classifiers [188]. Computational results proved that the overall recognition accuracy of the presented SVMs decision tree is larger than 90%. The studies presented here are performed within the AveSound project [11].

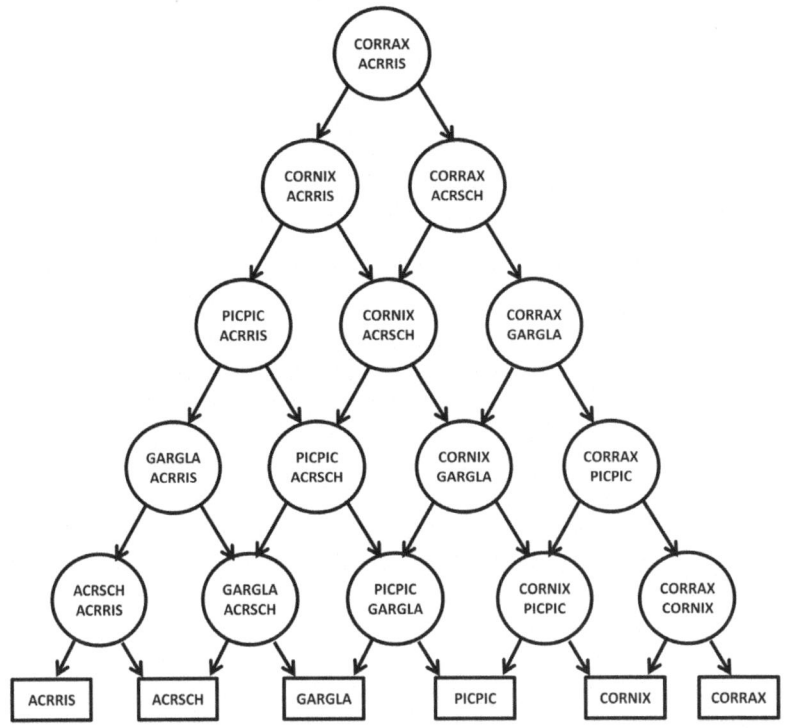

Fig. 6.7 The structure of the SVM decision tree used for recognizing bird species. Image from [71].

6.5.2 Detection of meat and bone meal

Since the emergence of the mad cow crisis in Europe, and all its socio-economic consequences, European Union regulatory agencies have undertaken many legal measures to ensure the safety and quality of feedstuffs for animals. One of the most important decisions was to ban meat and bone meal in feedstuffs destined to farm animals which are kept, fattened, or bred for the production of food. Controls are needed for verifying if meat and bone meal is instead used for accidental contamination or against the law. Therefore, the effective enforcement of this regulation requires accurate and efficient analytical methods capable of analyzing thousands of samples per year. Some methods have been developed for this purpose, which are reliable but tedious. The main problem is that visual observations and interpretations by an experienced analyst are needed, and this makes the process slower, expensive and subject to errors.

Near-infrared microscopy is an alternative method, which works well in discriminating the different ingredients found in compound feeds. Each particle in the feedstuffs is evaluated based on its chemical properties rather than appearance, reducing in this way the human subjectivity. Unfortunately, this method is slower than classical microscopy. Therefore other methods have been developed. One of these methods combines the advantages of spectroscopic and microscopic methods along with much faster sample analysis. An imaging spectrometer gathers spectral and spatial data simultaneously by recording sequential images of a predefined sample. The set of data obtained by this method (a collection of *spectra* in this case) can be used for training an SVM with the aim of defining a classifier able to discriminate between vegetable and meat and bone meal.

In this section, the training process of an SVM for these purposes [187] is presented. Spectra coming from 26 pure animal meals and spectra coming from 59 pure vegetable meals have been used for creating the training set. The animal and vegetable materials analyzed have been selected to span the diversity of materials mainly used for the formulation of compound feeds. In total, more than 267,000 spectra have been collected from pure animal and vegetable meals. All spectra have been kept as raw absorbance units.

In this application, samples belonging to two classes only have to be discriminated, and therefore only one SVM is needed. Different kernels are tested on the obtained set of data, and the results show that the best choice is the Gaussian kernel. The SVM parameters are optimized using the "grid search" method with a fixed calibration and validation set. The optimal parameter settings for C and σ are then selected as the values that give the maximum correct classification rate. When C is increased, the second term of the objective function dominates, forcing SVM toward a solution with the least training error, which decreases the amount of regularization. Moreover, a larger number of calibration samples are retained as support vectors, which increases the computation time of prediction. In this case, animal particles begin to be classified as vegetable, but no plant particles are misclassified.

In this particular area it is very important that all the samples are classified with a good precision. Indeed, a false detection of meat and bone meal can severely damage the reputation of honest and scrupulous farmers and manufacturers. Human analysis can be included in the process for verifying if there are false detections of meat and bone meal, but this would require additional expenses. For validating the trained SVM, a cross-validation technique is used (see Chapter 8). For this purpose, a testing set of 76,800 spectra is created in the same manner as the calibration samples, using the same imaging instrument.

The prediction of the data in the testing set is handled in two different ways. For details, please refer to [187]. During the first test, most of the animal particles are well detected and no vegetable particles are misclassified. During the second test, the detected animal particles correspond better with the true meat and bone meal particles, and no vegetable or background particles are misclassified.

6.6 MATLAB and LIBSVM

There is a MATLAB toolbox especially designed for SVMs. However, we will not discuss its potentialities in this chapter, and the interested reader can find additional information on this toolbox on the MATLAB Web site. This section is instead devoted to the free software LIBSVM (a LIBrary for Support Vector Machines). MATLAB is used just to generate instances that will be solved by using LIBSVM. This will give an example of how to interface two different software. The code in MATLAB we propose is simple and easy to modify for personal purposes.

LIBSVM is an integrated software for SVM classification and also regression and distribution estimation [43]. LIBSVM is distributed with the source code, so that it can be compiled and used on any platform. Executable files are also available for DOS and Windows users. It is composed of 4 procedures:

- svmtrain can be used for training an SVM by a certain training set and using different parameters.
- svmpredict can be used for predicting classifications by SVMs defined with the previous procedure.
- svmscale can be used for scaling the data. This procedure is highly recommended by the authors of LIBSVM for avoiding what they call "numerical difficulties" during the calculations. In fact, variables having a greater variability can dominate on the ones with smaller ranges of variability, and this may spoil the classification accuracy.
- svmtoy is a LIBSVM procedure which can be used for *playing* with SVMs. It has a graphic interface, where two-dimensional points can be drawn on a virtual plane and different classifications can be associated to them. The procedure provides graphical representations of SVMs modeling the drawn points. This can be a valuable exercise for checking the SVM classification skills in different situations, such as linear and nonlinear separable data.

In the following, it is shown how a training set can be generated and used for training an SVM. For generating the data, the MATLAB function generate is used. In this case, however, the data do not have to be used in the MATLAB environment. Hence, the data need to be stored in a text file formatted so that it can be read by the LIBSVM software.

The LIBSVM procedures are able to read text files formatted as follows. At least two text files need to be generated: one containing the samples of the training set and another one containing the samples of a testing test. These samples need to be listed row by row in the text files, so that each sample is represented on one single row. Each row starts with the identifier of the class the sample belongs to. If the samples are divided in two classes, the identifiers can be -1 and $+1$. After the identifier, all the components of the vector representing the sample need to be inserted. For each component, the component counter $\{1, 2, \ldots, n\}$ and its value are inserted and separated by the symbol ':'. If known, the class to which the sample belongs can be inserted also in the text file related to the testing test. In this way, svmpredict is able to verify how many unknown samples are classified correctly by the SVM. In Figure 6.8 a modified version of the MATLAB function generate (Figure 3.16) is given. It saves the generated data in the text file trainset.txt by using the functions fopen and fprintf. The function generate4libsvm assigns

```
%
% this function generates a random sets of data
% in the two-dimensional space and prints it in
% the text file "trainset.txt" formatted in the
% LIBSVM format
%
% input:
% n    - number of random samples to be generated
% eps  - predefined margin between samples separated by the line x = 0
%
% output:
% x - x coordinates of the samples
% y - y coordinates of the samples
%
% [x,y] = generate4libsvm(n,eps)

function [x,y] = generate4libsvm(n,eps)

    output = fopen('trainset.txt','w');

    for i = 1:n,

        random = rand();
        if random < 0.50,
            x(i) = -eps - rand();
        else
            x(i) =  eps + rand();
        end
        y(i) =  2.0*rand() - 1.0;
        if random < 0.50,
            fprintf(output,'-1   1:%f   2:%f\n',x(i),y(i));
        else
            fprintf(output,'+1   1:%f   2:%f\n',x(i),y(i));
        end

    end

    fclose(output);

end
```

Fig. 6.8 The MATLAB function generate4libsvm.

each sample of the type $(-x, y)$ to class -1 and each sample of type $(+x, y)$ to class $+1$.

A set of 100 samples has been generated by function generate4libsvm with eps $= 0.1$. The first samples contained in the text file are shown in Figure 6.9. Another set of 1000 samples have then been generated by the same function and imposing eps $= 0.0$. This second set is used as a testing set, and hence its name has been modified from trainset.txt to testset.txt after the generation. The two-dimensional points in the sets of data are generated in a way that their components range approximately in the set $[-1, 1] \times [-1, 1]$, depending on the eps value. For this reason, the procedure svmscale is not used in this example. Figure 6.10 provides the commands used for training and testing an SVM. The procedure svmtrain is used for training the SVM. The procedure has many parameters. If they are not specified, the default values are used for such parameters. In this example, the option '-t' is used for specifying one of the possible kernels that can be employed. The procedure svmpredict is then used for performing the classification of unknown samples by

```
-1  1:-0.600916   2:-0.341989
-1  1:-0.430370   2:0.260895
-1  1:-0.548602   2:0.182902
+1  1:0.173930    2:0.886739
-1  1:-0.301908   2:-0.606139
+1  1:0.561792    2:0.250258
+1  1:0.801817    2:-0.182400
-1  1:-0.165739   2:0.066201
-1  1:-0.415784   2:-0.242835
-1  1:-0.728737   2:-0.424297
-1  1:-0.571923   2:-0.329246
+1  1:0.219608    2:0.522800
-1  1:-0.381697   2:-0.523980
-1  1:-0.330673   2:-0.572741
-1  1:-0.409537   2:0.352594
+1  1:0.644180    2:-0.959050
+1  1:0.449036    2:-0.778927
-1  1:-0.230469   2:0.714791
+1  1:0.507112    2:0.338388
-1  1:-0.931043   2:-0.975567
-1  1:-1.054988   2:-0.882514
+1  1:0.398074    2:-0.844848
+1  1:0.852871    2:-0.896554
+1  1:0.483371    2:-0.019956
+1  1:0.168895    2:0.768616
+1  1:0.169426    2:-0.769558
-1  1:-0.867158   2:-0.823894
+1  1:0.287914    2:-0.596423
... ...
```

Fig. 6.9 The first rows of file `trainset.txt` generated by `generate4libsvm`.

using the trained SVM. This procedure needs two text files as input and one text file as output. The first one is `testset.txt`, where the samples to be classified are stored. The second one is `trainset.txt.model`, which is a text file generated by svmtrain where the parameters related to the SVM are saved. Finally, the output file `testresult.txt` will contain the classification of the unknown samples. The overall accuracy is 98%.

6.7 Exercises

This section presents some exercises related to SVMs. All the solutions are reported in Chapter 10.

```
LIBSVM> svmtrain -t 3 trainset.txt
*
optimization finished, #iter = 16
nu = 0.213405
obj = -14.075954, rho = -0.091571
nSV = 23, nBSV = 20
Total nSV = 23

LIBSVM> svmpredict testset.txt trainset.txt.model testresult.txt
Accuracy = 98.1% (981/1000) (classification)
```

Fig. 6.10 The DOS commands for training and testing an SVM by SVMLIB.

1. Let us suppose that a set of points in a three-dimensional space is defined as follows. The generic point of this set is the triplet

$$(A, B, C)$$

such that the components can have value 0 or 1. Let us suppose that all the points grouped in the class C^0 satisfy the rule:

$$A \quad \text{AND} \quad B \quad \text{AND} \quad C \quad = \quad 0,$$

whereas all points grouped in the class C^1 satisfy the rule:

$$A \quad \text{AND} \quad B \quad \text{AND} \quad C \quad = \quad 1.$$

State whether the two classes C^0 and C^1 are linearly separable.

2. As in the previous exercise, check if the two classes C^0 and C^1 are linearly separable, when the classes are defined as:

$$C^0 = \{(A, B, C) : \quad \text{NOT} \quad A \quad \text{AND} \quad B \quad = \quad 0\}$$
$$C^1 = \{(A, B, C) : \quad \text{NOT} \quad A \quad \text{AND} \quad B \quad = \quad 1\}$$

and when the classes are defined as:

$$C^0 = \{(A, B, C) : \quad (A \quad \text{OR} \quad B) \quad \text{AND} \quad (A \quad \text{AND} \quad C) \quad = \quad 0\}$$
$$C^1 = \{(A, B, C) : \quad (A \quad \text{OR} \quad B) \quad \text{AND} \quad (A \quad \text{AND} \quad C) \quad = \quad 1\}.$$

3. Suppose that a set of points and their classifications in two classes C^+ and C^- are specified as follows:

$$((0, 0), C^-), \quad ((0, 1), C^+), \quad ((1, 0), C^+), \quad ((1, 1), C^-).$$

State why the classes C^+ and C^- are not linearly separable.

4. Consider the set of points and their classification as described in Exercise 3. Transform the set of points by using the function

$$\Phi(x_1, x_2) = \begin{pmatrix} 1 \\ \sqrt{2}x_1 \\ \sqrt{2}x_2 \\ x_1^2 \\ x_2^2 \\ \sqrt{2}x_1x_2 \end{pmatrix}.$$

Check also if the set of points is linearly separable after the transformation.

5. Consider the set of points and their classification as described in Exercise 3. Formulate the primal optimization problem for finding the maximum margin classifier in the higher-dimensional space defined by the function $\Phi(x_1, x_2)$ in Exercise 4.

6. Reproduce the experiment discussed in Section 6.6 by using different kernel functions.
7. Considering the context of Section 6.1, prove that

$$M = \frac{2}{\sqrt{w^T w}}.$$

Chapter 7
Biclustering

7.1 Clustering in two dimensions

Clustering techniques aim at partitioning a given set of data into clusters. Chapter 3 presents the basic k-means approach and many variants to the standard algorithm. All these algorithms search for an optimal partition in clusters of a given set of samples. The number of clusters is usually denoted by the symbol k. As previously discussed in Chapter 3, each cluster is usually labeled with an integer number ranging from 0 to $k - 1$. Once a partition is available for a certain set of samples, the samples can then be sorted by the label of the corresponding cluster in the partition. If a color is then assigned to the label, a graphic visualization of the partition in clusters is obtained. This kind of graphic representation is used often in two-dimensional spaces for representing partitions found with biclustering methods.

A set of data can be represented through a matrix. The samples can be represented by m-dimensional vectors, where the components of these vectors represent the features used for describing each sample. All the vectors representing the samples can be grouped in a matrix

$$
A = \begin{pmatrix}
a_{11} & a_{12} & a_{13} & \cdots & a_{1n} \\
a_{21} & a_{22} & a_{23} & \cdots & a_{2n} \\
a_{31} & a_{32} & a_{33} & \cdots & a_{3n} \\
\cdots & \cdots & \cdots & \cdots & \cdots \\
a_{m1} & a_{m2} & a_{m3} & \cdots & a_{mn}
\end{pmatrix}.
$$

If a given set of data contains n samples which are represented by m features, then A is an $m \times n$ matrix. Each column of the matrix represents one sample, and it provides information on the expression of its m features. Each row represents a feature, and it provides the expression of that feature on the n samples of the set of data.

Standard clustering methods partition the samples in clusters, i.e., the columns of the matrix A are partitioned in clusters. Biclustering methods work instead simultaneously on the columns and the rows of the matrix A. Besides clustering the samples, even their features are partitioned in clusters. Two different partitions are therefore

A. Mucherino et al., *Data Mining in Agriculture*, Springer Optimization and Its Applications 34, 143
DOI: 10.1007/978-0-387-88615-2_7,

needed. The search of the two partitions is not performed independently, but rather the clusters of samples and the clusters of features are related. The concept of "bicluster" is introduced for this purpose. A *bicluster* is a collection of pairs of samples and features subsets $B = \{(S_1, F_1), (S_2, F_2), \ldots, (S_k, F_k)\}$, where k, as usual, is the number of biclusters [32]. Each bicluster (S_r, F_r) is formed by two single clusters: S_r is a cluster of samples, and F_r is a cluster of features.

The following conditions must be satisfied:

$$\bigcup_{r=1}^{k} S_r \equiv A, \qquad S_\zeta \cap S_\xi = \emptyset \quad 1 \le \zeta \neq \xi \le k,$$

$$\bigcup_{r=1}^{k} F_r \equiv A, \qquad F_\zeta \cap F_\xi = \emptyset \quad 1 \le \zeta \neq \xi \le k.$$

Note that the union of all the clusters S_r must be A because each sample, organized in columns in the matrix, must be contained in at least one cluster S_r. Similarly, the union of all the clusters F_r must be A as well. The only difference is that the features are organized on the rows of the matrix A. Note also that these same conditions are imposed on clusters when standard clustering is applied. Besides ensuring that each single sample or feature is contained in a cluster, they guarantee that all the clusters of samples and the clusters of features are disjoint.

The aim of biclustering techniques is to find a partition of the samples and of their features in biclusters (S_r, F_r). In this way, not only a partition of samples is obtained, but also the features causing this partition are identified. As for the standard clustering, the single clusters S_r and F_r can be labeled from 0 to $k-1$. Independently, the clusters S_r can be sorted by their own labels, and the same can be done for the clusters F_r. A color or a gray scale can be associated to each label, and a matrix of pixels can be created. On the rows of such matrix, the clusters F_r are ordered by their labels, and the clusters S_r are ordered on the columns. Even though this matrix is built considering the clusters S_r and F_r independently, it gives a graphic visualization of the biclusters (S_r, F_r). The matrix shows a checkerboard pattern where the biclusters can be easily identified. This pattern can be easily noticed, for instance, in Figure 7.4, related to the application of biclustering discussed in Section 7.4.1.

Biclustering is widely applied for partitioning gene expression data, and therefore some of the nomenclature in biclustering is similar to the one in gene expression analysis. In [159], a survey of biclustering algorithms for biological data is presented. Since biology is currently the main field of application of biclustering, this survey can be actually considered a survey on biclustering. It is updated to the year 2004, and hence it does not include recent developments, which are discussed in Section 7.2 of this chapter.

Following the definition, a bicluster is a pair of clusters (S_r, F_r), where S_r is a cluster of samples and F_r is a cluster of features. Since the samples and the features are organized in the matrix A as explained above, a bicluster can also be seen as a submatrix of A. A submatrix of an $m \times n$ matrix can be identified by the set of row indices and column indices it takes from A. For instance, if

$$A = \begin{pmatrix} 1 & 2 & 3 \\ 1 & 1 & 0 \\ 0 & -1 & 2 \end{pmatrix},$$

then the submatrix with the first and third row of A and the second and third column of A is

$$S_A = \begin{pmatrix} 2 & 3 \\ -1 & 2 \end{pmatrix}.$$

In the following, bicluster and submatrix of A will be used interchangeably.

Different kinds of biclusters can be defined. One might be interested in biclusters in which the corresponding submatrices of A have constant values. This requirement may be too strong in some cases, and it may work on non-noisy data only. Indeed, data from real-life applications are usually affected by errors, and a bicluster with constant values may not be possible to find. Formally, these kinds of biclusters are the ones in which

$$a_{ij} = \mu \quad \forall i, j: \quad a_i \in F_r \quad a^j \in S_r,$$

where μ is a real constant value. If the data contain errors, the following formalism can be used

$$a_{ij} = \mu + \eta_{ij} \quad \forall i, j: \quad a_i \in F_r \quad a^j \in S_r,$$

where η_{ij} is the noise associated to a real value μ of a_{ij}. The problem of finding biclusters with constant values can be formulated as an optimization problem in which the variance of the elements of the biclusters have to be minimized. If I_{S_r} is the set of column indices related to the samples $a^j \in S_r$, i.e., I_{S_r} contains all the j indices associated to S_r, and I_{F_r} is the set of row indices related to the features $a_i \in F_r$, then

$$f(S_r, F_r) = \sum_{i \in I_{S_r}} \sum_{j \in I_{F_r}} (a_{ij} - M)^2$$

evaluates the quality of the bicluster (S_r, F_r), where M is the average of all the elements in (S_r, F_r). If the data are not affected by errors, a *perfect* bicluster with constant values is such that $f(S_r, F_r) = 0$. Otherwise, minimizing the function $f(S_r, F_r)$ equals finding the bicluster which is closest to the optimal one. It is worth noting that every bicluster containing one row and one column is a perfect bicluster with constant values, since its only element a_{ij} equals M. In general, when the function $f(S_r, F_r)$ is optimized, constraints must take into account that the number of rows and columns of the submatrices representing the biclusters must be greater than a certain threshold.

Biclusters with constant row values and constant column values can also be of interest. If the row values in a bicluster are constant, then all the samples in the bicluster (and in S_r) have a constant subset of features (the ones in F_r). Inversely, if the columns have constant values, then the samples in S_r have all the features in F_r constant. In this case, different samples have different feature values, but all the feature values in the same sample are the same. A bicluster having constant rows satisfies the condition

$$a_{ij} = \mu + \alpha_i \quad \forall i, j : \quad a_i \in F_r \quad a^j \in S_r$$

or the condition

$$a_{ij} = \mu \alpha_i \quad \forall i, j : \quad a_i \in F_r \quad a^j \in S_r$$

where μ is a typical value within the bicluster and α_i is the adjustment for row $i \in I_{S_r}$. Similarly, a bicluster having constant columns satisfies the condition

$$a_{ij} = \mu + \beta_j \quad \forall i, j : \quad a_i \in F_r \quad a^j \in S_r$$

or the condition

$$a_{ij} = \mu \beta_j \quad \forall i, j : \quad a_i \in F_r \quad a^j \in S_r.$$

Even here, the presented conditions can be satisfied only if the data are not noisy, otherwise the noise parameters η_{ij} can be used, as in the previous example of biclusters with constant values.

The easiest way to approach the problem of finding biclusters with constant row values or constant column values is the following one. Let us suppose a bicluster with constant rows is contained in a matrix A and that the submatrix which corresponds to it is

$$S_A = \begin{pmatrix} 1\,1\,1\,1 \\ 2\,2\,2\,2 \\ 3\,3\,3\,3 \\ 4\,4\,4\,4 \end{pmatrix}.$$

Since all the values on the rows are constant, the mean among all these values corresponds to any of the row values. If each row is normalized by the mean of all its values, then the following matrix is obtained

$$\hat{S}_A = \begin{pmatrix} 1/1\,1/1\,1/1\,1/1 \\ 2/2\,2/2\,2/2\,2/2 \\ 3/3\,3/3\,3/3\,3/3 \\ 4/4\,4/4\,4/4\,4/4 \end{pmatrix} = \begin{pmatrix} 1\,1\,1\,1 \\ 1\,1\,1\,1 \\ 1\,1\,1\,1 \\ 1\,1\,1\,1 \end{pmatrix},$$

which corresponds to a bicluster with constant values. Therefore, the row and the columns normalization can allow the identification of biclusters with constant values on the rows or on the columns of the matrix A by transforming these biclusters into constant biclusters.

Biclusters have coherent values when the generic element of the corresponding submatrix can be written as

$$a_{ij} = \mu + \alpha_i + \beta_j \quad \forall i, j : \quad a_i \in F_r \quad a^j \in S_r.$$

Particular cases of coherent biclusters are biclusters with constant rows ($\beta_j = 0$), or biclusters with constant columns ($\alpha_i = 0$), or biclusters with constant values ($\alpha_i = \beta_j = 0$). This kind of bicluster can be represented by submatrices such as

$$S_A = \begin{pmatrix} \mu + \alpha_1 + \beta_1 & \mu + \alpha_1 + \beta_2 & \cdots & \mu + \alpha_1 + \beta_m \\ \mu + \alpha_2 + \beta_1 & \mu + \alpha_2 + \beta_2 & \cdots & \mu + \alpha_2 + \beta_m \\ \cdots & \cdots & \cdots & \cdots \\ \mu + \alpha_n + \beta_1 & \mu + \alpha_n + \beta_2 & \cdots & \mu + \alpha_n + \beta_m \end{pmatrix}.$$

The whole submatrix S_A can be built using the value μ and the two vectors $\alpha \equiv (\alpha_1, \alpha_2, \ldots, \alpha_n)$ and $\beta \equiv (\beta_1, \beta_2, \ldots, \beta_m)$.

The following proves that a generic element a_{ij} of a submatrix S_A can be obtained from means among the rows, the columns and all the elements of the matrix. The mean among the elements of the ith row of S_A is

$$M_i = \mu + \alpha_i + \frac{1}{m} \sum_{k=1}^{m} \beta_k,$$

whereas the mean among the elements of the jth column of S_A is

$$M_j = \mu + \frac{1}{n} \sum_{k=1}^{n} \alpha_k + \beta_j.$$

Moreover, the mean of all the elements of the matrix S_A is

$$M = \mu + \frac{1}{n} \sum_{k=1}^{n} \alpha_k + \frac{1}{m} \sum_{k=1}^{m} \beta_k.$$

From simple computations, it results that

$$M_i + M_j - M = \mu + \alpha_i + \beta_j = a_{ij}. \tag{7.1}$$

Therefore, the generic element of a coherent bicluster can be written as the mean of its rows, plus the mean of its columns, minus the mean of the whole submatrix. If the data are affected by errors, then equation (7.1) may not be satisfied. The residue $r(a_{ij})$ associated to an element a_{ij} is then defined as

$$r(a_{ij}) = a_{ij} - M_i - M_j + M$$

and consists of the difference between the value a_{ij} and the value obtained applying equation (7.1). A perfect (not affected by noise) coherent bicluster would have all the residues $r(a_{ij})$ equal to zero. Thus, the following function is able to evaluate the coherency of biclusters:

$$H(S_r, F_r) = \frac{1}{nm} \sum_{i=1}^{n} \sum_{j=1}^{m} \left[r(a_{ij}) \right]^2 .$$

Coherent biclusters can be located in the matrix A by minimizing this objective function.

As shown in this section, the problem of finding a bicluster or a partition in biclusters can be formulated as an optimization problem. The easiest way to solve it is through an exhaustive search among all the possible biclusters. This can be affordable only if the considered set of data contains a small number of samples and features. When this is not the case, optimization methods need to be used. In Section 1.4, some standard methods for optimization are presented. However, usually the optimization methods used for biclustering are tailored to the particular problem to solve [66, 83].

7.2 Consistent biclustering

In this section, the notion of consistent biclustering is introduced. This part of the chapter makes a large use of mathematical symbols: the symbology utilized follows. As already observed, the set of clusters S_r and the set of clusters F_r represent two partitions of the samples and of the features of a set of data. Each cluster S_r or F_r has a certain center. Since we have to deal with two different partitions (samples and features), let us denote the center of the generic cluster S_r with the symbol c_r^S and the center of the generic cluster F_r with the symbol c_r^F. The center c_r^S refers to the rth cluster of the samples. Since it is the average of samples represented by m-dimensional vectors, c_r^S is an m-dimensional vector. These vectors can be organized into an $m \times k$ matrix C_S, where the centers are stored column by column, just as the samples in the matrix A. The same can be done in correspondence of the clusters F_r and their centers. The generic center c_r^F refers to the rth cluster of features. A matrix C_F can be defined where such centers are organized column by column. C_F is an $n \times k$ matrix, since each feature is represented by an n-dimensional vector. Since the matrices C_S and C_F contain averages, their elements are the average expressions of the corresponding samples and features. It is clear that the nomenclature "average expression" comes from the studies on gene expression data. An average expression can be evaluated by a non-negative number: we will suppose in the following that all the centers have non-negative values.

Matrices are widely used in biclustering: A contains the set of data to partition in biclusters; C_S and C_F contain the centers of the clusters S_r and F_r, respectively. a_{ij} refers to the ith feature of the jth sample. A sample can be referred to as a^j: the j as superscript means that the j refers to the column index of the matrix A. Similarly, a_i refers to the ith row of the matrix, i.e., to the ith feature. The same symbology can be used for elements in C_S and C_F. c_{ir}^S refers to the ith component of the center of the cluster S_r; c_{jr}^F refers to the jth component of the center of the cluster F_r.

As already pointed out, the two single clusters in a bicluster (S_r, F_r) are related. Actually, once a partition in clusters of the samples is provided, a corresponding partition in clusters of the features can be obtained. Vice versa, a partition in clusters S_r can be obtained from the clusters F_r. Let us suppose then that the clusters S_r are known. In this case, each sample or column a^j is assigned to a certain cluster. The centers of all the clusters S_r are also known and contained in the matrix C_S column by column. The generic element c_{ir}^S of the matrix represents the average expression of the ith feature in the rth cluster, among all the samples in S_r. Let $\hat{r}$ be the cluster in which the ith feature is most expressed. In mathematical formulas, $\hat{r}$ can be defined as the index such that the following condition is satisfied:

$$a_i \in F_{\hat{r}} \quad \Longleftrightarrow \quad c_{i\hat{r}}^S > c_{i\xi}^S \quad \forall \xi \in \{1, 2, \ldots, k\} \quad \xi \neq \hat{r}. \tag{7.2}$$

Intuitively, it is reasonable to assign the feature a_i to the cluster $F_{\hat{r}}$. If the condition (7.2) is applied for all the indices $i \in \{1, 2, \ldots, k\}$ and all the features a_i are assigned to the corresponding clusters $F_{\hat{r}}$, a partition in clusters F_r is obtained from a previous partition in clusters S_r.

The same procedure can be applied for obtaining a partition of the samples when a partition of the features is known. The following rule can be used for assigning a sample a^j to a certain cluster $\hat{S}_r$:

$$a^j \in \hat{S}_{\hat{r}} \quad \Longleftrightarrow \quad c_{j\hat{r}}^F > c_{j\xi}^F \quad \forall \xi \in \{1, 2, \ldots, k\} \quad \xi \neq \hat{r}. \tag{7.3}$$

If this rule is applied for each j, a new partition in clusters $\hat{S}_r$ is obtained from the partition in clusters F_r. Note that a symbol is used for discriminating the generic cluster S_r and the generic cluster $\hat{S}_r$. Indeed, S_r is the generic cluster used for finding a partition in clusters F_r of the features, whereas $\hat{S}_r$ represents the partition in clusters obtained from the clusters F_r. Two different notations for S_r and $\hat{S}_r$ are used because these two partitions of samples can be different in general. Even though S_r generated F_r and F_r generated $\hat{S}_r$, there are no reasons why S_r and $\hat{S}_r$ should correspond. If they correspond, then the partition in biclusters (S_r, F_r) is called *consistent*.

It is important to note that not all the sets of data admit a consistent partition in biclusters. This may happen because there may not be a statistical evidence that a sample or a feature belongs to a certain cluster. If a consistent partition in biclusters exists for a certain set of data, then it is said to be *biclustering-admitting*. When it is not the case, samples or features are usually deleted from the set of data for letting it become biclustering-admitting. In this case, it is important to delete the least possible in order to preserve the information in the set of data. This procedure is known as *feature selection*.

The requirement of consistency can be weak in some cases. Let us suppose that a partition in clusters S_r is available, and that a partition in clusters F_r is obtained from it. Each feature is therefore assigned to the cluster $F_{\hat{r}}$ such that $c_{i\hat{r}}^S$ has the largest value in the vector c_i^S. Let us suppose now that the following condition holds:

$$\min_{\xi \neq \hat{r}} \{c_{i\hat{r}}^S - c_{i\xi}^S\} \leq \varepsilon \tag{7.4}$$

where ε is a small number. In this case, small changes in the data can bring different partitions of the features in the clusters F_r. Indeed, small variations of the samples bring variations of the centers of the clusters S_r, and this can bring a different feature to be more expressed. The following example should clarify this concept.

Let us suppose that the data are partitioned in two biclusters only. S_1 and S_2 are known, as well as their centers c_1^S and c_2^S. The features are also partitioned into two clusters F_1 and F_2. Each feature is assigned to one of the two clusters depending on their average expressions in the corresponding clusters S_r. Therefore, the generic feature a_i is assigned to F_1 if $c_{i1}^S > c_{i2}^S$, and vice versa. Let us suppose for instance that $c_{i1}^S = 5.9$ and $c_{i2}^S = 6.1$. Then, a_i is assigned to F_2. However, the condition (7.4) holds with $\alpha \geq 0$. This means that it is not evident statistically that a_i belongs to F_2. Indeed, let us suppose that another sample is added to the set of data, and that it is assigned to cluster S_1. The center of S_1 hence changes, and in particular its ith component changes. If the feature a_i is more expressed in this sample, the average c_{i1}^S can increase. Since it is an average and it considers all the samples in the same cluster, it cannot change dramatically, even though the new sample might be different from the others. However, in the considered example, the feature a_i might be assigned to a different cluster after the new sample is added. If indeed c_{i1}^S is now equal to 6.2, then $c_{i1}^S > c_{i2}^S$, and the feature a_i is assigned to F_1.

In order to overcome this kind of problem, conditions stronger than consistent biclustering are introduced in [176]. A biclustering is called an *additive consistent biclustering* with parameter α or an *α-consistent biclustering* if the following two relations holds

$$a_i \in \hat{F}_{\hat{r}} \iff c_{i\hat{r}}^S > \alpha_j^F + c_{i\xi}^S \quad \forall \xi \in \{1, 2, \ldots, k\} \quad \xi \neq \hat{r} \qquad (7.5)$$

$$a^j \in \hat{S}_{\hat{r}} \iff c_{j\hat{r}}^F > \alpha_i^S + c_{j\xi}^F \quad \forall \xi \in \{1, 2, \ldots, k\} \quad \xi \neq \hat{r} \qquad (7.6)$$

where each α_j^F and α_i^S are positive numbers. It is easy to prove that an α-consistent biclustering is a consistent biclustering, but not the inverse. Indeed, if the conditions (7.5) and (7.6) are satisfied with $\alpha_j^F > 0$ and $\alpha_i^S > 0$, then they keep being satisfied with $\alpha_j^F = 0$ and $\alpha_i^S = 0$. Inversely, let us suppose that $c_{i\hat{r}}^S > \alpha_j^F + c_{i\xi}^S$ for all the ξ different from $\hat{r}$, in correspondence with some feature a_i and with $\alpha_j^F = 0$. If α_j^F is successively modified and it becomes positive, then the condition may not be satisfied anymore. The quantity $\alpha_j^F + c_{i\xi}^S$ becomes larger, and therefore the quantity $c_{i\hat{r}}^S$ may not be greater than it anymore.

Similar to α-consistent biclustering is the β-consistent biclustering. A biclustering is called a *multiplicative consistent biclustering* with parameter β or a *β-consistent biclustering* if the following two relations holds

$$a_i \in \hat{F}_{\hat{r}} \iff c_{i\hat{r}}^S > \beta_j^F c_{i\xi}^S \quad \forall \xi \in \{1, 2, \ldots, k\} \quad \xi \neq \hat{r} \qquad (7.7)$$

$$a^j \in \hat{S}_{\hat{r}} \iff c_{j\hat{r}}^F > \beta_i^S c_{j\xi}^F \quad \forall \xi \in \{1, 2, \ldots, k\} \quad \xi \neq \hat{r} \qquad (7.8)$$

where $\beta_j^F > 1$ and $\beta_i^S > 1$. As before, a β-consistent biclustering is a consistent biclustering.

7.3 Unsupervised and supervised biclustering

Biclustering is a technique for clustering on two dimensions. On the first dimension, the samples contained in a set of data are taken into account. Standard clustering methods work on this dimension only. On the second dimension, moreover, biclustering considers the features that are used for representing the samples. The simultaneous clustering of samples and features allows one to partition the data in clusters where similar samples are contained, and to find out the features that cause these similarities.

Biclustering can be performed by solving one of the optimization problems discussed in Section 1.4. In this way, the partition of the samples and the partition of the features are searched simultaneously. Biclustering can also be performed by using methods for standard clustering coupled with the concepts introduced in the previous section. For instance, the k-means algorithm can be applied for partitioning a given set of samples. Then, the conditions (7.2) can be used for finding a correspondent partition in clusters of the features. In this way, the biclusters can be defined. Besides the partition of the samples, the partition of their features allows one to identify the ones that generate the current partition of the samples.

However, the partition found in biclusters might not be consistent. From the partition in clusters of the features, a partition in clusters, the samples can be obtained using the conditions (7.3). As already pointed out, the obtained partition of the samples can be equal or not to the starting partition, i.e., to the partition found by the k-means algorithm in this example. If they correspond, the biclustering is consistent, otherwise it is not. In the latter case, some features be can deleted from the set of data in order to let the biclustering become consistent. The feature selection process is not easy, and the consistent biclustering can be found only if the set of data is biclustering-admitting.

Clustering techniques are referred to as techniques for unsupervised classifications, because they are used when there is not any previous knowledge about the data. Biclustering can be also supervised, because the information from a training set can be actually used. If a training set is available, a set of data is available that is already partitioned in different classes. In this case, a partition algorithm such as k-means is not needed, because the data are already partitioned. Then, a partition of the features can be obtained applying the conditions (7.2). At this point, a set of biclusters is defined, which is able to provide information on the features that caused the classification of the samples given by the training set. As before, this information is accurate if the biclustering is consistent, otherwise there is not a strong statistical evidence that a feature belongs to one cluster or another.

The problem of finding a consistent biclustering, once a partition of the samples is given, can be formulated as an optimization problem (see Section 1.4). Before formulating the optimization problem, let us introduce some notations. Let F be an $m \times k$ matrix whose elements can have value 0 or 1 only. The generic f_{ir} element has value 1 if the feature a_i belongs to the cluster F_r, and 0 otherwise. By using this matrix, the condition of consistency can be written as follows. Suppose that the clusters S_r are known. Suppose that the clusters F_r are built by using the conditions (7.2). Then, the clustering in biclusters (S_r, F_r) is consistent if S_r is obtained when the conditions (7.3) are applied. Equivalently, the following conditions must hold:

$$\frac{\sum_{i=1}^{m} a_{ij} f_{i\hat{r}}}{\sum_{i=1}^{m} f_{i\hat{r}}} > \frac{\sum_{i=1}^{m} a_{ij} f_{i\xi}}{\sum_{i=1}^{m} f_{i\xi}}, \quad \forall \hat{r}, \xi \in \{1, 2, \ldots, k\}, \hat{r} \neq \xi, j \in S_{\hat{r}}. \tag{7.9}$$

Let us introduce now the binary vector x of length m whose generic element x_i is 1 if the feature a_i is taken into account, and 0 otherwise. The condition (7.9) on a subset of features can be written as follows:

$$\frac{\sum_{i=1}^{m} a_{ij} f_{i\hat{r}} x_i}{\sum_{i=1}^{m} f_{i\hat{r}} x_i} > \frac{\sum_{i=1}^{m} a_{ij} f_{i\xi} x_i}{\sum_{i=1}^{m} f_{i\xi} x_i}, \quad \forall \hat{r}, \xi \in \{1, 2, \ldots, k\}, \hat{r} \neq \xi, j \in S_{\hat{r}}. \tag{7.10}$$

As already pointed out, when deleting features in order to find a consistent biclustering, the minimum possible features have to be removed. The problem of choosing a subset of features that is as large as possible and such that the corresponding biclustering is consistent can be formulated as an optimization problem. The function to maximize is

$$f(x) = \sum_{i=1}^{m} x_i \tag{7.11}$$

while subject to the constraints (7.10). In the optimization field, this problem is called *fractional 0-1 programming problem*. Its solution provides an efficient selection of the features to take into account. This optimization problem can be solved by using a suitable method for global optimization (Section 1.4), but it is usually quite difficult to manage. Therefore, ad hoc methods have been developed. Details about these methods can be found in [32, 176].

The solutions of the formulated optimization problem allow one to obtain consistent biclusterings where the maximum number of features is considered. Similarly, the following optimization problem provides α-consistent biclusterings:

$$\max_{x} f(x)$$

subject to

$$\frac{\sum\limits_{i=1}^{m} a_{ij} f_{i\hat{r}} x_i}{\sum\limits_{i=1}^{m} f_{i\hat{r}} x_i} > \alpha_j + \frac{\sum\limits_{i=1}^{m} a_{ij} f_{i\xi} x_i}{\sum\limits_{i=1}^{m} f_{i\xi} x_i}, \quad \forall \hat{r}, \xi \in \{1, 2, \ldots, k\}, \hat{r} \neq \xi, j \in S_{\hat{r}}.$$

This other optimization problem provides instead β-consistent biclusterings:

$$\max_x f(x)$$

subject to

$$\frac{\sum\limits_{i=1}^{m} a_{ij} f_{i\hat{r}} x_i}{\sum\limits_{i=1}^{m} f_{i\hat{r}} x_i} > \beta_j \frac{\sum\limits_{i=1}^{m} a_{ij} f_{i\xi} x_i}{\sum\limits_{i=1}^{m} f_{i\xi} x_i}, \quad \forall \hat{r}, \xi \in \{1, 2, \ldots, k\}, \hat{r} \neq \xi, j \in S_{\hat{r}}.$$

7.4 Applications

Biclustering techniques are nowadays mainly applied to the field of biology, and in particular for the analysis of microarray data. In Section 7.4.1 we will discuss in detail this kind of application and we will report the experiments presented in [32], where supervised biclustering has been applied. Moreover, even other applications of biclustering have emerged in the literature. Biclustering is used for collaborative filtering, where the aim is to identify subgroups of customers with similar preferences or behaviors toward a subset of products [55, 228, 244]. In information retrieval and text mining [60], biclustering can be successfully used to identify subgroups of documents with similar properties relative to subgroups of attributes, such as words or images. In [103], biclustering has been used for analyzing electoral data and, in [142], it has been used for studying the exchanges of foreign currencies. To the best of our knowledge, biclustering has never been used before for solving problems related to agriculture. However, as we will explain in Section 7.4.2, it is our opinion that biclustering techniques can be successfully applied to agricultural-related data mining problems.

7.4.1 Biclustering microarray data

Microarrays in biology are used for studying the expression of genes under different conditions. Genes in humans, for instance, have different expression levels in presence of diseases. Finding the set of genes that have similar expression levels in the

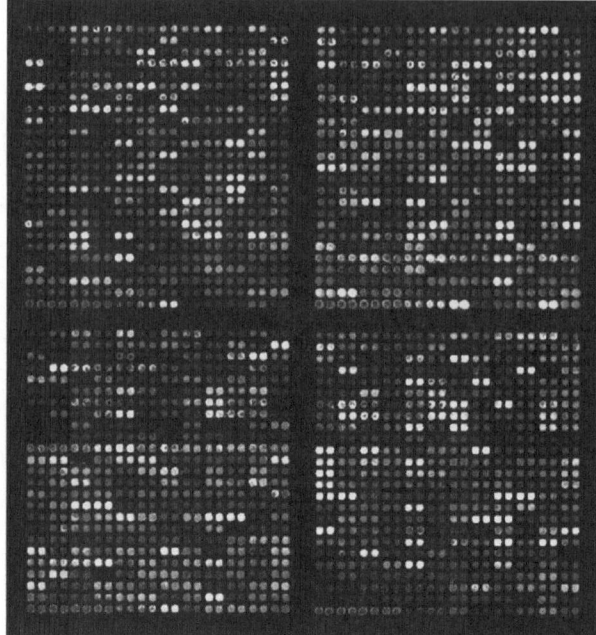

Fig. 7.1 A microarray.

presence of a certain disease can help understanding the disease itself and how the body reacts to it. Microarray data are organized as in a matrix: each row of the matrix is related to a gene, and each column is related to a different condition. Therefore, the generic element of a microarray gives the expression level of the gene, specified by the current row, under the condition specified by the current column. The expression levels are usually visualized by a matrix of colors ranging from light green to red. In black and white pictures, this range of colors corresponds to a gray scale from white to black. Figure 7.1 shows a microarray.

The expression levels obtained by a microarray can be placed in a $m \times n$ numerical matrix A. The samples contained in this matrix are organized column by column: each of them represents an experimental condition through the expression levels of all the considered genes. The features used for describing such samples are hence the expression levels of the genes. Each row of A contains all the measured expression levels of the same gene under the different experimental conditions.

Biclusters in the matrix A can reveal genetic pathways that can be used, for instance, for identifying the genes with different expression levels in presence of a disease. A bicluster of samples and features groups a subset of similar conditions that are caused by a subset of genes having similar expression levels. The meaning of the term "similar" depends on the kind of considered bicluster. For instance, biclusters can have constant values, on the whole bicluster or only on its rows or columns, or it can be a bicluster with coherent values.

Another way for finding biclusters in the matrix A is to look for a consistent biclustering of the data as explained in Section 7.2. Let us suppose that the samples (the experimental conditions in this application) are already classified in clusters. Then, the rule (7.2) can be used for finding a partition in clusters of the features, i.e., a partition in clusters of the genes. In this way, biclusters containing conditions and genes can be identified, and the genes causing certain conditions can be located. It is important to note that the correlation between conditions and genes is statistically evident only if the partition found in biclusters is consistent. For this reason, the best way to find such partition is to solve the optimization problem (7.11)–(7.10). In this way, the features that cause the biclustering not to be consistent are removed.

In [32, 176], this technique has been applied to a well-researched microarray data set containing samples from patients diagnosed with *acute lymphoblastic leukemia* (ALL) and *acute myeloid leukemia* (AML) diseases [89]. The original set of data has been divided in two parts: a part used as training set and another used as validation set. Hence, the training set used contains 27 samples classified as ALL and 11 sample classified as AML; the validation set contains 20 ALL samples and 14 AML samples. A consistent biclustering is obtained by following a methodology described in [32], which is based on the optimization of the problem (7.11)–(7.10). After that, the samples of the validation set are subsequently classified choosing for each of them the class with the highest average feature expression: 3439 features for class ALL and 3242 features for class AML have been selected. The obtained classification contains only one error: one AML-sample was classified into the ALL class. The obtained partition in biclusters is shown in Figure 7.2.

The same methodology has also been applied to the *Human Gene Expression* (HuGE) Index data set [112]. The purpose of the HuGE project is to provide a comprehensive database of gene expressions in normal tissues of different parts of the human body and to highlight similarities and differences among the organ systems [111]. The data set consists of 59 samples from 19 distinct tissue types. It was obtained using oligonucleotide microarrays capturing 7070 genes. The samples were obtained from 49 human individuals: 24 males with median age of 63 and 25 females with median age of 50. Each sample came from a different individual except for the first 7 BRA (brain) samples that were from different brain regions of the same individual and 5th LI (liver) sample, which came from that individual as well. The list of considered tissue types with their abbreviations and the number of samples for each of them is given in Figure 7.3. Figure 7.4 presents the partition in biclusters obtained by applying the same methodology as above. The distinct block-diagonal pattern of the heatmap evidences the high quality of the obtained feature classification.

7.4.2 Biclustering in agriculture

There are currently no applications in the agricultural field for biclustering techniques. The reason might be the fact that biclustering techniques are used only in recent years, in which they have been mainly applied to gene expression analysis.

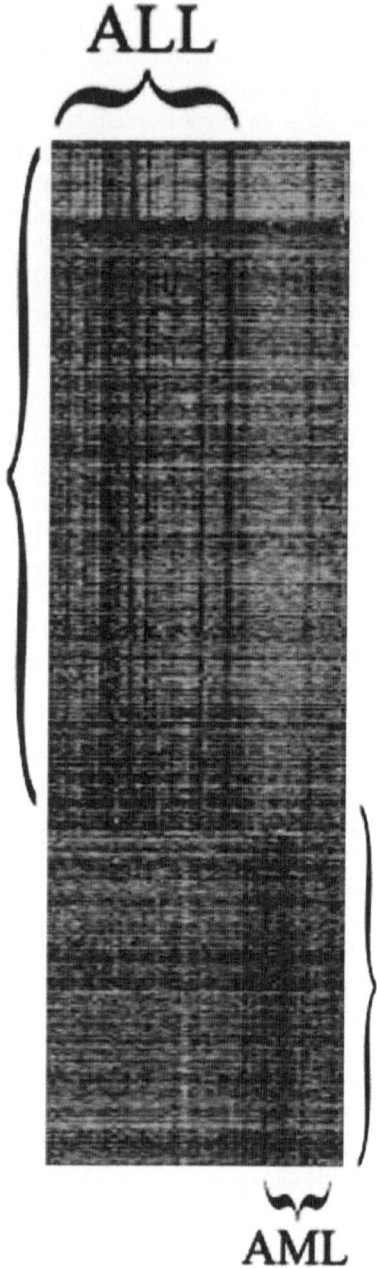

Fig. 7.2 The partition found in biclusters separating the ALL samples and the AML samples.

Tissue type	Abbreviation	Number of samples
Blood	BD	1
Brain	BRA	11
Breast	BRE	2
Colon	CO	1
Cervix	CX	1
Endometrium	ENDO	2
Esophagus	ES	1
Kidney	KI	6
Liver	LI	6
Lung	LU	6
Muscle	MU	6
Myometrium	MYO	2
Ovary	OV	2
Placenta	PL	2
Prostate	PR	4
Spleen	SP	1
Stomach	ST	1
Testes	TE	1
Vulva	VU	3

Fig. 7.3 Tissues from the HuGE Index set of data.

In fact, biclustering was introduced in the literature in 1972 by Hartigan [103], but only later, in 2000, Cheng and Church took the idea and applied it to expression data [47]. Another reason for the non-use of biclustering in agriculture may be the complexity of the method. As usual, scientists who are expert in fields different from numerical analysis and computer science tend to use easier solutions. This is one of the reasons why methods such as k-means are applied more than neural networks or support vector machines in applied fields.

However, it is our opinion that biclustering may provide good results if applied to agricultural problems. Let us take as example the problem considered in Section 3.5.1, where wine fermentation problems are predicted by a k-means approach. In this example, each sample is represented as a vector having as components some compounds measured in the wine during the fermentation process. The goal is to predict wine fermentation problems that may occur using information about the compounds measured not later than 3 days after the start of the fermentation process. The clustering algorithm used provides a partition of the samples but no considerations are made about the compounds that are responsible for these partitions. Biclustering might also provide this kind of information. If the feature is known, a particular compound in this case that is associated to a cluster of samples, then such samples are similar because of that feature. In this application, besides discovering patterns that signal fermentation problems, the compounds that are more responsible for such problems can be located. This may help the work of the enologist when his intervention is required to correct the fermentation process.

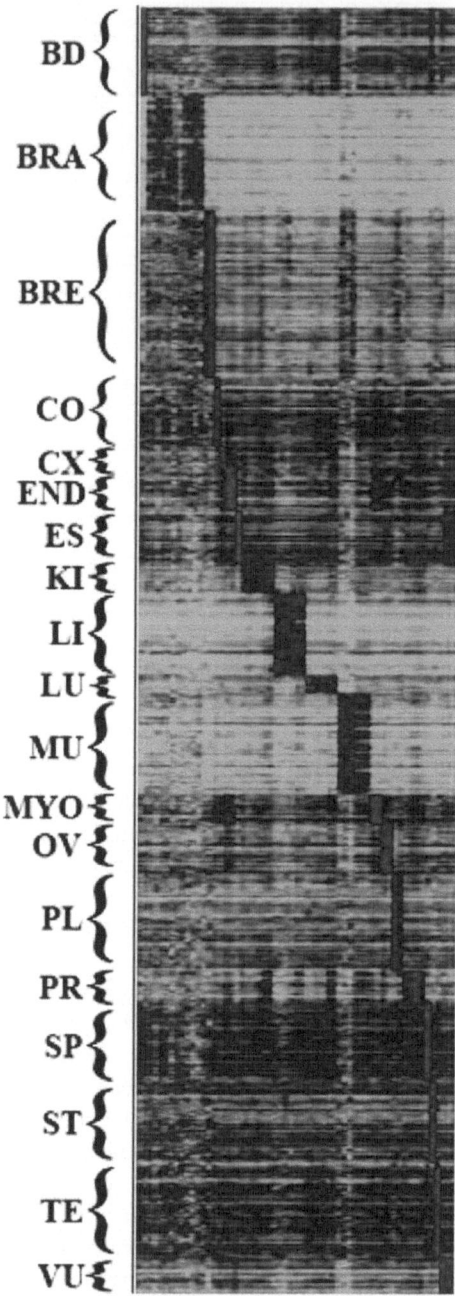

Fig. 7.4 The partition found in biclusters of the tissues in the HuGE Index set of data.

Biclustering can be applied even to other applications discussed in the other chapters of the book. In particular, when a training set is available, and classification techniques can be used, then a partition in biclusters of the data can be found before the classification technique is applied. This can be done using the rule (7.2). When the biclusters are found, each class in the original training set is associated to a cluster of features. This allows one to find out which are the features responsible for grouping a subset of samples in a certain class. In order to be sure that each feature is actually assigned to the right class, the partition in biclusters has to be consistent. The consistency can be checked by applying the rule (7.3) and checking if the original classification in the training set is found again. In the case the partition is not consistent, then some of the features need to be discarded. This task could be done by hand if the classification problem is not so large. Otherwise, the optimization problem (7.11)–(7.10) needs to be solved.

Note that, once the samples in a testing set have been classified by using a classification technique, the rule (7.3) can be applied to it and another partition in biclusters can be found. The classification technique tries to reproduce the classification in the training set on unclassified samples. Therefore, choosing a certain class, the corresponding bicluster in the training set and the one in the testing set should be similar. This may also be used for validating the data mining technique used.

7.5 Exercises

In this section some exercises related to biclustering are presented.

1. Consider the matrix
$$A = \begin{pmatrix} 1 & 2 & 3 & -4 & 5 \\ 1 & 1 & 0 & 0 & 1 \\ 0 & 1 & 2 & 2 & 0 \\ -1 & 3 & 1 & 0 & 2 \\ 3 & -1 & 1 & 2 & 1 \end{pmatrix}.$$

   Locate a bicluster with constant row values having dimension 2×2.

2. Consider 6 samples in a three-dimensional space:
$$x_1 = (7, 0, 0), \quad x_2 = (5, 0, 0), \quad x_3 = (0, 1, 0),$$
$$x_4 = (0, 3, 0), \quad x_5 = (0, 0, 1), \quad x_6 = (0, 0, 5).$$

   Suppose that they are assigned to 3 clusters as follows:
$$x_1 \in S_1, \quad x_2 \in S_1, \quad x_3 \in S_2, \quad x_4 \in S_2, \quad x_5 \in S_3, \quad x_6 \in S_3.$$

   By using the rule (7.2), find a partition of the features used for representing the three-dimensional points. Then, define a partition of the points in biclusters.

3. Verify that the partition in biclusters obtained in the previous exercise is consistent.

4. Consider 4 samples in a three-dimensional space:

$$x_1 = (1, 2, 3), \quad x_2 = (2, 3, 4), \quad x_3 = (3, 4, 2), \quad x_4 = (4, 5, 1).$$

Suppose that

$$x_1 \in S_1, \quad x_2 \in S_1, \quad x_3 \in S_2, \quad x_4 \in S_2.$$

Find a partition in biclusters by using the rule (7.2) and check if the biclustering is consistent.

5. Provide an example of partition in biclusters of a given set of data which is α-consistent but not consistent for a certain α value.

Chapter 8
Validation

8.1 Validating data mining techniques

This book presents details for some of the most frequently used data mining techniques in the field of agriculture. As pointed out in Chapter 1, data mining techniques can be mainly divided into clustering and classification techniques. Clustering techniques are used when there is not any previous knowledge about the data, and hence a partition in clusters grouping similar data is searched. When a training set is available, classification techniques can be applied. In such cases, the training set is exploited for classifying data of unknown classification. The training set can be exploited in two ways: it can be used directly for performing the classification, or it can be used for setting up the parameters of a model which fits the data.

Chapter 3 presents the most frequently used clustering algorithm, the k-means algorithm, and many of its variants. Samples in a set of data are partitioned into clusters; each cluster groups a subset of samples very similar to one another. The similarities between the samples are measured using a distance function. Each cluster contains the samples closest to the center of the cluster. An error function monitoring the distances between the samples and the centers is used to evaluate the quality of a given partition in clusters. Chapter 7 introduces the simultaneous partition of the samples and their features in biclusters. In this case, the quality of the biclusters is evaluated using error functions as well, where the variance in the elements of a bicluster is measured. These error functions depend on the kinds of biclusters that are searched.

The classification techniques discussed in this book are the k-nearest neighbor (Chapter 4), the artificial neural networks (Chapter 5), the support vector machines (Chapter 6), and the supervised biclustering (Chapter 7). All these techniques require the use of a training set. k-nearest neighbor exploits such a training set directly for classifying samples with unknown classification. An unclassified sample is compared to similar samples in the training set, and the classification is assigned in accordance with the ones such similar samples have. As before, the similarities between samples are measured through distance functions. Artificial neural networks consist of a set

A. Mucherino et al., *Data Mining in Agriculture*, Springer Optimization and Its Applications 34, 161
DOI: 10.1007/978-0-387-88615-2_8,
© Springer Science + Business Media, LLC 2009

of neurons performing simple tasks and connected to each other in a structure that resembles the human brain. A neural network can be trained using the information available in a training set. During this phase, they are supposed to learn from the data and generalize from them. Once trained, a neural network should be able to classify unknown samples because of the information extracted during the learning phase from the known samples of the training set. Similarly, support vector machines learn from a training set how to classify unknown data. They are linear classifiers and can be extended to nonlinear cases. The basic assumption is that a classifier able to separate two distinct classes of samples with a larger margin is a better classifier. Finally, supervised biclustering uses a training set of samples for simultaneously classifying in biclusters the samples themselves and even their features. Therefore, not only the samples are categorized in classes, but even their features, so that the features responsible for the classification of a class can be identified.

In the clustering techniques discussed in this book, an error function is usually used for finding the best partition in clusters or biclusters of the data. Such error function gives an evaluation of the quality of a given partition: the lower is the error function value, the higher is the quality. This can be considered as an evaluation of the quality of the solution. However, even when the error function has a small value, the obtained partition may not be accurate. For instance, let us suppose that the k-means algorithm is used with different values for the k parameter. For a given k, the error function values show the best partition among a set of partitions in k clusters. Unfortunately, if k changes, and k_1 and k_2 are for instance used, then the error function values cannot be used for comparing the partitions in k_1 clusters to the partitions in k_2 clusters. Therefore, sometimes validation techniques are needed when clustering methods are used. Reference [98] presents a survey of validation techniques applied to clustering methods. This survey takes into account even clustering methods that are not presented in this book.

The situation is different when dealing with classification techniques. In the k-nearest neighbor approach, an unknown sample is classified considering the classification of its neighbors in the training set. The accuracy of the classification depends on the value chosen for the parameter k, and some k values may be good for some types of applications and not as good for other types of applications. Although the method provides a simple and often effective classification, unfortunately, the accuracy of the classification needs verification. In the neural network approach, an error function is defined for monitoring how the network fits the data during the learning phase. This error function evaluates the mean error occurring when the network is used for classifying the samples of a training set. Once the network is trained, and eventually also pruned, it can be used for classifying unknown samples. Even in this case, the network is not able to provide an estimation for the accuracy of the classification, and therefore the results need to be validated in a different way. In general, all the classification techniques using a training dataset are able to estimate the accuracy of their classification on the known data only.

Therefore, it is important to validate the classifiers used in the classification process. Validation techniques can be used for this purpose. Usually, the available training set is divided at least in two parts. The first part is actually used as training set and the second part is used for validation purposes. The latter part is usually called *validation* or *testing* set. Both names, in general, can refer to the set of known samples used for evaluating the quality of the classifications. In some cases, however, validation and testing sets are actually two different objects. As an example, during the learning phase of a neural network, the parameters of the network are improved step by step and they converge toward the optimal values. Therefore, at each iteration, the parameters can be used for classifying samples different from the ones in the training set. This allows one to check if the parameters are converging to optimal values, or if there is overfitting, during the learning phase. The set of samples used in this case is usually referred to as the validation set. Once the network has been trained, then a set of known samples can be used to check the quality of the classifications obtained by the network. This last set of known samples is referred to as the testing set.

In the following sections, three validation techniques are presented and for each of them an example in MATLAB® is provided. For simplicity, regression models and the simple k-nearest neighbor rule are validated on a random set of samples in a two-dimensional space in MATLAB. For more details about these techniques, the reader may consider Andrew Moore's lecture that can be found on the Internet [167].

8.2 Test set method

The training set contains the information needed for performing the classification of unknown samples. It consists in a set of pairs grouping samples and their corresponding classifications. All the other samples which are not contained in the training set have an unknown classification, and hence they cannot be used for validation purposes.

The *test set* method is based on the following idea. Since only the samples in the training set have a known classification, the idea is to split the training set in two parts: a part which is actually used as a training set, and another part used for the validation. In general, 70% of the data can be used as a training set, and the remaining (30%) can be used for the validation process.

Let us suppose that the k-nearest neighbor rule is used for classifying a set of unknown samples. To validate the effectiveness of this rule, 30% of the training set is classified using the remaining 70% of the training set. Since samples in both cases are taken from the training set, their classification is known, and therefore the classification obtained by the k-nearest neighbor rule can be validated. Similarly, a neural network or a support vector machine can be trained using 70% of the training set, and then the accuracy of the classification provided by the trained network or support vector machine can be evaluated on the remaining 30% of the training set.

8.2.1 An example in MATLAB

In this example, a linear regression model is validated by using the test set method. It is supposed that a set of points in a two-dimensional space is available, and that it is needed to model these points by linear regression. These points can represent measurements of a certain process that it is known to be linear. The available set of points is used as a training set: the general rule governing the process needs to be discovered from this set. Once the regression model has been found, it should be able to approximate with an acceptable accuracy the points of the training set, and it should also be able to generalize to other unknown points.

In order to validate the quality of the regression model, the test set method can be applied. Following this method, the original training set has to be divided in two parts. Let us suppose the training set contains the following 10 points:

$$(1, 4), \quad (2, 2), \quad (3, 3), \quad (4, 1.7), \quad (5, 1)$$
$$(6, 1.2), \quad (7, 1.5), \quad (8, 1.9), \quad (9, 2.3), \quad (10, 2.7).$$

Three of these points (30%) can be used for validating the model, while the other seven points (70%) are used as a training set for finding the model. One issue can be how to decide the points to place in the validation set and the ones to place in the actual training set. This separation can be done in a totally random way, but there might be cases in which this can lead to problems. Let us consider, for instance, that the three points having the smallest x value are used as a validation set, whereas the others are used as a training set. The following MATLAB code has been used for performing the validation and generating Figure 8.1.

```
x = [1 2 3 4 5 6 7 8 9 10];
y = [4 2 3 1.7 1 1.2 1.5 1.9 2.3 2.7];
x1 = x(4:10);
y1 = y(4:10);
x2 = x(1:3);
y2 = y(1:3);
plot(x1,y1,'ks','MarkerSize',16,'MarkerEdgeColor','k','MarkerFaceColor',
     [.49 1 .63])
hold on
plot(x2,y2,'ko','MarkerSize',16,'MarkerEdgeColor','k','MarkerFaceColor',
     [.87 1 .23])
c = polyfit(x1,y1,1);
xx = 0:0.1:12;
yy = polyval(c,xx);
plot(xx,yy,'k')
err = abs(y(1) - polyval(c,x(1)))
```

The x vector is initialized with the x coordinates of the whole set of points, and the y vector is initialized with their y coordinates. In the vectors x1 and y1 are then placed the points that are actually used for computing the regression model. In x2 and y2 are instead placed the remaining points, the ones that are used for the validation. In this example, the compact symbologies 1:3 and 4:10 are used for considering vectors whose first component is 1 (or 4), whose last component is 3 (or 10) and having distance between any consecutive components equal to 1 (for details see Appendix A). These points separated in this way are then printed by using the function plot. Note that many options are used for controlling the symbols used when the points are drawn. In particular, the points in the validation set are marked by circles, and the points in the training set are marked by squares.

The function polyfit is able to find the coefficient of the linear function that better approximates the points in x1 and y1 (see Section 2.4). The specified degree for the polynomial is 1, because a linear model is searched. The output of the function `polyfit` is placed in the vector c, which is soon used as input in the function `polyval` that evaluates the polynomial in the x coordinates stored in xx. The vector xx is created so that it contains all the x coordinates in x. The vector yy generated by `polyval` contains the corresponding y coordinates. The vectors xx and yy are finally given as input to the function `plot`.

Figure 8.1 shows that the found linear regression does not give a good approximation of the points placed in the validation set. For instance, the error `err` computed on the point $(x(1),y(1))$ has value 3.59. This is not a small error, if it is compared to the coordinates of the points. This error is larger than 3 times the difference between two consecutive x components. If the points (x1,y1) and (x2,y2) are instead chosen in the following way

```
x1 = x([1 2 4 5 7 8 10]);
y1 = y([1 2 4 5 7 8 10]);
x2 = x([3 6 9]);
y2 = y([3 6 9]);
```

then the accuracy grows. Figure 8.2 shows the new-found linear regression. In this case, the whole set of points is represented better by the chosen 70% of the original training set. This brings to a reduction of the overall error on the points in the validation set. The largest error is here due to the points (x6,y6) and it corresponds to 0.85. In general, more than one random division of the training set could be considered and the test set method applied for each of these divisions.

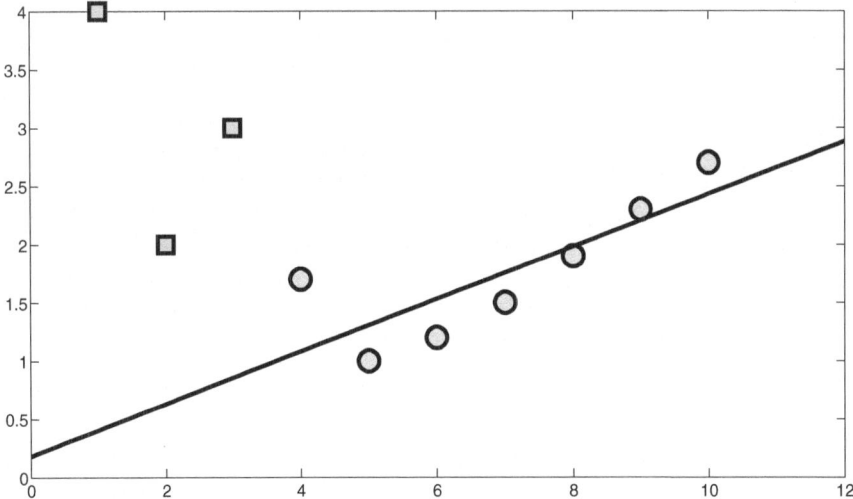

Fig. 8.1 The test set method for validating a linear regression model.

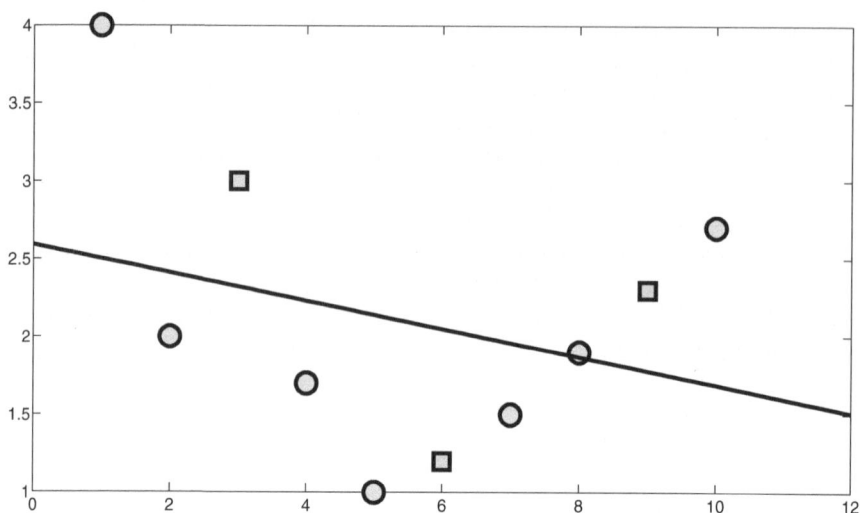

Fig. 8.2 The test set method for validating a linear regression model. In this case, a validation set different from the one in Figure 8.1 is used.

8.3 Leave-one-out method

There are two disadvantages in using the test set method for validation. First, a consistent part of the training set is actually not used as a training set, but it is used as a validation set. Second, the validation set is generally randomly extracted from the original training set, and it may not be a good representative of the whole set. Therefore, the original training set has to be reduced for applying this method, and it may be a problem if there is not much data available. Furthermore, the validation set may not provide an accurate validation. For instance, if only one sample of a certain class is contained in the validation set, which the accuracy of classifications in this class is evaluated only on such a sample, and this is statistically irrelevant.

The *leave-one-out* method overcomes these problems. As the name suggests, the validation is performed by leaving only one sample out of the training set: all the samples except the one left out are used as a training set, and the classification method is validated on the sample left out. If this procedure is performed only once, then the result would be statistically irrelevant as well. The procedure is indeed performed as many times as the number of samples in the training set, that one by one are taken out of the training set. The overall accuracy of the classifications of the samples left out gives an evaluation of the classification method.

8.3.1 An example in MATLAB

A quadratic regression model is validated in the example discussed in this section. Let us suppose that the same set of points used in the example in Section 8.2.1

is available, but this time it is known that the model fitting these points has to be quadratic. In practice, the parabola that better fits the points is searched. As before, the available set of points can be used as a training set for finding the quadratic regression model which is able to approximate the points in the training set and even unknown points.

The leave-one-out method is used for evaluating the quality of several quadratic models that can be generated from the set of points. In particular, each model is created by using the whole training set except only one point, which is later used for the validation. The following MATLAB code can be used for building one of these quadratic models leaving out the point (x1,y1):

```
x = [1 2 3 4 5 6 7 8 9 10];
y = [4 2 3 1.7 1 1.2 1.5 1.9 2.3 2.7];
x1 = x;
y1 = y;
x1(1) = [];
y1(1) = [];
x2 = x(1);
y2 = y(1);
plot(x1,y1,'ks','MarkerSize',16,'MarkerEdgeColor','k','MarkerFaceColor',
    [.49 1 .63])
hold on
plot(x,y,'ko','MarkerSize',16,'MarkerEdgeColor','k','MarkerFaceColor',
    [.87 1 .23])
c = polyfit(x1,y1,2);
xx = 0:0.1:12;
yy = polyval(c,xx);
plot(xx,yy,'k')
err = abs(y(1) - polyval(c,x(1)))
```

It is supposed that the original training set is the same used in Section 8.2.1, containing points whose x and y coordinates are stored in x and y, respectively. In x1 and y1 are specified the points that are part of the training set. In the code, x1 and y1 are initially set equal to x and y, and then the first component of both of them is deleted. The instruction x1(1) = [] actually removes the component 1 of x1, since it assigns to x1(1) an empty matrix []. The validation set contains in this case only one point, whose x and y coordinates are stored in x2 and y2. Figure 8.3(a) is generated by the two calls to the function plot, separated by the instruction hold on.

The MATLAB function polyfit is used for creating the quadratic regression model. It receives as inputs the actual training set through the vectors x1 and y1, and the degree of the approximating polynomial, which is 2 in this example. The provided output consists of the obtained polynomial coefficients, stored in the vector c. In order to draw the polynomial, a vector xx is defined and the function polyval is called, similarly as in the example showed in Section 8.2.1. Another call of the function plot finally draws the quadratic regression in Figure 8.3(a).

The error occurring when the point left out, (x(1),y(1)) in this case, is compared to the corresponding point (x(1),polyval(c,x(1))) is 0.75. Following the leave-one-out method, the same procedure has to be repeated leaving out all the points of the training set, one by one. Figure 8.3(b) shows the quadratic regression obtained leaving out the point (x(4),y(4)). In Figure 8.4(a) the point (x(7),y(7)) is left out, and in Figure 8.4(b) the point (x(10),y(10)) is left out. The obtained errors are 0.01 when (x(4),y(4)) is left out, 0.11 when (x(7),y(7)) is left out, and 0.44

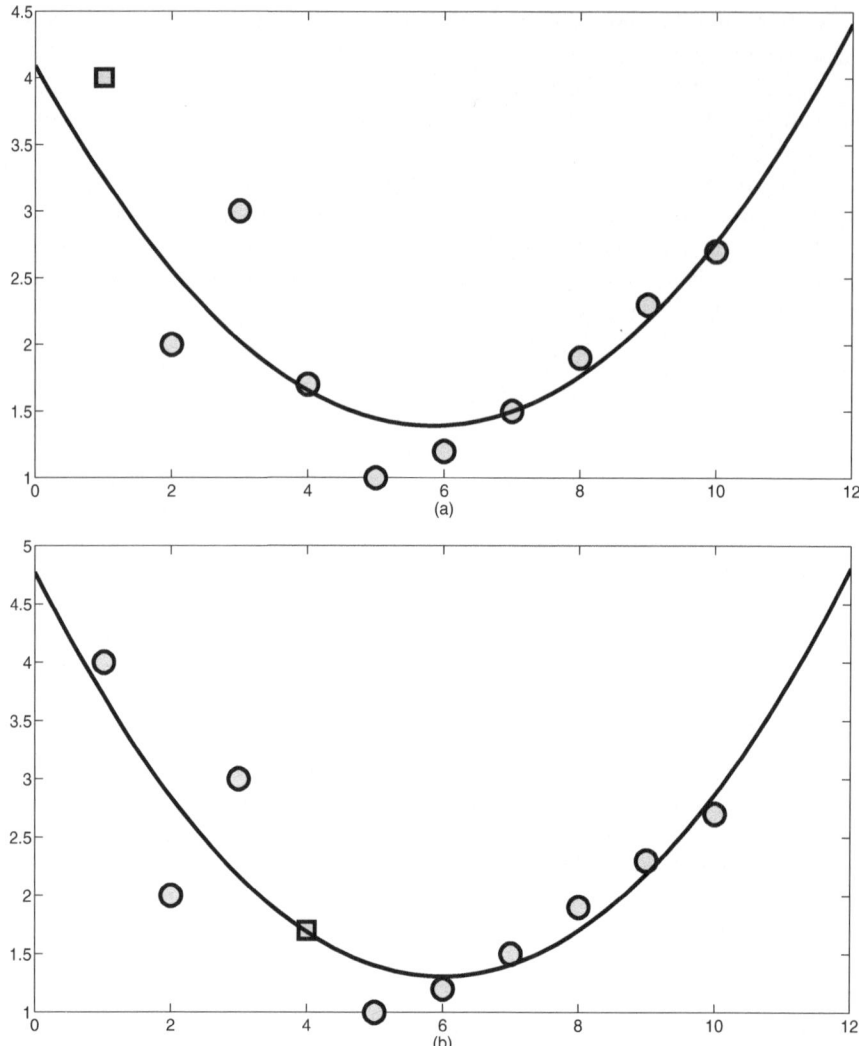

Fig. 8.3 The leave-one-out method for validation. (a) The point $(x(1),y(1))$ is left out; (b) the point $(x(4),y(4))$ is left out.

when $(x(10),y(10))$ is left out. The errors on the other points of the training set, when left out, have similar values. Therefore, in general, this regression model can be considered sufficiently accurate, since such errors are quite small.

8.4 k-fold method

As previously observed, the test set method may not be very efficient as a validation method because the validation set takes data from the training set and because

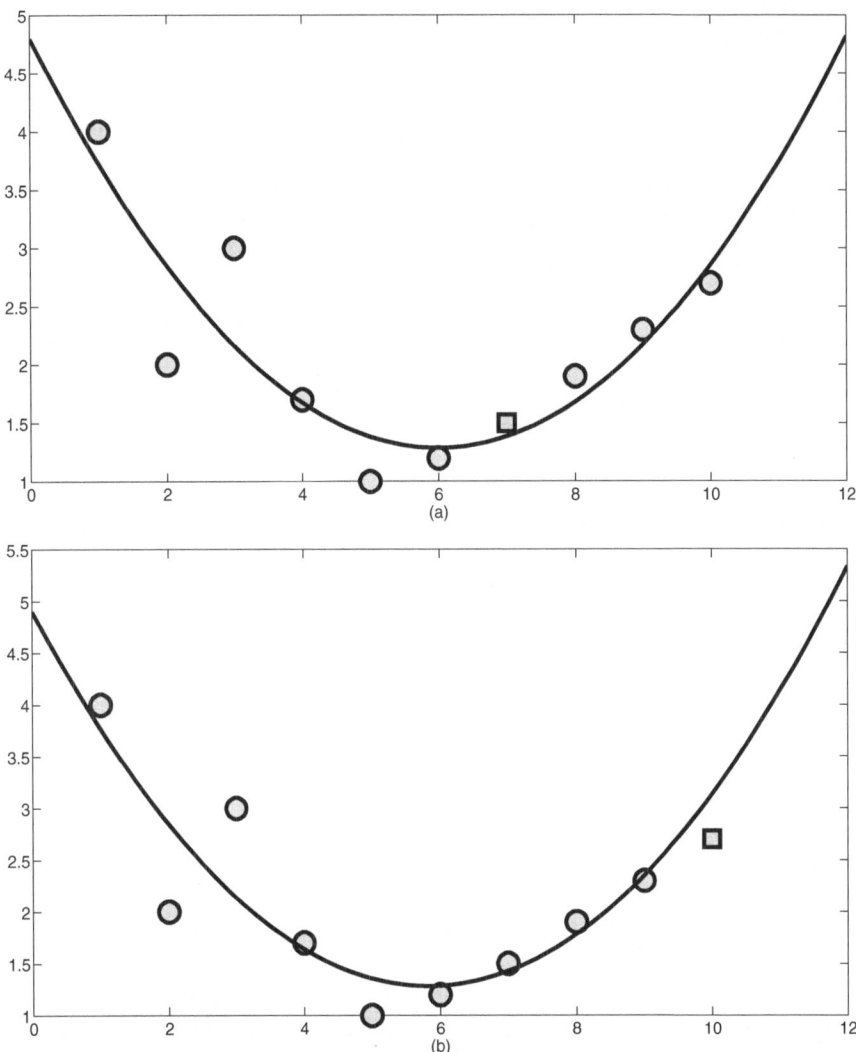

Fig. 8.4 The leave-one-out method for validation. (a) The point (x(7),y(7)) is left out; (b) the point (x(10),y(10)) is left out.

these data may not be a good representative of the original set. These problems are overcome if the leave-one-out method is instead used. In this case, indeed, only one sample is taken out of the training set at a time, and hence the amount of data actually used as a training set is not reduced. Moreover, all the samples, one by one, are also used for testing the accuracy of the classification, overcoming the problem of using a validation set that may not be a good representative of the whole set of data. The leave-one-out method seems to be the optimal choice, but it actually introduces another issue. This issue is related to the computational cost of the validation method.

If the training set contains n samples, then, following the leave-one-out method, the used classification method needs to be trained and applied n times. If n is large enough, this can be computationally demanding.

The optimal choice between the speed of the test set method and the reliability of the leave-one-out method is the *k-fold* method. In this method, the samples are partitioned in k groups. Then, for each of these k groups, the classification method is performed using as a training set the original set without the samples contained in one of these groups. After that, the group left out from the training set is used as a validation set. Note that if $k = n$, then the k-fold method corresponds to the leave-one-out method. If $k = 4$, then one iteration of the k-fold method, in which about 25% of the training set is devoted to the validation, is similar to the test set method. Therefore, the choice of a value for the parameter k is very important as it provides the trade-off between accuracy and computational speed.

8.4.1 An example in MATLAB

In this example, an application of the k-nearest neighbor method is validated by using the k-fold method. The example is carried out in the MATLAB environment and the code used for performing it is the following one:

```
[x,y] = generate(100,0.2);
[class] = hmeans(100,x,y,2);
plotp(100,x,y,class);
xA = x(1:50);
yA = y(1:50);
xB = x(51:100);
yB = y(51:100);
classA = class(1:50);
classB = class(51:100);
[class] = knn(50,xA,yA,2,50,xB,yB,classB);
plotp(50,xA,yA,class)
[class] = knn(50,xB,yB,2,50,xA,yA,classA);
plotp(50,xB,yB,class)
```

The MATLAB functions used in this example have been discussed in the previous chapters and their source codes are available in the book. The function generate is used for creating a random set of points in a two-dimensional space. One hundred points are generated, and they are randomly separated in two subgroups having a margin equal to 0.2 (see Section 3.6 and Figure 3.16). The chosen margin is quite wide, so that a clustering method is able to discover easily this pattern in the data. In particular, the function hmeans is used for partitioning the points in two parts. The partition is stored through the vector class, whose components can have value 1 or 2 (Section 3.6, Figure 3.20). This set of points and its partition are used as a training set for the application of the k-nearest neighbor. The call to the function plotp generates Figure 8.5.

The k-fold method is used for validating the application of the k-nearest neighbor method in which the training set is the one generated above. For simplicity, the parameter k in k-fold is set to 2, and therefore the training set is divided in 2 parts only. The division in 2 parts can be performed randomly, or the strategy used here

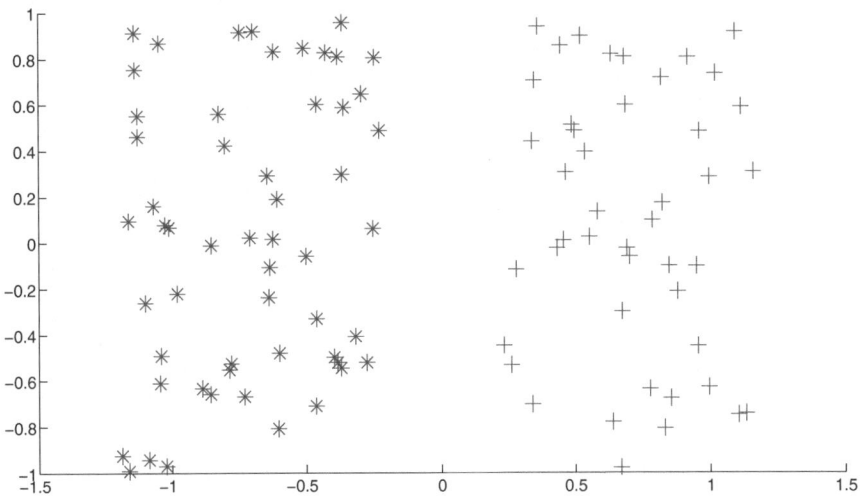

Fig. 8.5 A set of points partitioned in two classes.

can be implemented. xA and yA are defined so that they contain the first 50 points stored in x and y; xB and yB, instead, are defined so that they contain the last 50 points stored in x and y. The vectors classA and classB are defined similarly. The *k*-nearest neighbor method must be applied twice. The training set is specified by xB, yB and classB and the points stored in xA and yA are classified. Successively, the training set is specified by xA, yA and classA and the points stored in xB and yB are instead classified. The function plotp is used for plotting the points in xA and yA, where the vector class is the one just obtained by the function kNN. The obtained result is shown in Figure 8.6(a). Successively, the function plotp is used again for printing the points xB and yB marked in accordance with the classification given by the *k*-nearest neighbor. This other plot is shown in Figure 8.6(b). If Figure 8.5 and Figures 8.6(a) and 8.6(b) are compared, it is easy to see that the points are correctly classified by the *k*-nearest neighbor in both the cases. This classification method on this simple example is therefore validated.

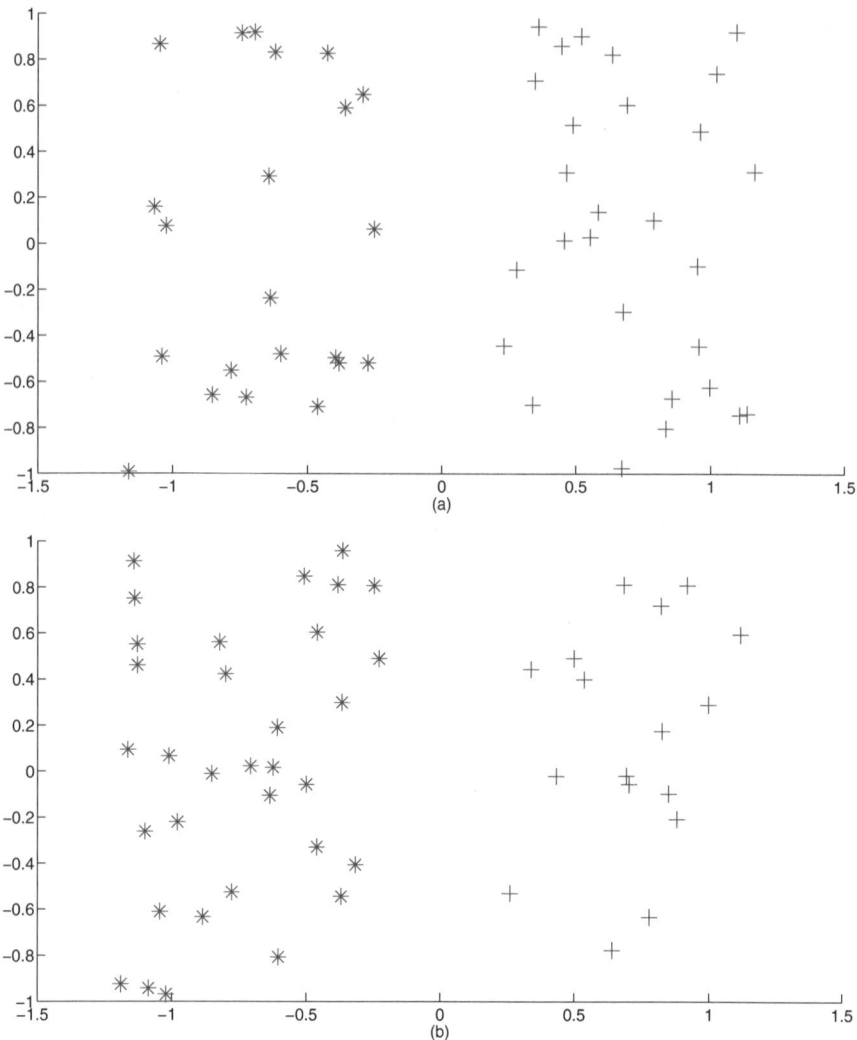

Fig. 8.6 The results obtained applying the k-fold method. (a) Half set is considered as a training set and the other half as a validation set; (b) training and validation sets are inverted.

Chapter 9
Data Mining in a Parallel Environment

9.1 Parallel computing

In this section, we give a very brief introduction to parallel computing, with the aim of giving to the reader the basic knowledge needed to understand the parallel version of some of the data mining techniques discussed in this book. A very simple example of a parallel algorithm is presented in Section 9.2. A parallel version of the k-means algorithm, the k-nearest neighbor decision rule, and the training phases of a neural network and a support vector machine are presented in Section 9.3.

When there is the need to analyze a large amount of data, the parallel computing paradigm can be used to fulfill these tasks and also reduce both the computational time and the memory requirement. A parallel environment is a machine or a set of machines in which more processors can simultaneously work on the same task. When working in a parallel environment, the computational time needed for carrying a standard algorithm out is sped up, because it is performed *in parallel* on more processors. The basic idea is to split the problem at hand into smaller subproblems that can be solved on different processors simultaneously. Each processor can also have a private memory in which it can store its own data. This reduces the memory requirement on each single processor.

The simplest and cheapest way to build a parallel machine is to interconnect single personal computers by a network and make them work together. The obtained parallel machine is also called a *Beowulf* cluster of computers. Each computer of the cluster can keep working independently from the others, but they can also work in parallel on a single task. These clusters of computers belong to the group of the MIMD parallel architectures, where MIMD stands for multiple instruction multiple data. As already mentioned, the basic idea in parallel computing is to divide a certain problem into smaller subproblems. On MIMD computers, each subproblem is solved independently from the others on different processors. The instructions are multiple, and therefore each subproblem can be solved by using an algorithm which is completely different from the ones implemented on the other processors. The data are also multiple, and therefore each processor can refer to a set of data different from

A. Mucherino et al., *Data Mining in Agriculture*, Springer Optimization and Its Applications 34,   173
DOI: 10.1007/978-0-387-88615-2_9,
© Springer Science + Business Media, LLC 2009

the ones the other processors refer to. Thus, each processor on a MIMD computer can work independently from the others. It is very important that the computational load on each processor is as balanced as possible. In other words, it is important that the computational cost for solving each subproblem is similar on each processor. If this aim is reached, the time for solving a problem in parallel on a machine with p processors could be the time for solving it on a sequential machine divided by p. However, this result is quite difficult to reach. Indeed, the subproblems in which a problem can be split are usually dependent from each other. For instance, some variable computed while solving one of these subproblems could be needed for the solution of another subproblem. For this reason, the processors often need to exchange data among them, and this operation may have a relevant computational cost. A good parallel algorithm is the one in which the computational load is well distributed among the processors and the number of synchronizations among the processors is limited.

Nowadays, Beowolf clusters of computers are much used. They work as a MIMD computer in which each processor can be located on a different personal computer. In particular, a Beowulf cluster is a MIMD parallel computer with distributed memory. In fact, each personal computer is equipped with its own processor and its own memory. Each processor can then access its own memory only, and not the memories of the others. Therefore, when the processors need to synchronize and exchange data, they actually need to communicate. Some processor may need to send its data to another or to all the processors. Some other processor may need to receive and save these data on its own memory. All these communications have a computational cost, which depends on the particular Beowulf cluster.

It is worth noting that other parallel computers with different architectures exist. For instance, MIMD computers can also have a shared memory. In this case, all the processors of the parallel machine refer to the same memory. Therefore, this memory must be big enough for containing all the data needed for all the processors. Moreover, the processors read and write on the same memory, and then the data a processor access can be modified even by another processor. This makes the development of algorithms for MIMD computers with shared memory more complex. On the other hand, there is no need of communications, and therefore there is no computational cost for communications in this case. Figure 9.1 compares the MIMD computers with distributed and shared memory.

Another kind of parallel machine is the SIMD computer, where SIMD stands for single instruction multiple data. As before, the data are multiple, and hence all the

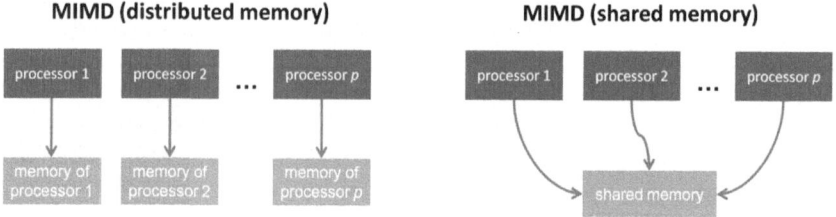

Fig. 9.1 A graphic scheme of the MIMD computers with distributed and shared memory.

processors of the parallel machine can work on different data. The instructions are single, though. This means that all the processors carry out the same instructions on different data. Differently from the MIMD computers, the processors are always synchronized in this case. A detailed classification of the parallel machines can be found in [77]. In this context, a single personal computer is referred to as a SISD machine, where SISD stands for single instruction single data. Recently, hybrid machines using both MIMD and SIMD architectures have also been developed. Details can be found in [217]. Finally, it is important to note that parallel computing recently evolved into the so-called *grid computing* [79]. The main difference between standard parallel computing and grid computing stands in the fact that many remote computers having difference properties (such as the CPU clock) are used simultaneously in grid computing. This brings consequences which are not discussed in this book.

Let us consider a simple problem and let us try to develop a parallel algorithm for solving this problem. Given an integer and positive number N, it could be desirable to know if such number is prime or not. The easiest way for finding out if N is prime is to try all the possible divisions of N by all the integer numbers smaller than N. The totality of these divisions can then be split among the p processors of a parallel machine so that each processor can try different divisors in parallel. In theory, the processors can work without knowing anything about the other processors. When all the processors have finished their job, p answers are available. Each processor indeed can provide as output the divisibility or not of N considering only its part of all possible divisors. If all the processors give as an answer "not divisible," then N is prime. If at least one processor gives as an answer "divisible," then N is not prime.

A way for improving this parallel algorithm can be the following one. While the processors work simultaneously, one of them may find an integer $\bar{n}$ such that N is divisible by $\bar{n}$. This would mean that N is not prime, and then the parallel algorithm can already provide its output: N is not prime. All the next divisions that all the processors might perform are completely useless, because the output is already known. In a sequential algorithm, this situation can be simply handled by a loop such as `while` or `repeat..until`. In this case, instead, only the processor which finds $\bar{n}$ can stop making divisions in this way. How to tell the other processors that continuing work is useless? The processors need to communicate.

On MIMD computers with distributed memory, all the processors have a private memory. On each memory, a variable can be stored and used as a sort of signal for the divisibility. This variable can have value 1 when no divisors have been found, and 0 otherwise. Naturally, if one processor changes the signal variable to 0, the others cannot see the change and stop working. For overcoming this problem, the processors can periodically exchange their own signal variable. If one of them at least is 0, all the processors can stop working and all of them can provide as output "the number is not prime." Instead, if all the processors finish working on their divisions, they exchange the signal variables and they find out that none of them is 0, then the global (or parallel) output is: "the number is prime."

Organizing the communications among the processors of a parallel computer in an efficient way is a crucial point for the success of a parallel procedure. Let us suppose a variable, say a, needs to be computed somewhere during the algorithm, and that the variable is needed later for computing other variables by all the processors. The variable a should then be located in the memory of each processor, because it is needed for performing a part of the instructions. There is therefore the need to let the processors communicate for exchanging data. An example may be that a single processor computes a and then it sends a to another processor that receives this information. Another example is that a single processor sends a to all the other processors. This kind of communication is called *broadcast*, because one processor sends its data to all the others involved in the computation. Let us suppose now that another variable, say b, is needed in all the memories because it has to be used in some part of the algorithms performed on each processor. Another communication may be activated for sending b to all the other processors by the time they need it. However, communications among the processors require time, and the time saved in the computation must pay off the time spent in the communication process. When the computation of a or b is not very expensive, it is preferable to let all the processors compute such variables in order to avoid communications, which might require more time.

Since clusters of computers are currently more frequently used than other parallel architectures, we will focus in the following on algorithms to be implemented on these kinds of parallel machines. On these machines, each processor can work on a different algorithm by using its own data, which are stored on its own memory. Communications are needed during the execution of the parallel algorithm. The message passing interface (MPI) [168] provides interfaces for allowing processors to communicate to each other. It was originally designed to be used on clusters of computers. It is based on C and Fortran programming languages and it is available on almost all platforms and operating systems. MPI consists of a library of C functions and Fortran subroutines needed for making the processors communicate. Basic communications are implemented, such as for sending or receiving variables from one processor to another. Other functions or subroutines allow more than two processors to communicate to one another and to perform simultaneously predetermined tasks.

9.2 A simple parallel algorithm

In order to show the basic idea behind the development of a computational procedure in a parallel environment, a simple parallel algorithm is presented in this section. The aim is to compute the distances between one sample A and all the samples contained into a certain set S and to identify the closest sample in S to A. This parallel procedure can be used as a sub-procedure when dealing with some of the data mining methods discussed in the previous chapters. For instance, a possible implementation of the k-NN method with $k = 1$ could use this parallel sub-procedure.

The parallel algorithm is developed for working on a MIMD parallel machine. There are therefore p processors having a different memory and working on different tasks at the same time. They need to communicate among them for exchanging data.

This algorithm can be mainly divided into two parts: the computation of all the distances between A and the samples in S; and the identification of the minimum distance found. During the first part, the samples in S can be divided among the p processors so that each of them has the same computational load. The sample A must be in the memory of all the processors, whereas only a part of the set S must be in the memory of every single processor. Locally, during the first part of the algorithm, each processor can then compute the distances between A and all the samples in S allocated to this processor. In this phase, each processor is completely independent, and it does not need to exchange data with the others.

The second part of the algorithm consists of finding the minimum distance which has been computed before. The minimum of a set of real numbers must therefore be computed. These numbers are divided into the memories of the p processors working simultaneously, because each processor could save each computed distance only in its own memory. Exchange of data is then needed, but it must be reduced to the minimum, in order to decrease the number of communications and in this way the time needed for carrying the algorithm out. An efficient method for computing this minimum distance is to let the processors work alone for computing the minimum among the distances each of them has in memory. After that, the processors need to communicate with each other. Each of them can send to one predetermined processor the minimum distance on its own memory: in this way, one processor has the partial minimum of the distances related to each processor. At this point, this processor can compute the minimum among all the partial minima and therefore find the desired algorithm output. If the final output is needed in all the processor memories for performing other tasks, there are two strategies that can be used. Once one processor has computed the global minimum distance, it can send to the others this information. This requires time for one communication of one processor with all the others. If the computation of the final output is very fast, as in this case, a more efficient strategy can be the following: when the processors exchange their partial minimum distances, they can communicate so that each of them has in memory all these partial distances. In this way, all of them can compute the global minimum and have it in their own memories. Since the processors work together on the same data, the parallelism seems not to be exploited during this phase. However, the time in which they work this way is smaller than the time needed for letting the processors communicate for receiving the result from another processor. A sketch of the parallel algorithm is given in Figure 9.2. Note that there are more complex strategies for exchanging the partial distances among the processors and for computing the minimum distance. For instance, the processors can communicate as in a tree-like scheme and compute in parallel the minimum distance in a more efficient way.

9.3 Some data mining techniques in parallel

In this section, we present a parallel version of some of the data mining techniques discussed in the previous chapters.

```
A = one sample
S = set of samples
equally divide the samples in S among the p processors
for each processor (in parallel)
    for all the samples s in S and in the processor memory
        compute distance between s and A
        save each distance in the vector dist
    end for
    partDist = minimum distance value in dist
    send partDist to all the other processors
    receive partDist from the other processors
    minDist = minimum among all the partDist
end for
```

Fig. 9.2 A parallel algorithm for computing the minimum distance between one sample and a set of samples in parallel.

9.3.1 k-means

k-means is a data mining technique for clustering. A detailed description of the technique is provided in Chapter 3. In the same chapter different improvements and variants of the technique are also discussed. A sketch of the standard k-means algorithm is provided in Figure 3.2. An application in C implementing a simple variant of the algorithm is presented in Appendix B and is referred to as h-means algorithm (see Figure 3.9). The aim of the algorithm is to find a suitable partition of a set of samples in clusters. The main tasks to be performed during the algorithm are: the computation of all the distances between samples and the centers of the clusters, and the computation of the centers. These tasks can be carried out in parallel in order to speed the clustering algorithm up. In [119, 191, 222, 250] some parallel versions of the k-means algorithm can be found. In the following we will present a parallel h-means algorithm based on the ideas presented in [119].

The set of data to be partitioned can be divided among the memories of the processors involved in the computation. If p is the number of processors, and n is the size of the set of data, every processor can store on its own memory approximately $n_p = n/p$ samples. This allows reduction of the quantity of memory that must be devoted to storing the data on each processor memory. Each single processor can then work only on the samples allocated to this processor. Each cluster has a center. The current k centers of the clusters can be stored in the memories of all the processors, because they are frequently needed during the algorithm.

The computation of the centers of the clusters can be performed in parallel as follows. Each processor contains on its own memory a part of the samples to partition. Hence it does not have all the information for computing the centers, because the samples it has can belong to different clusters. A center can be simply computed by calculating the arithmetic mean among all the samples belonging to the same cluster. Precisely, the sum of all the samples in the same cluster must be computed and the obtained value must be divided by the number of samples. This simple

```
for each processor k_p in {0, 1, ..., p - 1} (in parallel)
    compute the partial sums of the samples belonging to the same cluster
    send the partial sums to the other processors
    receive the partial sums from the other processors
    for each cluster
        sum the partial sums
        divide the total sum by the number of samples in the current cluster
    end for
end for
```

Fig. 9.3 A parallel algorithm for computing the centers of clusters in parallel.

task must be split in p sub-tasks in order to compute such centers in parallel. The main idea is that all the processors exploit all the information they have, while the number of communications is kept minimal. The samples in each processor can in general belong to each of the k clusters. Hence, only the partial sum of the samples belonging to the same cluster can be computed on each processor. During this phase, no communications are needed. When these partial sums are computed, the processors need to exchange them for computing the centers. Each processor has to send its partial sum to the others and needs to receive the partial sums from all the other processors. After the communication phase, each processor can sum the p partial sums and then divide the result by the number of samples contained in each of the k clusters. In this way, each processor has the k cluster centers. A sketch of this parallel algorithm is given in Figure 9.3.

At each step of the h-means algorithm, the distances between each sample and the centers of the clusters are computed and the sample is assigned to the cluster having the closest center. Since all the processors know the centers, this task can be performed independently on each processor on the samples stored in the local memories. There is no need of communications at all, and this makes this phase of the h-means algorithm very efficient in parallel. When all samples are re-assigned, new centers need to be computed. At this point, the parallel procedure discussed above can be reused. A communication phase is included in such procedure, but it is the only one needed during an entire iteration of the h-means algorithm. Therefore, this parallel version of the h-means algorithm can be efficiently implemented on parallel computers. A sketch of the parallel h-means algorithm is given in Figure 9.4.

9.3.2 k-NN

k-NN is a method for classification (see Chapter 4). It classifies unknown samples by checking the classification of the k-nearest known samples contained in a given training set. The basic k-NN algorithm is provided in Figure 4.2. k-NN is a very simple algorithm, but it can be computational expensive if the training set and the set of samples to be classified are large. Parallel computing can help in speeding the

```
equally divide the samples among the p processors
for each processor kp in {0, 1, ..., p − 1} (in parallel)
    randomly assign each sample in processor kp to one cluster
end for
compute the cluster centers in parallel
while the centers are not stable
    for all the samples Samples(i) in processor kp
        compute the distances between Samples(i) and all the centers
        find k' such that the k'-th center is the closest to Sample(i)
        assign Sample(i) to the cluster k'
    end for
    recompute the centers of the clusters in parallel
end while
```

Fig. 9.4 A parallel version of the h-means algorithm.

algorithm up. Since the basic algorithm is very simple, there are parallel versions of it which are simple as well.

The most simple parallel solution is presented in [84]. Given an unknown sample, it must be compared to all the samples contained in the training set, but it is not compared to any of the other unknown samples. Hence, if the set of unknown samples is divided among the p processors involved into a parallel computation, and the training set is replicated on each of them, each processor can work independently from each other. The standard k-NN algorithm can be performed on each processor by using the whole training set and only a part of the set of samples to be classified. This is highly efficient, because no communications are needed during the computation at all. A sketch of this parallel k-NN algorithm is given in Figure 9.5.

```
equally divide the unknown samples among the p processors
for each processor kp in {0, 1, ..., p − 1} (in parallel)
    for all the unknown samples UnSample(i) in processor kp
        for all the known samples Sample(j)
            compute the distance between UnSamples(i) and Sample(j)
        end for
        find the k smallest distances and check the related classification
        assign UnSample(i) to the class which appears more frequently
    end for
end for
```

Fig. 9.5 A parallel version of the k-NN algorithm.

If the training set is much larger than the set of unknown samples, then this parallel algorithm may not be so effective. This is a very rare situation though, where a lot of known data are available for classifying few unknown samples. In any case, for reducing the computational time, large training sets can be reduced by using one of the techniques presented in Section 4.2. If the training set is still too large to be stored in all the processor memories, it might be split among the p processors. The basic k-NN could be carried out locally on each processor, but a communication phase would be needed before an unknown sample could be classified. For this reason, if there are not problems related to the memory space on each processor, the algorithm in Figure 9.5 is the most efficient one. Other studies on parallel versions of the k-NN algorithm are presented in [7].

9.3.3 ANNs

Neural networks can be used for solving classification problems (see Chapter 5). They are inspired by studies on the human brain. The multilayer perceptron is a neural network in which the neurons are organized in layers. The input signal is fed to the network through the input layer and then such signal propagates layer by layer. The neurons of the output layer provide the network output. Layer by layer, the neurons on the same layer manage the data they receive simultaneously and then they send their output to the neurons of the following layer. A general scheme of the multilayer perceptron is given in Figure 5.1.

There is an inherent parallelism in neural networks [206]. Neurons belonging to the same layer work in parallel, and hence they can be distributed on different processors. When the signal passes from one layer to another, each neuron in the first layer must send its information to all the neurons on the second layer. During this phase, then, processors need to communicate with each other for receiving information from neurons working on other processors. Every time a certain input is given to the network, the processors need to communicate a number of times equal to the number of layers before obtaining the corresponding output. The quantity of communications is therefore high, and therefore neural networks can be efficiently used in parallel environments if the number of neurons on each layer is sufficiently large. For small networks, instead, this kind of parallelism would not be so efficient. Indeed, the computational cost in terms of operations would be much smaller than the computational cost in terms of communications.

Another kind of parallelism can also be introduced for neural networks. The training process of a neural network can be formulated as a global optimization problem where the function to be optimized is the function (5.3) (see Section 5.2). As already pointed out, global optimization problems can be difficult to solve and methods designed for solving them may give as solution a point which actually is only one of the local optima. For more details refer to Section 1.4. When meta-heuristic algorithms are used, different executions of the algorithms can lead to different solutions, because the algorithm is probabilistically driven. In such cases, the algorithm is performed more than once for solving the exact same problem, and the best solution obtained over a certain number of trials is considered to be the global optimal solution.

The parallelism can then be introduced at another level. The training phase of the neural network can be considered as sequential and not parallel. However, more training phases can be performed in parallel on a parallel computer. On each processor, one training phase can start by using different initial parameters (such as the

```
nt = number of trials (must be divisible by p)
for i = 1 to nt step i = i + p
   for each processor (in parallel)
      set a seed for the random number generator
      generate randomly the initial neuron weights
      minimize function (5.3) by heuristic optimization
      save the solution in the local memory
   end for
end for
each processor sends its solutions to the others
each processor receives the solutions from the others
each processor identifies the best solution
```

Fig. 9.6 A parallel version of the training phase of a neural network.

seed of the random number generator in heuristic methods or the initial values of the neuron weights). When the training process is finished on each processor, they can communicate and exchange the solution they found. At this point, the best solution found by the p processors can be considered as the solution of the parallel algorithm. Moreover, the best solution might be only stored as the best "parallel" solution found so far, while the processors start another training phase in parallel. At the end of this other phase, the processors need to exchange their solutions another time. The final solution would be in this case the best solution among the best ones obtained during the previous parallel training phase and the new generated solutions on the p processors. A sketch of a possible parallel version of the training process of a neural network is given in Figure 9.6. Note that this kind of parallelism can be applied to other problems where different attempts are carried out using different initial conditions. An example could be the general resolution of a global optimization problem using a meta-heuristic method.

9.3.4 SVMs

Support vector machines are used for finding linear and nonlinear classifiers. They are based on the idea that the best classifier is the one maximizing the margin between the support vectors (see Chapter 6). As already pointed out before, the support vectors satisfy a very interesting property which can be exploited for using SVMs on parallel computers. The support vectors are able to redefine the same exact classifier obtained by using an entire training set. If the support vectors are known, the SVM can be trained by using just them and discarding all the other samples in the training set. The problem is that the support vectors are identified only when the classifier is defined. However, samples in the training set which are not support vectors can be discarded, in order to improve the performances of the training process.

A parallel training process for SVMs is presented in [91]. In these studies, the training set is divided among the p processors involved in the computation. The subsets of the training set are used for training smaller SVMs in parallel on each processor. Each of the found SVMs are actually not good classifiers, because they are based only on the data in random representatives of the whole training set. However, the interesting result is given by the support vectors identified with the classifiers on each processor. Indeed, non-support vectors of a subset have a good chance to be non-support vectors of the whole training set. All the non-support vectors are then eliminated from the subsets, couples of subsets are merged and other SVMs are trained in parallel using these new subsets. The general scheme of this strategy has a tree structure, where at the top there are the initial subsets and at the root there is the last subset. The SVM trained by this last subset provides the final classifier and uses only the samples which are support vectors in all the previous SVMs on the tree. The final support vectors can be tested for global convergence by feeding the result back to the top of the tree. Figure 9.7 shows the tree scheme used.

A sketch of the parallel algorithm is given in Figure 9.8. In this algorithm, it is supposed that the number of initial SVMs is equal to the number of processors p. However, the initial SVMs can also be greater than the number of processors, and for instance two SVMs can be trained on the same processor at the first step. This does not exploit the parallelism abilities of the parallel computer, but divides the original

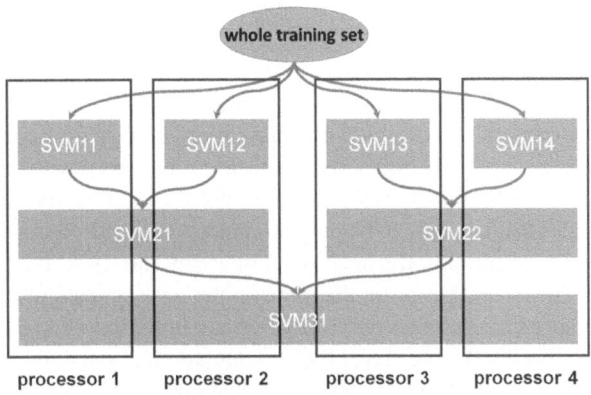

Fig. 9.7 The tree scheme used in the parallel training of a SVM.

```
equally divide the training set among the p processors
Set(i) = subset of the training set assigned to the i-th processor
for each processor p
    for i to log₂ p
        train a SVM by using Set(i)
        locate the support vectors
    end for
    merge support vectors following the tree scheme
    update all the subsets Set(i)
end for
```

Fig. 9.8 A parallel version of the training phase of a SVM.

problem into smaller problems with a consequent reduction of the complexity. In the case of an algorithm following the scheme in Figure 9.7, $\log_2 p$ steps are needed, as the tree scheme suggests. When these kinds of schemes are used, the parallelism cannot be completely exploited. In fact, at the first step (top of the tree) all the processors train different SVMs, but, from the second step on, at least two processors work on the same problem. At the root of the tree, only one SVM must be trained, and this cannot be done in parallel. Either only one processor can work on that, or all the processors can work on the same problem. The computational cost is exactly the same in both cases. In the last one, the solution though is present in all the memories of the processors, and no communications have been performed.

9.4 Parallel computing and agriculture

Currently, there are no parallel computing applications in data mining and agriculture. However, the growing amount of data collected from agricultural-related activities and the need for analyzing these data in an efficient and fast way will force researchers to use parallel computing. As discussed in this chapter, indeed, parallel computing allows one to perform a certain task in parallel on different processors working simultaneously. The advantage in this is the possibility to perform the same task in a shorter time.

Let us take as example the application we discussed in Section 3.5.2. Apples running on a conveyor are analyzed with the aim of discriminating between good and bad apples for marketing. Pictures are taken in real time, and a k-means approach is exploited in the analysis. In big industries or farms, the speed with which a certain task is performed is very important. If more apples can be analyzed in a shorter time, more apples can be ready to be put onto market earlier. In this application, the speed needed for performing the analysis is given by the speed the apples run on the conveyor. The faster the speed, the more apples are analyzed in the same time. However, this speed can increase only until a certain limit. Indeed, the pictures taken from the apples need to be processed by a computational system implementing the k-means approach. This process requires time that must be shorter than the time the considered apple is still on the conveyor and reachable by a robot arm which puts it with the other good or bad apples. It is obvious then how parallel computing can help to let this process become more efficient.

Chapter 10
Solutions to Exercises

In this chapter, all the solutions of the exercises appearing at the ends of the chapters of this book are presented. Each following section contains the solutions related to one chapter.

10.1 Problems of Chapter 2

1 The variability of the components of the points

$$(1, -1), \quad (3, 0), \quad (2, 2)$$

has to be computed. Let us consider the x components. They can assume values 1, 3 and 2, and therefore the range of variability of x is 2. The value 2 comes from the difference between the largest and the smallest values the x component can have. Similarly, the variability of the y component can be computed and it is equal to 3.

2 The following MATLAB® instructions generate a random set of points in a two-dimensional space lying on the line $y = x$. Then, the principal component analysis is applied in order to reduce the dimension of the set of points to 1:

```
>> x = rand(1,20);
>> y = x;
>> A = cov(x,y);
>> [v,d] = eig(A);
>> d

d =

        0        0
        0    0.1619

>> x1 = v(1,2)*x + v(2,2)*y;
>> y1 = v(1,1)*x + v(2,1)*y;
>> var_y1 = max(y1) - min(y1)

var_y1 =

    0
```

A. Mucherino et al., *Data Mining in Agriculture*, Springer Optimization and Its Applications 34, 185
DOI: 10.1007/978-0-387-88615-2_10,
© Springer Science + Business Media, LLC 2009

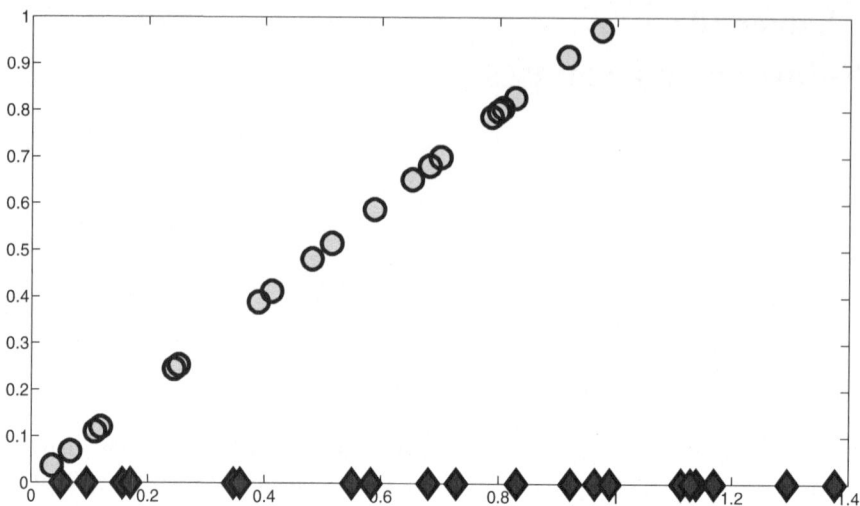

Fig. 10.1 A set of points before and after the application of the principal component analysis.

3 If the variables used in the previous exercise are still in the memory of the MATLAB environment, then the following instructions can be used for creating the Figure 10.1, as required by the exercise:

```
>> plot(x,y,'ko','MarkerSize',10,'MarkerEdgeColor','k','MarkerFaceColor',
     [.49 1 .63])
>> hold on
>> plot(x1,y1,'kd','MarkerSize',10,'MarkerEdgeColor','k','MarkerFaceColor',
     [.49 0 .63])
```

4 The equation of the unique line passing through the two points

$$(x_1, y_1) = (1, 0), \quad (x_2, y_2) = (0, -2)$$

needs to be computed. The general equation of a line l is

$$y = ax + b.$$

In this very easy case, the equation of the l can be easily obtained imposing the passage of the line through the points as follows:

$$(x_1, y_1) \in l \Longrightarrow y_1 = ax_1 + b \Longrightarrow 0 = a + b$$
$$(x_2, y_2) \in l \Longrightarrow y_2 = ax_2 + b \Longrightarrow -2 = 0 + b$$

Then,

$$\begin{cases} a = 2 \\ b = -2 \end{cases}.$$

Let us check if the line l of equation

$$y = 2x - 2$$

passes through the given points. Since

$$x_1 = 1 \implies y_1 = ax + b = 2 \cdot 1 - 2 = 0$$
$$x_2 = 0 \implies y_2 = ax + b = 2 \cdot 0 - 2 = -2$$

the passage is verified.

5 The following instructions draw the line which is the solution of Exercise 4 (see Figure 10.2):

```
>> x = [1 0];
>> y = [0 -2];
>> plot(x,y)
>> hold on
>> plot(x,y,'ko','MarkerSize',10,'MarkerEdgeColor','k','MarkerFaceColor',
        [.49 1 .63])
```

6 The only parabola passing through the points

$$(x_1, y_1) = (0, 1), \quad (x_2, y_2) = (1, 2), \quad (x_3, y_3) = (-1, 3)$$

has to be computed. The general equation of the Newton polynomial is

$$y = f(x_1) + \sum_{i=2}^{n+1} f[x_1, \ldots, x_i] \prod_{j=1}^{i-1} (x - x_j).$$

In this case, the Newton polynomial can be written as:

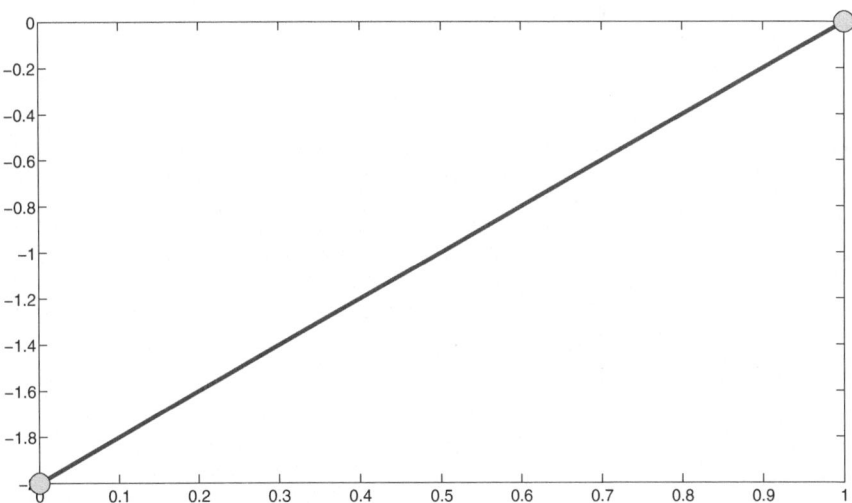

Fig. 10.2 The line which is the solution of Exercise 4.

$$y = f(x_1) + f[x_1, x_2](x - x_1) + f[x_1, x_2, x_3](x - x_1)(x - x_2).$$

Two divided differences have to be computed for finding the equation of the parabola. The first one is

$$f[x_1, x_2] = \frac{y_2 - y_1}{x_2 - x_1} = \frac{2 - 1}{1 - 0} = 1.$$

The second one needs the computation of the divided difference $f[x_2, x_3]$, because

$$f[x_1, x_2, x_3] = \frac{f[x_2, x_3] - f[x_1, x_2]}{x_3 - x_1}.$$

Since

$$f[x_2, x_3] = \frac{y_3 - y_2}{x_3 - x_2} = \frac{3 - 2}{-1 - 1} = -\frac{1}{2},$$

the needed divided difference is

$$f[x_1, x_2, x_3] = \frac{-\frac{1}{2} - 1}{-1 - 0} = \frac{3}{2}.$$

By substituting the divided differences in the Newton polynomial, the following equation is obtained:

$$y = 1 + x + \frac{3}{2}x(x - 1).$$

The passage of the given points is satisfied by the obtained equation:

$$(x_1, y_1) \Longrightarrow 1 = 1 + 0 + \frac{3}{2}0(0 - 1) = 1$$

$$(x_2, y_2) \Longrightarrow 2 = 1 + 1 + \frac{3}{2}1(1 - 1) = 2$$

$$(x_3, y_3) \Longrightarrow 3 = 1 - 1 - \frac{3}{2}1(-1 - 1) = 3.$$

Therefore, the obtained equation is actually a parabola passing from the given points.

7 A figure in which the points

$$(4, 2), (2, 2), (1, 4), (0, 0), (-1, 3)$$

and the join-the-dots function interpolating such points are displayed needs to be created. The MATLAB instructions for performing this exercise are the following ones:

```
>> x = [4 2 1 0 -1];
>> y = [2 2 4 0 3];
>> plot(x,y)
>> hold on
>> plot(x,y,'ko','MarkerSize',10,'MarkerEdgeColor','k','MarkerFaceColor',
        [.49 1 .63])
```

What is obtained is shown in Figure 10.3.

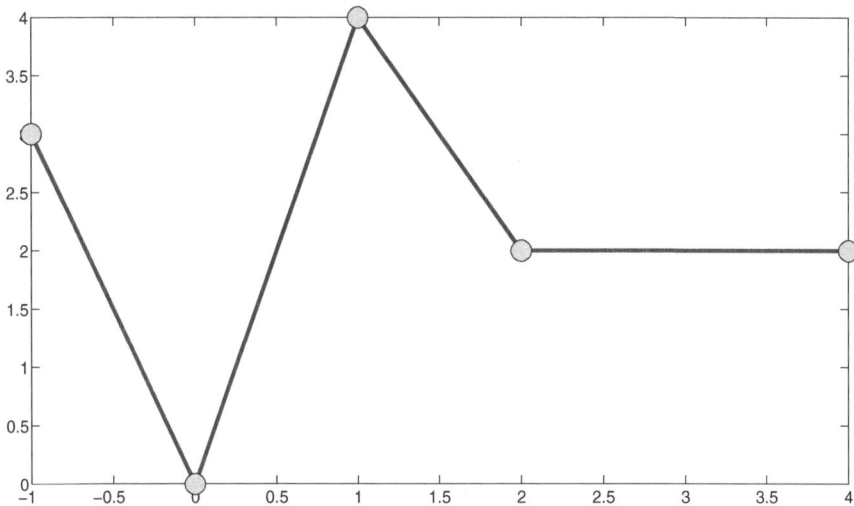

Fig. 10.3 The solution of Exercise 7.

8 Considering the same points given in Exercise 7 and supposing that the join-the-dots function is replaced by a quadratic regression function, then the exercise can be solved by the following MATLAB instructions:

```
>> x = [4 2 1 0 -1];
>> y = [2 2 4 0 3];
>> plot(x,y,'ko','MarkerSize',10,'MarkerEdgeColor','k','MarkerFaceColor',
       [.49 1 .63])
>> hold on
>> c = polyfit(x,y,2);
>> xx = min(x)-1:0.1:max(x)+1;
>> yy = polyval(c,xx);
>> plot(xx,yy)
```

What obtained is shown in Figure 10.4.

9 In this exercise, the linear and quadratic regression functions approximating the points

$$(1, 2), (2, 3), (1, -1), (-1, 3), (1, -2), (0, -1)$$

have to be computed in MATLAB. Figure 10.5 shows the result obtained by using the following instructions in the MATLAB environment:

```
>> x = [1 2 1 -1 1 0];
>> y = [2 3 -1 3 -2 -1];
>> plot(x,y,'ko','MarkerSize',10,'MarkerEdgeColor','k','MarkerFaceColor',
       [.49 1 .63])
>> hold on
>> c = polyfit(x,y,1);
>> xx = min(x)-1:0.1:max(x)+1;
>> yy = polyval(c,xx);
>> plot(xx,yy)
>> c = polyfit(x,y,2);
>> yy = polyval(c,xx);
>> plot(xx,yy,'m:')
```

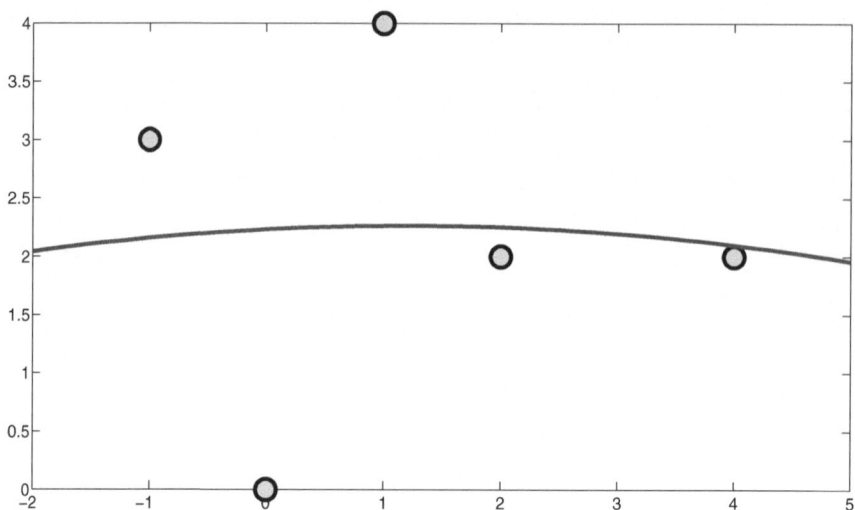

Fig. 10.4 The solution of Exercise 8.

10 In the previous exercise, the linear and quadratic regression functions related to a set of 6 points are computed. If it is supposed that each point (x, y) is approximated with the corresponding point $(x, f(x))$ of the linear regression f, then the mean arithmetic error on these 6 points can be computed by using the following MATLAB code:

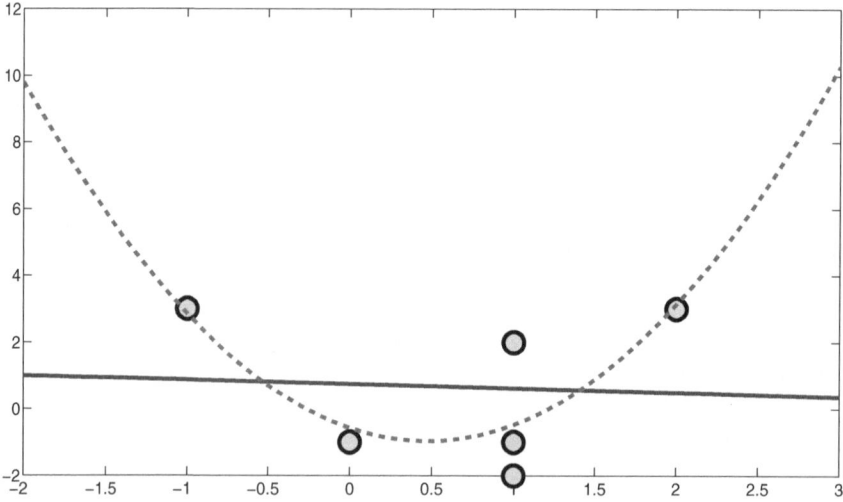

Fig. 10.5 The solution of Exercise 9.

```
>> err = 0; for i = 1:6, err = err + abs(y(i) - polyval(c,x(i))); end
>> err = err/6

err =

    2
```

10.2 Problems of Chapter 3

1 The aim of the exercise is to partition a small set of points by using the standard k-means algorithm. Let us assign a label to each considered point:

$$x_1 = (-1, -1), \quad x_2 = (-1, 1), \quad x_3 = (1, -1),$$
$$x_4 = (1, 1), \quad x_5 = (7, 8), \quad x_6 = (8, 7).$$

As suggested by the exercise, the 1st, 3rd and 5th samples are initially assigned to class 1, and the 2nd, 4th and 6th samples are initially assigned to class 2:

$$x_1 \to 1 \quad x_2 \to 2 \quad x_3 \to 1 \quad x_4 \to 2 \quad x_5 \to 1 \quad x_6 \to 2.$$

Let us compute the centers of these two clusters:

$$c_1 = \frac{x_1 + x_3 + x_5}{3} = \frac{(-1, -1) + (1, -1) + (7, 8)}{3} = \left(\frac{7}{3}, 2\right)$$

$$c_2 = \frac{x_2 + x_4 + x_6}{3} = \frac{(-1, 1) + (1, 1) + (8, 7)}{3} = \left(\frac{8}{3}, 3\right).$$

Following the k-means algorithm, for each point x_i, the distances $d(x_i, c_1)$ and $d(x_i, c_2)$ must be computed and the point has to be assigned to the cluster corresponding to the nearest center. Let us start from the first point x_1:

$$d(x_1, c_1) = 4.48 \quad d(x_1, c_2) = 5.43.$$

Since $d(x_1, c_1) < d(x_1, c_2)$, the point x_1 is closer to the center of the cluster 1, and therefore it is not moved to the other one. Let us consider now the second point:

$$d(x_2, c_1) = 3.48 \quad d(x_2, c_2) = 4.18.$$

The closest center is the one of the cluster 1, whereas x_2 is currently assigned to cluster 2. Then, the point x_2 is moved from cluster 2 to cluster 1. Following the algorithm, the new centers of the two clusters need to be recomputed when there is a change. In fact, the two clusters do not contain the same points anymore, and hence their centers have changed. The new partition is

$$x_1 \to 1 \quad \mathbf{x_2 \to 1} \quad x_3 \to 1 \quad x_4 \to 2 \quad x_5 \to 1 \quad x_6 \to 2$$

and the new centers are

$$c_1 = \frac{x_1 + x_2 + x_3 + x_5}{4} = \frac{(-1, -1) + (-1, 1) + (1, -1) + (7, 8)}{4} = \left(\frac{3}{2}, \frac{7}{4}\right)$$

$$c_2 = \frac{x_4 + x_6}{2} = \frac{(1, 1) + (8, 7)}{2} = \left(\frac{9}{2}, 4\right).$$

By considering the centers just computed, let us keep checking the distances starting from the point x_3:

$$d(x_3, c_1) = 2.80 \quad d(x_3, c_2) = 6.10.$$

The point x_3 results to be in the right cluster, hence it is not moved. The next point is x_4:

$$d(x_4, c_1) = 0.90 \quad d(x_4, c_2) = 4.61.$$

In this case, x_4 needs to be moved from cluster 2 to cluster 1:

$$x_1 \to 1 \quad x_2 \to 1 \quad x_3 \to 1 \quad \mathbf{x_4 \to 1} \quad x_5 \to 1 \quad x_6 \to 2,$$

and therefore new centers are computed:

$$c_1 = \frac{x_1 + x_2 + x_3 + x_4 + x_5}{5} = \frac{(-1, -1) + (-1, 1) + (1, -1) + (1, 1) + (7, 8)}{5}$$

$$= \left(\frac{7}{5}, \frac{8}{5}\right)$$

$$c_2 = x_6 = (8, 7).$$

The next point to consider is x_5, and its distances from the centers just recomputed are checked:

$$d(x_5, c_1) = 8.50 \quad d(x_5, c_2) = 1.41.$$

Since x_5 is closer to c_2, it is moved to cluster 2:

$$x_1 \to 1 \quad x_2 \to 1 \quad x_3 \to 1 \quad x_4 \to 1 \quad \mathbf{x_5 \to 2} \quad x_6 \to 2$$

and new centers are computed:

$$c_1 = \frac{x_1 + x_2 + x_3 + x_4}{4} = \frac{(-1, -1) + (-1, 1) + (1, -1) + (1, 1)}{4} = (0, 0)$$

$$c_2 = \frac{x_5 + x_6}{2} = \frac{(7, 8) + (8, 7)}{2} = \left(\frac{15}{2}, \frac{15}{2}\right).$$

The last point of the set that needs to be checked is

$$d(x_6, c_1) = 10.63 \quad d(x_6, c_2) = 0.71,$$

and it is closer to the center of the cluster it is currently assigned to, and hence it is not moved. All the points have been checked at least once, and during this phase the centers changed several times. The centers are therefore not stable yet, and the algorithm needs to restart checking the points from the first one:

$$d(x_1, c_1) = 1.41 \quad d(x_1, c_2) = 12.02.$$

The point x_1 is not moved. All the other points are not moved as well:

$$d(x_2, c_1) = 1.41 \quad d(x_2, c_2) = 10.70$$
$$d(x_3, c_1) = 1.41 \quad d(x_3, c_2) = 10.70$$
$$d(x_4, c_1) = 1.41 \quad d(x_4, c_2) = 9.19$$
$$d(x_5, c_1) = 10.63 \quad d(x_5, c_2) = 0.71$$
$$d(x_6, c_1) = 10.63 \quad d(x_6, c_2) = 0.71.$$

Since all the points have been checked and none of them changed cluster, the centers are finally stable and the k-means algorithm can terminate.

2 In this exercise, the set of points

$$x_1 = (1, 0), \quad x_2 = (1, 2), \quad x_3 = (2, 0),$$
$$x_4 = (0, 1), \quad x_5 = (1, -3), \quad x_6 = (2, 3), \quad x_7 = (3, 3)$$

has to be partitioned in two clusters using the basic k-means algorithm. The initial partition in clusters is

$$x_1 \to 1 \quad x_2 \to 2 \quad x_3 \to 1 \quad x_4 \to 2 \quad x_5 \to 1 \quad x_6 \to 2 \quad x_7 \to 1.$$

The current centers of the clusters are

$$c_1 = \frac{x_1 + x_3 + x_5 + x_7}{4} = \frac{(1, 0) + (2, 0) + (1, -3) + (3, 3)}{4} = \left(\frac{7}{4}, 0\right)$$

$$c_2 = \frac{x_2 + x_4 + x_6}{3} = \frac{(1, 2) + (0, 1) + (2, 3)}{3} = (1, 2).$$

Following the k-means algorithm, all the points from x_1 to x_7 have to be considered and their distances from the centers of the clusters have to be checked. In this example, all the points from x_1 to x_6 do not need to be moved, because the computed distances are

$$d(x_1, c_1) = 0.75 \quad d(x_1, c_2) = 2.00$$
$$d(x_2, c_1) = 2.13 \quad d(x_2, c_2) = 0.00$$
$$d(x_3, c_1) = 0.25 \quad d(x_3, c_2) = 2.23$$
$$d(x_4, c_1) = 2.02 \quad d(x_4, c_2) = 1.41$$
$$d(x_5, c_1) = 3.09 \quad d(x_5, c_2) = 5.00$$
$$d(x_6, c_1) = 3.01 \quad d(x_6, c_2) = 1.41.$$

Then, the point x_7 is moved to the cluster 2, because the distances from the centers are

$$d(x_7, c_1) = 3.25 \quad d(x_7, c_2) = 2.24.$$

The new partition is therefore

$$x_1 \to 1 \quad x_2 \to 2 \quad x_3 \to 1 \quad x_4 \to 2 \quad x_5 \to 1 \quad x_6 \to 2 \quad \mathbf{x_7 \to 2,}$$

and the new centers are

$$c_1 = \frac{x_1 + x_3 + x_5}{3} = \frac{(1, 0) + (2, 0) + (1, -3)}{4} = \left(\frac{4}{3}, -1\right)$$

$$c_2 = \frac{x_2 + x_4 + x_6 + x_7}{4} = \frac{(1, 2) + (0, 1) + (2, 3) + (3, 3)}{4} = \left(\frac{3}{2}, \frac{9}{4}\right).$$

The centers changed, and hence another iteration of the algorithm has to be performed. These are the distances of all the points in the set from the new two centers:

$$
\begin{array}{ll}
d(x_1, c_1) = 1.05 & d(x_1, c_2) = 2.30 \\
d(x_2, c_1) = 3.02 & d(x_2, c_2) = 0.56 \\
d(x_3, c_1) = 1.20 & d(x_3, c_2) = 2.30 \\
d(x_4, c_1) = 2.40 & d(x_4, c_2) = 1.95 \\
d(x_5, c_1) = 2.03 & d(x_5, c_2) = 5.27 \\
d(x_6, c_1) = 4.06 & d(x_6, c_2) = 0.90 \\
d(x_7, c_1) = 4.33 & d(x_7, c_2) = 1.67.
\end{array}
$$

Since all the points are closer to the centers of the cluster to which they belong, none of them is moved. The algorithm then stops.

3 In this exercise, the set of points

$$x_1 = (-1, -1), \quad x_2 = (-1, 1), \quad x_3 = (1, -1),$$
$$x_4 = (1, 1), \quad x_5 = (7, 8), \quad x_6 = (8, 7),$$

must be partitioned in two clusters using the h-means algorithm. The centers of the initial partition in clusters are (see Exercise 1):

$$c_1 = \left(\frac{7}{3}, 2\right), \quad c_2 = \left(\frac{8}{3}, 3\right).$$

In the h-means algorithm, all the distances from the points and the centers c_1 and c_2 are computed and each point is moved to the cluster with closest center. Even though some point can migrate from a cluster to another, the centers are updated only after all the points have been checked. Let us compute all the distances:

$$
\begin{array}{ll}
d(x_1, c_1) = 4.48 & d(x_1, c_2) = 5.43 \\
d(x_2, c_1) = 3.48 & d(x_2, c_2) = 4.18 \\
d(x_3, c_1) = 3.28 & d(x_3, c_2) = 4.33 \\
d(x_4, c_1) = 1.67 & d(x_4, c_2) = 2.60 \\
d(x_5, c_1) = 7.60 & d(x_5, c_2) = 6.62 \\
d(x_6, c_1) = 7.76 & d(x_6, c_2) = 6.67.
\end{array}
$$

According to these distances, the new partition of the points becomes:

$$x_1 \to 1 \quad \mathbf{x_2} \to \mathbf{1} \quad x_3 \to 1 \quad \mathbf{x_4} \to \mathbf{1} \quad \mathbf{x_5} \to \mathbf{2} \quad x_6 \to 2.$$

The new centers are:

$$c_1 = \frac{x_1 + x_2 + x_3 + x_4}{4} = \frac{(-1, -1) + (-1, 1) + (1, -1) + (1, 1)}{4} = (0, 0)$$
$$c_2 = \frac{x_5 + x_6}{2} = \frac{(7, 8) + (8, 7)}{2} = \left(\frac{15}{2}, \frac{15}{2}\right).$$

This is the same partition obtained at the end of the solution of Exercise 1: this is the optimal partition of the points. Note that the same partition has been obtained by computing the centers only twice by using the h-means algorithm, whereas they have been computed 4 times when the k-means algorithm has been applied.

4 The k-means algorithm can find 4 different partitions in clusters having the same error function value (3.1) if, for instance, the following input is provided:

$$(-1, -1), (-1, 1), (1, -1), (1, 1).$$

5 An example of 8 points on a Cartesian plane that can be partitioned by k-means in 2 different ways that correspond to the same error function value (3.1) is the following one:
$$(-1, 1), (0, 1), (1, 1),$$
$$(-1, 0), (1, 0),$$
$$(-1, -1), (0, -1), (1, -1).$$

6 The set of points

$$x_1 = (-1, -1), \quad x_2 = (-1, 1), \quad x_3 = (1, -1),$$
$$x_4 = (1, 1), \quad x_5 = (7, 8), \quad x_6 = (8, 7).$$

is initially assigned to the clusters 1, 2 and 3 as follows:

$$x_1 \to 1 \quad x_2 \to 2 \quad x_3 \to 1 \quad x_4 \to 2 \quad x_5 \to 1 \quad x_6 \to 2.$$

Note that the cluster 3 is currently empty. According to the k-means+ algorithm, the cluster 3 can be filled by the point that currently is the farthest from its center. Let us compute the distances from each point to the corresponding center:

$$d(x_1, c_1) = 4.48$$
$$d(x_2, c_2) = 4.18$$
$$d(x_3, c_1) = 3.28$$
$$d(x_4, c_2) = 2.60$$
$$d(x_5, c_1) = 7.60$$
$$d(x_6, c_2) = 6.67.$$

The farthest point is x_5: the new partition of the points is therefore the following one:

$$x_1 \to 1 \quad x_2 \to 2 \quad x_3 \to 1 \quad x_4 \to 2 \quad \mathbf{x_5} \to \mathbf{3} \quad x_6 \to 2.$$

The current centers are

$$c_1 = (0, 1), \quad c_2 = \left(\frac{8}{7}, 3\right), \quad c_3 = (7, 8).$$

Let us check the distances of the points from these 3 centers:

$$d(x_1, c_1) = 1.00 \quad d(x_1, c_2) = 4.54 \quad d(x_1, c_3) = 12.04$$
$$d(x_2, c_1) = 2.24 \quad d(x_2, c_2) = 2.93 \quad d(x_2, c_3) = 10.63.$$

According to the algorithm, x_2 is moved to the cluster 1, and the updated centers need to be computed before proceeding. The new partition is

$$x_1 \to 1 \quad \mathbf{x_2 \to 1} \quad x_3 \to 1 \quad x_4 \to 2 \quad x_5 \to 3 \quad x_6 \to 2$$

and the new centers are

$$c_1 = \left(-\frac{1}{3}, -\frac{1}{3}\right), \quad c_2 = \left(\frac{9}{2}, 4\right), \quad c_3 = (7, 8).$$

Let us continue checking the other points:

$$d(x_3, c_1) = 1.49 \quad d(x_3, c_2) = 6.10 \quad d(x_3, c_3) = 10.82$$
$$d(x_4, c_1) = 1.89 \quad d(x_4, c_2) = 4.61 \quad d(x_4, c_3) = 9.22.$$

The point x_4 is then moved to cluster 1. The partition is now

$$x_1 \to 1 \quad x_2 \to 1 \quad x_3 \to 1 \quad \mathbf{x_4 \to 1} \quad x_5 \to 3 \quad x_6 \to 2$$

and the centers are

$$c_1 = (0, 0), \quad c_2 = (8, 7), \quad c_3 = (7, 8).$$

Let us continue checking the points until the last one:

$$d(x_5, c_1) = 10.63 \quad d(x_5, c_2) = 1.41 \quad d(x_5, c_3) = 0.00$$
$$d(x_6, c_1) = 10.63 \quad d(x_6, c_2) = 0.00 \quad d(x_6, c_3) = 1.41.$$

x_5 and x_6 are not moved. Another iteration of the algorithm starts:

$$d(x_1, c_1) = 1.41 \quad d(x_1, c_2) = 12.04 \quad d(x_1, c_3) = 12.04$$
$$d(x_2, c_1) = 1.41 \quad d(x_2, c_2) = 10.82 \quad d(x_2, c_3) = 10.63$$
$$d(x_3, c_1) = 1.41 \quad d(x_3, c_2) = 10.63 \quad d(x_3, c_3) = 10.82$$
$$d(x_4, c_1) = 1.41 \quad d(x_4, c_2) = 9.22 \quad d(x_4, c_3) = 9.22$$
$$d(x_5, c_1) = 10.63 \quad d(x_5, c_2) = 1.41 \quad d(x_5, c_3) = 0.00$$
$$d(x_6, c_1) = 10.63 \quad d(x_6, c_2) = 0.00 \quad d(x_6, c_3) = 1.41.$$

None of the points are moved, none of the clusters are empty, and therefore the k-means+ algorithm can stop.

7 The set of points

$$x_1 = (-1, -1), \quad x_2 = (-1, 1), \quad x_3 = (1, -1),$$
$$x_4 = (1, 1), \quad x_5 = (7, 8), \quad x_6 = (8, 7)$$

are initially assigned to 3 clusters as in the previous exercise. The cluster 3 is empty, and since x_5 is the point which is the farthest from its center (see previous exercise), it is chosen for filling the empty cluster. Then the current partition in clusters is

$$x_1 \to 1 \quad x_2 \to 2 \quad x_3 \to 1 \quad x_4 \to 2 \quad x_5 \to 3 \quad x_6 \to 2$$

and the centers of the clusters are

$$c_1 = (0, 1), \quad c_2 = \left(\frac{8}{7}, 3\right), \quad c_3 = (7, 8).$$

According to the h-means+ algorithm, all the distances from the points and the centers have to be checked and the centers must be updated only when all the points have been checked. The distances are

$$\begin{aligned}
&d(x_1, c_1) = 1.00 \quad d(x_1, c_2) = 4.54 \quad d(x_1, c_3) = 12.04 \\
&d(x_2, c_1) = 2.24 \quad d(x_2, c_2) = 2.93 \quad d(x_2, c_3) = 10.63 \\
&d(x_3, c_1) = 1.00 \quad d(x_3, c_2) = 4.00 \quad d(x_3, c_3) = 10.82 \\
&d(x_4, c_1) = 2.24 \quad d(x_4, c_2) = 2.01 \quad d(x_4, c_3) = 9.22 \\
&d(x_5, c_1) = 11.40 \quad d(x_5, c_2) = 7.70 \quad d(x_5, c_3) = 0.00 \\
&d(x_6, c_1) = 11.31 \quad d(x_6, c_2) = 7.94 \quad d(x_6, c_3) = 1.41.
\end{aligned}$$

Because of the distances obtained, x_2 is moved to cluster 1, and x_6 is moved to cluster 3. The new partition is then

$$x_1 \to 1 \quad \mathbf{x_2 \to 1} \quad x_3 \to 1 \quad x_4 \to 2 \quad x_5 \to 3 \quad \mathbf{x_6 \to 3}$$

and the corresponding centers are

$$c_1 = \left(-\frac{1}{3}, -\frac{1}{3}\right), \quad c_2 = (1, 1), \quad c_3 = \left(\frac{15}{2}, \frac{15}{2}\right).$$

All the distances are checked another time:

$$\begin{aligned}
&d(x_1, c_1) = 0.94 \quad d(x_1, c_2) = 2.83 \quad d(x_1, c_3) = 12.02 \\
&d(x_2, c_1) = 1.49 \quad d(x_2, c_2) = 2.00 \quad d(x_2, c_3) = 10.70 \\
&d(x_3, c_1) = 1.49 \quad d(x_3, c_2) = 2.00 \quad d(x_3, c_3) = 10.70 \\
&d(x_4, c_1) = 1.89 \quad d(x_4, c_2) = 0.00 \quad d(x_4, c_3) = 9.19 \\
&d(x_5, c_1) = 11.10 \quad d(x_5, c_2) = 9.22 \quad d(x_5, c_3) = 0.71 \\
&d(x_6, c_1) = 11.10 \quad d(x_6, c_2) = 9.22 \quad d(x_6, c_3) = 0.71.
\end{aligned}$$

None of the points changed cluster, and then the h-means+ can stop.

This exercise also requires to compare the partition obtained in this exercise to the one obtained in the previous one. The two partitions are different, and this shows that the k-means(+) and h-means(+) algorithms can provide different solutions. In particular, the error function (3.1) has value 5.34 in this partition, and value 5.64 in the one of the previous exercise. Therefore, in this case, the h-means+ algorithm provided a better partition.

8 The following MATLAB code can be used for generating Figure 10.6.

```
x = [-1 -1 1 1 7 8];
y = [-1 1 -1 1 8 7];
class = [1 1 1 1 2 2];
plotp(6,x,y,class)
```

9 The possible code for the MATLAB function hmeans implementing the h-means algorithm in the two-dimensional space follows.

```
%
% this function performs a h-means algorithm
% on a two-dimensional set of data
%
% input:
% n - number of samples
% x - x coordinates of the samples
% y - y coordinates of the samples
% k - number of classes
%
% output:
% class - classes to which each sample belongs
%
% [class] = hmeans(n,x,y,k)

function [class] = hmeans(n,x,y,k)
```

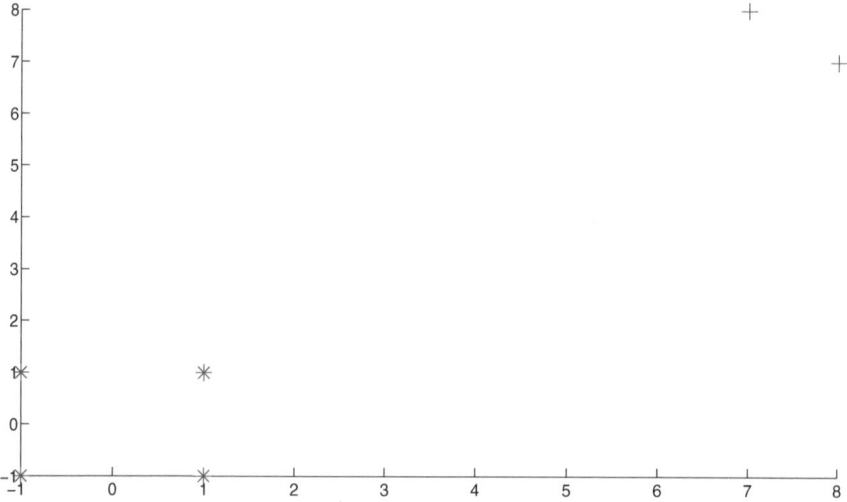

Fig. 10.6 The set of points of Exercise 1 plotted with the MATLAB function `plotp`. Note that 3 of these points lie on the x or y axis of the Cartesian system.

```
% initializing the clusters

for i = 1:n,
  class(i) = int16(k*rand());
  if class(i) == 0,
    class(i) = k;
  end
end

% computing the cluster centers

[cx,cy] = centers(n,x,y,k,class);

stable = 1;   % unstable

while stable == 1,

  % computing the distances between samples (x,y) and centers (cx,cy)
  for i = 1:n,
    mindist = 10.e+100;
    minindex = 0;
    for j = 1:k,
      dist = (x(i) - cx(j))^2 + (y(i) - cy(j))^2;
      dist = sqrt(dist);
      if dist < mindist,
        mindist = dist;
        minindex = j;
      end
    end
    % changing cluster
    class(i) = minindex;
  end

  % checking the cluster centers

  [cxnew,cynew] = centers(n,x,y,k,class);

  stable = 0;
  for j = 1:k,
    if abs(cxnew(j) - cx(j)) > 1.e-6 | abs(cynew(j) - cy(j)) > 1.e-6,
      stable = 1;
    end
  end

  % preparing for the next iteration
  for j = 1:k,
    cx(j) = cxnew(j);   cy(j) = cynew(j);
  end

end % while

end
```

10 The simple proof of the equivalence follows. We have that

$$\begin{aligned}
||x_{j_1} - x_{j_2}||^2 &= ||x_{j_1} - c_i||^2 + ||x_{j_2} - c_i||^2 - 2||x_{j_1} - c_i|| \\
&\quad \cdot ||x_{j_2} - c_i|| \cos(x_{j_1} - c_i, x_{j_2} - c_i) \\
&= ||x_{j_1} - c_i||^2 + ||x_{j_2} - c_i||^2 - 2(x_{j_1} - c_i)(x_{j_2} - c_i).
\end{aligned}$$

Then the quantity

$$\sum_{j_1 \in S_i} \sum_{j_2 \in S_i} ||x_{j_1} - x_{j_2}||^2$$

is equal to

$$\sum_{j_1 \in S_i} \sum_{j_2 \in S_i} \left(||x_{j_1} - c_i||^2 + ||x_{j_2} - c_i||^2 \right) - 2 \sum_{j_1 \in S_i} \sum_{j_2 \in S_i} (x_{j_1} - c_i)(x_{j_2} - c_i).$$

The last term is zero, since

$$\sum_{j_1 \in S_i} \sum_{j_2 \in S_i} (x_{j_1} - c_i)(x_{j_2} - c_i) = \sum_{j_1 \in S_i} \left((x_{j_1} - c_i) \sum_{j_2 \in S_i} (x_{j_2} - c_i) \right)$$

and

$$\sum_{j_2 \in S_i} (x_{j_2} - c_i) = \sum_{j_2 \in S_i} x_{j_2} - |S_i|c_i = |S_i|c_i - |S_i|c_i = 0.$$

Thus,

$$\sum_{j_1 \in S_i} \sum_{j_2 \in S_i} ||x_{j_1} - x_{j_2}||^2 = \sum_{j_1 \in S_i} \sum_{j_2 \in S_i} \left(||x_{j_1} - c_i||^2 + ||x_{j_2} - c_i||^2 \right)$$

$$= 2|S_i| \sum_{j_1 \in S_i} ||x_j - c_i||^2,$$

which implies the equality.

10.3 Problems of Chapter 4

1 The 1-NN rule has to be applied for classifying the points $x_1 = (2, 1), x_2 = (-3, 1)$ and $x_3 = (1, 4)$ in the two classes C^+ and C^- by using the training set:

$$\{\{T_1 = (-1, -1), C^-\}, \{T_2 = (-1, 1), C^-\}, \{T_3 = (1, -1), C^+\}, \{T_4 = (1, 1), C^+\}\}.$$

Following the 1-NN rule, the points have to be classified in accordance with the classification of their closest point in the training set. Let us consider the first point x_1:

$$d(x_1, T_1) = 3.61, \ d(x_1, T_2) = 3.00, \ d(x_1, T_3) = 2.23, \ d(x_1, T_4) = 1.00.$$

Since the nearest point to x_1 in the training set is T_4, the point is classified in the same way as T_4:

$$x_1 \in C^+.$$

Following the same procedure, the other two points x_2 and x_3 can be classified with the same rule:

$$d(x_2, T_1) = 2.83, \quad d(x_2, T_2) = 2.00, \quad d(x_2, T_3) = 4.47,$$
$$d(x_2, T_4) = 4.00 \Longrightarrow x_2 \in C^-$$

$$d(x_3, T_1) = 5.38, \quad d(x_3, T_2) = 3.61, \quad d(x_3, T_3) = 5.00,$$
$$d(x_3, T_4) = 3.00 \Longrightarrow x_3 \in C^+.$$

2 In this exercise, the points

$$x_1 = (7, 8), \quad x_2 = (0, 0), \quad x_3 = (0, 2), \quad x_4 = (4, -2)$$

have to be classified in the classes C_A and C_B by using as training set the set of points:

$$\{T_1 = (0, 1), T_2 = (-1, -1), T_3 = (1, 1)\} \in C_A,$$
$$\{T_4 = (-2, -2), T_5 = (2, 2)\} \in C_B.$$

The 1-NN rule is applied:

$$d(x_1, T_1) = 9.90, \ d(x_1, T_2) = 12.04, \ d(x_1, T_3) = 9.22,$$
$$d(x_1, T_4) = 13.45, \ d(x_1, T_5) = 7.81$$
$$d(x_2, T_1) = 1.00, \ d(x_2, T_2) = 1.41, \ d(x_2, T_3) = 1.41,$$
$$d(x_2, T_4) = 2.83, \ d(x_2, T_5) = 2.83$$
$$d(x_3, T_1) = 1.00, \ d(x_3, T_2) = 3.16, \ d(x_3, T_3) = 1.41,$$
$$d(x_3, T_4) = 4.47, \ d(x_3, T_5) = 2.00$$
$$d(x_4, T_1) = 5.00, \ d(x_4, T_2) = 5.10, \ d(x_4, T_3) = 4.24,$$
$$d(x_4, T_4) = 6.00, \ d(x_4, T_5) = 4.47.$$

According to the distance values obtained, the unknown points are classified as follows:

$$x_1 \in C_B \quad x_2 \in C_A \quad x_3 \in C_A \quad x_4 \in C_A.$$

3 In this exercise, the points

$$x_1 = (5, 1), \quad x_2 = (-1, 4),$$

must be classified into the classes C_A and C_B by using the points:

$$\{T_1 = (0, 1), T_2 = (-1, -1), T_3 = (1, 1)\} \in C_A,$$
$$\{T_4 = (-2, -2), T_5 = (2, 2)\} \in C_B.$$

The 3-NN rule is applied:

$$d(x_1, T_1) = 5.00, \ d(x_1, T_2) = 6.32, \ d(x_1, T_3) = 4.00,$$
$$d(x_1, T_4) = 7.62, \ d(x_1, T_5) = 3.16$$
$$d(x_2, T_1) = 3.16, \ d(x_2, T_2) = 5.00, \ d(x_2, T_3) = 3.61,$$
$$d(x_2, T_4) = 6.08, \ d(x_2, T_5) = 3.61.$$

Both the points x_1 and x_2 are classified as belonging to the class C_A.

4 The following training set allows different classification for the point $\hat{x} = (1, 1)$ if the k-NN rule is applied with k equal to 1 or 3. The set of points contains:

$$x_{A1} = (1, 0), \quad x_{A2} = (3, 0),$$
$$x_{B1} = (0, 0), \quad x_{B2} = (-1, 0), \quad x_{B3} = (0, 2),$$

and they are classified in the classes C_A and C_B according to their subscripts. The point $\hat{x}$ is classified as belonging to class C_A if k is 1 and it is classified as belonging to class C_B if k is 3. Let us compute the distances between $\hat{x}$ and all the points in the training set:

$$d(\hat{x}, x_{A1}) = 1.00, \quad d(\hat{x}, x_{A2}) = 2.24,$$
$$d(\hat{x}, x_{B1}) = 1.41, \quad d(\hat{x}, x_{B2}) = 2.24, \quad d(\hat{x}, x_{B3}) = 1.41.$$

The nearest point to $\hat{x}$ is x_{A1}. If the 1-NN rule is then applied, $\hat{x}$ is classified as x_{A1}, i.e., it is assigned to the class C_A. If the 3-NN rule is instead used, the three nearest neighbors of $\hat{x}$ are x_{A1}, x_{B1} and x_{B3}. Since two of them belong to the class C_B and only one to the class C_A, the unknown point $\hat{x}$ is classified as the majority of its neighbors. In this case, then, $\hat{x}$ is assigned to the class C_B.

5 The training set and the unknown sample that satisfies the requirements of Exercise 3 can be plotted by the MATLAB function `plotp`. Figure 10.7 shows the training set and the point given as solution of Exercise 4.

6 The classification problem proposed in Exercise 1 can be easily solved by using the MATLAB environment and the function `knn`. A list of instructions in MATLAB follows:

```
>> ntrain = 4;
>> xtrain = [-1 -1 1 1];
>> ytrain = [-1 1 -1 1];
>> ctrain = [1 1 2 2];
>> x = [2 -3 1];
```

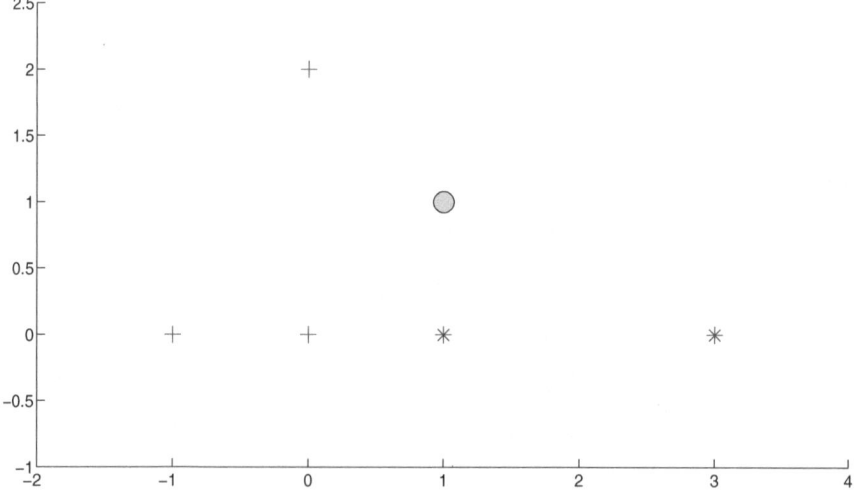

Fig. 10.7 The training set and the unknown point that represents a possible solution to Exercise 4.

```
>> y = [1 1 4];
>> class = knn(3,x,y,2,ntrain,xtrain,ytrain,ctrain)

class =

    2    1    1
```

7 In this exercise, a training set has to be randomly created and the corresponding condensed and reduced set have to be computed. In MATLAB, the following instructions can be used for this purpose:

```
>> [x,y] = generate(200,0.1);
>> [class] = hmeans(200,x,y,2);
>> [ntcnn,xtcnn,ytcnn,ctcnn] = condense(200,x,y,class,2);
>> ntcnn

ntcnn =

    11

>> [ntrnn,xtrnn,ytrnn,ctrnn] = reduce(200,x,y,class,2);
>> ntrnn

ntrnn =

    9
```

As shown, the condensed training set has only 11 points, and the reduced training set has only 9 points. The original training set was created with 200 points.

8 The figures required by the exercise can be generated using the function `plotp`. If the variables used in the previous exercise in MATLAB are still in memory, then the following instructions can be used:

```
>> plotp(200,x,y,class)
>> plotp(ntcnn,xtcnn,ytcnn,ctcnn)
>> axis([-1.5 1.5 -1 1])
>> plotp(ntrnn,xtrnn,ytrnn,ctrnn)
>> axis([-1.5 1.5 -1 1])
```

The first call to the function `plotp` generates Figure 10.8. The other two calls create Figures 10.9(a) and 10.9(b).

9 The solution of the exercise can be found by using the following instructions in MATLAB. It is supposed that the variables x, y and class used in the Exercise 7 are still in memory.

```
>> ntrain = 200;
>> xtrain = x;
>> ytrain = y;
>> ctrain = class;
>> [x,y] = generate(500,0);
>> [class] = knn(500,x,y,2,ntrain,xtrain,ytrain,ctrain);
>> plotp(500,x,y,class)
```

The call to the function `plotp` generates Figure 10.10.

10 If it is supposed that all the variables used in Exercise 7 are still in memory, such as the condensed and reduced subsets, then the following code can be used:

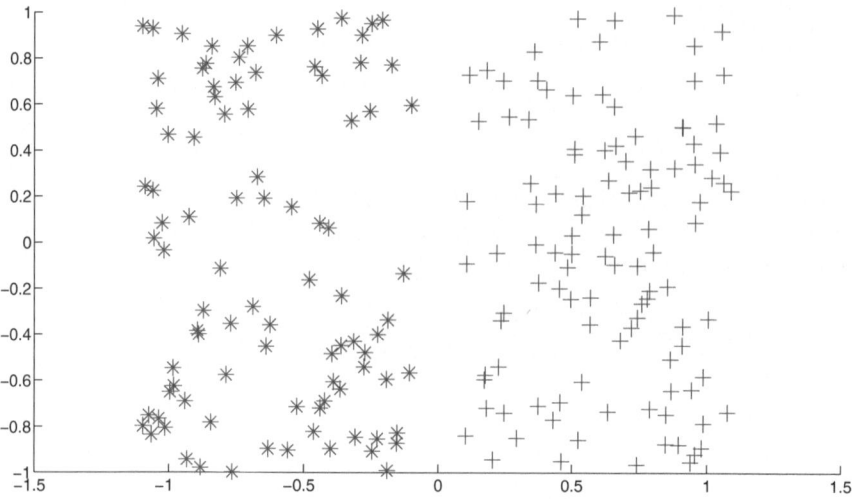

Fig. 10.8 A random set of 200 points partitioned in two clusters.

```
>> [class] = knn(500,x,y,2,ntcnn,xtcnn,ytcnn,ctcnn);
>> plotp(500,x,y,class)
>> [class] = knn(500,x,y,2,ntrnn,xtrnn,ytrnn,ctrnn);
>> plotp(500,x,y,class)
```

The two calls to the function `plotp` generate Figure 10.11.

10.4 Problems of Chapter 5

1 A multilayer perceptron having one input neuron, two hidden neurons on only one hidden layer and one output neuron has the structure shown in Figure 10.12. For the labels assigned to each neuron and weight, refer to the figure. The network has to be trained so that it is able to model the equation

$$y = 2x.$$

For simplicity, the function O_j assigned to each active neuron is the identity function, which can be expressed by the equation $y = x$. For training the network, let us consider a subset of couples of independent variables x and dependent variables y satisfying the equation $y = 2x$. For instance, the points

$$(1, 2), \quad (-1, -2), \quad (2, 4)$$

satisfy the equation. Let us start considering the first point: $(1, 2)$. A network trained as required should be able to provide 2 when 1 is fed. When $x = 1$ is fed, this signal is sent from the input neuron A to both the neurons of the hidden layer, B and C.

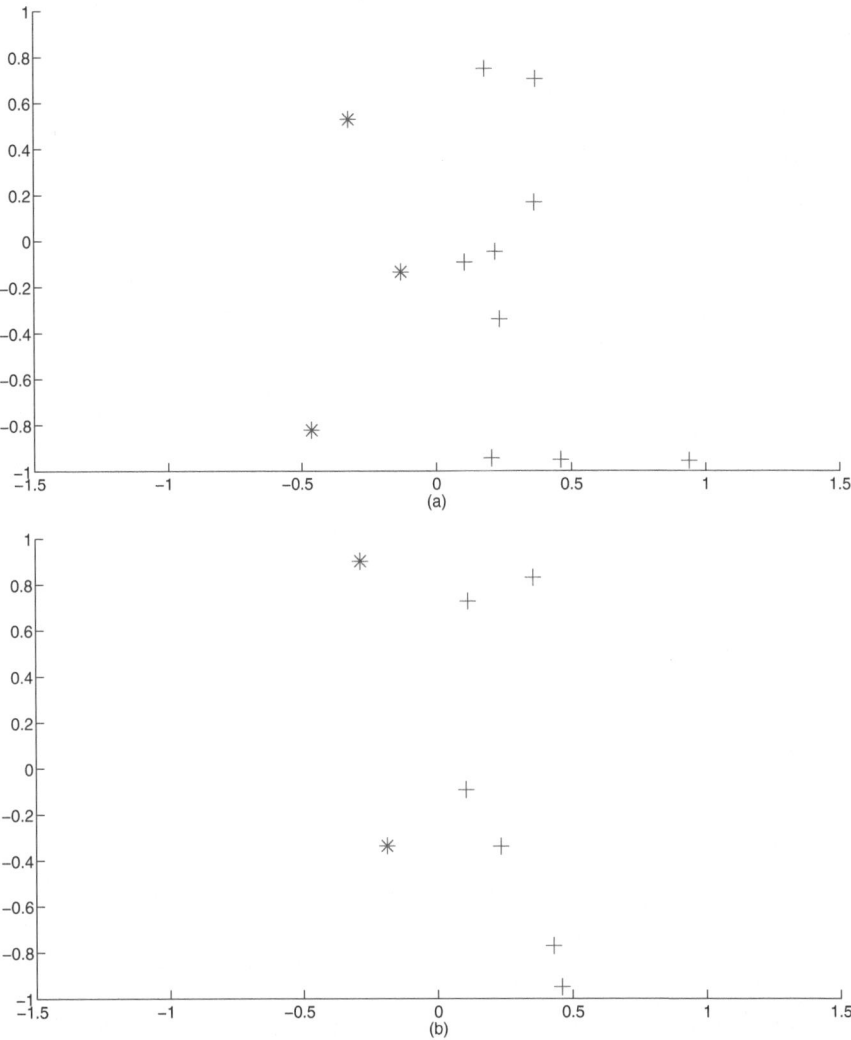

Fig. 10.9 The condensed and reduced set obtained in Exercise 7: (a) the condensed set corresponding to the set in Figure 10.8; (b) the reduced set corresponding to the set in Figure 10.8.

These two neurons compute their activation levels using the weights assigned to the links connecting them to the input neuron. In general, the activation level in B is $w_{11}x$ and the activation level in C is $w_{12}x$. Hence, in this case, the activation in B is w_{11} and the activation in C is w_{12}. The function O_j is the identity function, and therefore the neurons B and C do not modify the activation values, which are sent as they are to the output neuron. The activation level in D is $w_{11}w_{21} + w_{12}w_{22}$. As before, O_j is the identity function, and hence this is the final output provided by the network. Since the network has to mimic the equation $y = 2x$, the following

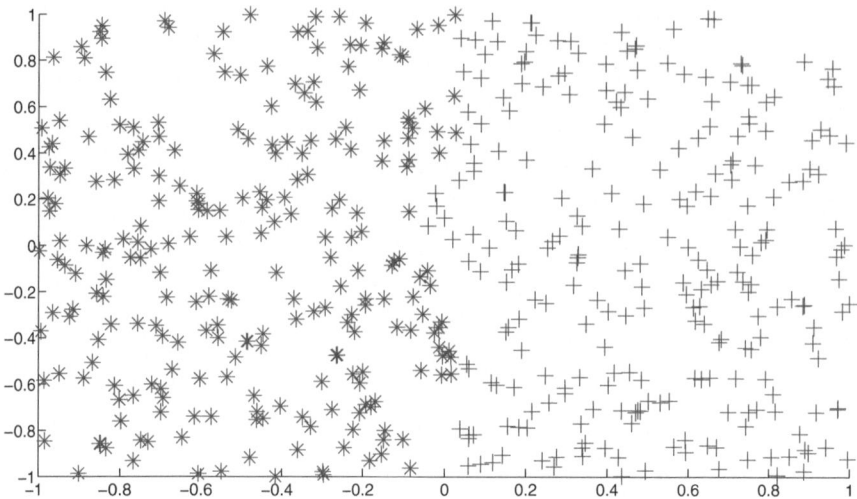

Fig. 10.10 The classification of a random set of points by using a training set of 200 points.

condition has to be satisfied:

$$w_{11}w_{21} + w_{12}w_{22} = 2. \tag{10.1}$$

If the point $(-1, -2)$ is considered, and -1 is fed to the network, the output from the network is -2 if the condition

$$-(w_{11}w_{21} + w_{12}w_{22}) = -2$$

is satisfied. Similarly, if $(2, 4)$ is considered, the condition

$$2(w_{11}w_{21} + w_{12}w_{22}) = 4$$

is obtained. Note that all these conditions depend on each other, and hence only one of them can be considered and the others discarded. If other points are considered, and other conditions obtained, they would be dependent on these ones. Let us take in account the condition (10.1). There are 4 unknown weights in only one condition, and therefore there is an infinite number of combinations of the 4 weights that satisfy such condition. For instance the weights

$$w_{11} = 1, \quad w_{21} = 1, \quad w_{12} = 2, \quad w_{22} = 1$$

satisfy the condition (10.1). The network with these weights works as the equation $y = 2x$.

2 It is needed to prove that a multilayer perceptron having one input neuron, two hidden neurons on only one hidden layer and one output neuron having the structure in Figure 10.12 cannot model the equation $y = 2x + 1$ exactly. In the previous exercise,

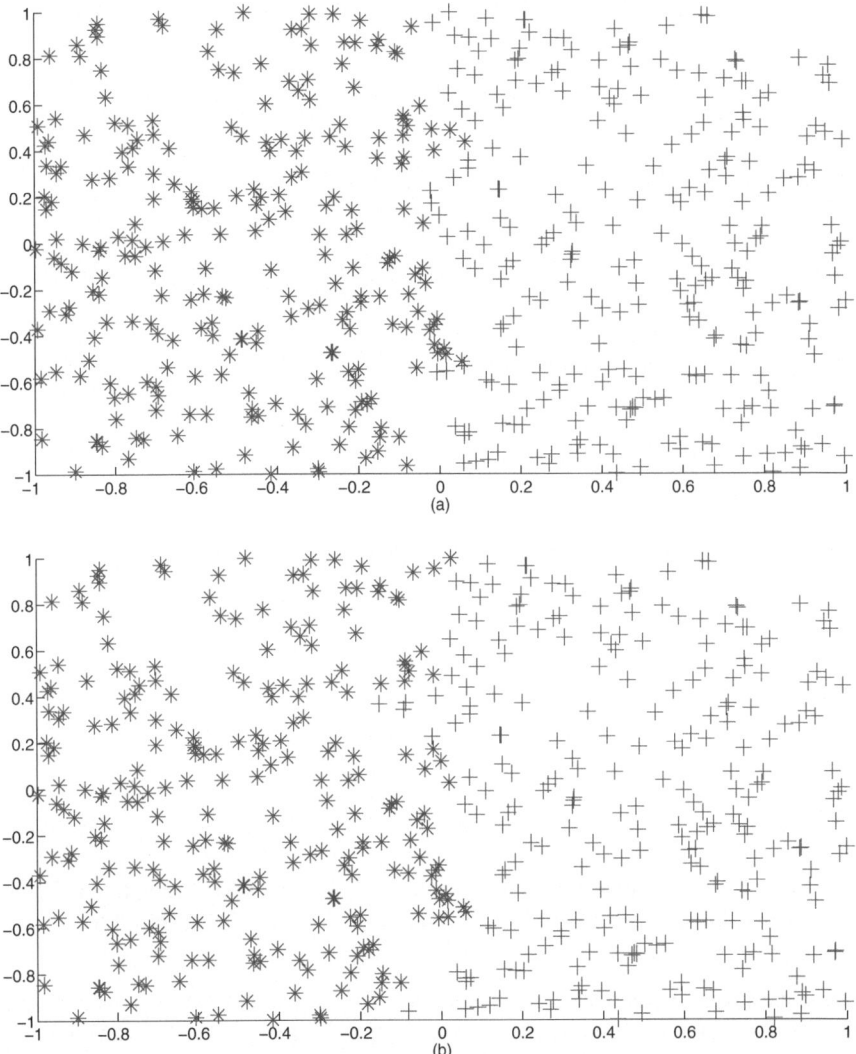

Fig. 10.11 The classification of a random set of points by using (a) the condensed set of the set in Figure 10.8; (b) the reduced set of the set in Figure 10.8.

the network has been fed with different points satisfying the equation $y = 2x$. Let us consider now the generic point satisfying the equation $y = 2x + 1$:

$$(x, 2x + 1).$$

Let us feed x to the network. The activation level in B is $w_{11}x$ and the activation level in C is $w_{12}x$. The function O_j is the identity function, and then these two activation levels are sent as they are to the output neuron. In D, the activation level

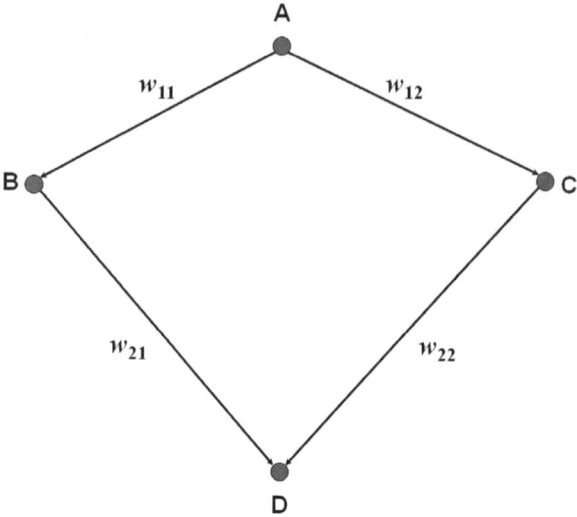

Fig. 10.12 The structure of the network considered in Exercise 1.

is $w_{11}w_{21}x + w_{12}w_{22}x$. Therefore, the following condition has to be satisfied if the network has to approximate the equation $y = 2x + 1$:

$$(w_{11}w_{21} + w_{12}w_{22})\, x = 2x + 1.$$

It follows that:

$$(w_{11}w_{21} + w_{12}w_{22} - 2)\, x = 1,$$

and this implies that the weights must depend on x for satisfying the equation. There are no possible choices for the weights that satisfy the condition for all the x, and for this reason this network cannot model the equation $y = 2x + 1$ exactly.

3 A multilayer perceptron having one hidden layer with 2 neurons has to be trained for the AND classification problem. Given two logical variables, X and Y, X AND Y must be the answer of the classification rule. As known, the AND logical operator works in accordance with the following table.

X	Y	X AND Y
True	True	True
True	False	False
False	True	False
False	False	False

In the exercise, the logical value 'true' is indicated by 0, and the logical value 'false' is indicated by 1. In this way, the previous table can be written in terms of 0 and 1.

X	Y	X AND Y
0	0	0
0	1	1
1	0	1
1	1	1

The network is trained so that, when X and Y are fed, the corresponding X AND Y value is given as output. The network has two input neurons, one corresponding to X and the other corresponding to Y, and it has only one output value, where X AND Y is provided. The hidden neurons on one hidden layer are 2. The structure of this network is in Figure 10.13: refer to the figure for the labels given to the neurons and the weights.

Let us feed the network with a generic couple (X,Y). The signal containing X starts from the neuron A and reaches the neuron C. The activation level of the neuron C is then $w_{11}X$. Similarly, the signal containing Y starts from the neuron B and reaches the neuron D. The activation level of the neuron D is then $w_{12}Y$. Successively, both neurons C and D send their signal to the input neuron E. The activation level on E is

$$w_{11}w_{21}X + w_{12}w_{22}Y.$$

Therefore, the network is able to provide the following results:

X	Y	Network output
0	0	0
0	1	$w_{12}w_{22}$
1	0	$w_{11}w_{21}$
1	1	$w_{11}w_{21} + w_{12}w_{22}$

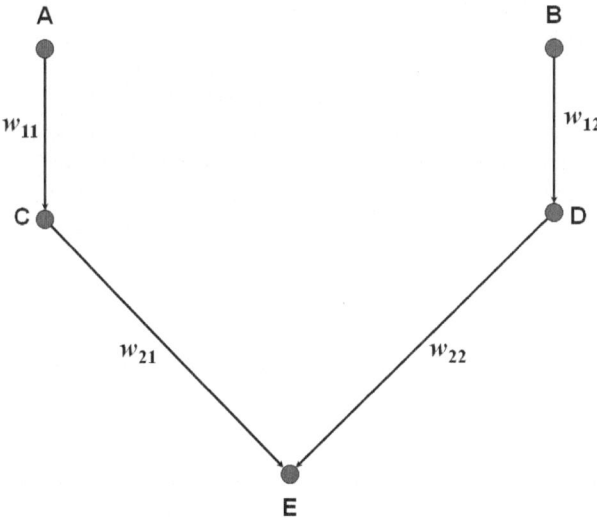

Fig. 10.13 The structure of the network considered in Exercise 3.

The network works as the AND classifier if all the weights are set to 1 and the function

$$O_j = \begin{cases} 0 \longrightarrow 0 \\ 1 \longrightarrow 1 \\ 2 \longrightarrow 1 \end{cases}$$

is associated to the neuron E.

4 The network considered in this exercise has the same structure as the one in Exercise 3. Its structure is provided in Figure 10.13. All the weights are set to 1, and the sigmoid function

$$O_j = \text{sigmoid}(x) = \frac{1}{1 + e^{-x}}$$

is associated to the output neuron. Let us feed the network with $(6, 1)$. The signal containing 6 starts from the neuron A and arrives at the neuron C unaltered. Similarly, the signal containing 1 starts from the neuron B and arrives at the neuron D unaltered. These signals start from the neurons C and D and arrive at E. The activation level in E is the weighted sum of the received signals, and therefore it is $6 + 1 = 7$. Associated to E is the sigmoid function, and hence the output value of the network is

$$\text{sigmoid}(7) = \frac{1}{1 + e^{-7}}.$$

If instead $(-1, -1)$ is fed to the network, the output value of the network is

$$\text{sigmoid}(-2) = \frac{1}{1 + e^2}.$$

5 In this exercise, the considered network has the same structure as the one in Figure 10.13. All the weights are equal to 2 and the logistic function is associated to the neuron E. When the signal propagates from one neuron to another it is doubled in value. Since there is only one hidden layer, the original signal is sent from the input layer to the hidden layer, and then from the hidden layer to the output neuron. In total, therefore, the original signal is amplified four times when it passes the network. When the neuron E receives its inputs, it sums them and applies the logistic function to the result. Thus, if $(1, 1)$ is fed to the network, then the output provided by the network is

$$\text{logistic}(4 + 4) = \frac{1}{1 + e^{-\frac{8}{2}}}.$$

The same result can be obtained when $(0, 2)$ is fed:

$$\text{logistic}(0 + 8) = \frac{1}{1 + e^{-\frac{8}{2}}}.$$

6 Two networks having the same structure as shown in Figure 10.13 are considered. The first network has all the weights equal to 1 and the sigmoid function associated to the output neuron. The second one has all the weights equal to 2 and the logistic

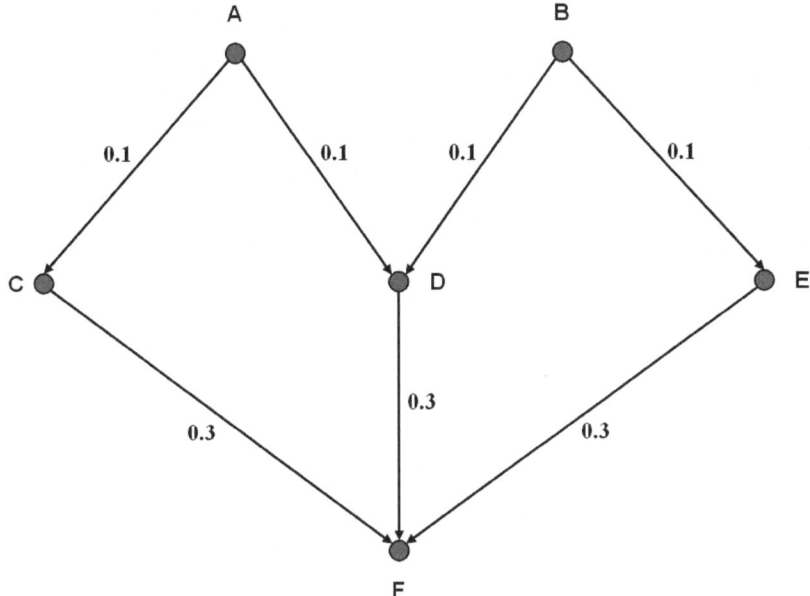

Fig. 10.14 The structure of the network considered in Exercise 7.

function associated to the output neuron. The first network can have the hidden layer removed without changing the its outputs, because the weights related to the hidden neurons are equal to 1 and no functions are associated to them. Such neurons actually do not have any effect.

7 The structure of the network considered in this exercise is shown in Figure 10.14. The weights on the links are assigned as specified in the figure. Let us feed the network with an arbitrary input $(1, 2)$. The signal containing 1 propagates from A and the signal containing 2 propagates from B. In C the signal is 0.1, in D it is 0.2, and it is 0.1 in E. The signal in the output neuron is 0.12. It is easy to verify that, if the link between A and C is removed, then the neuron C remains inactive. Similarly, E remains inactive if the link between B and E is removed. If only one of the other links is removed, no neurons remain inactive.

8 A network having the features required by the exercise is given in Figure 10.15.

10.5 Problems of Chapter 6

1 The set of points

$$(A, B, C)$$

whose components can have 0 or 1 as value are separated in the two classes

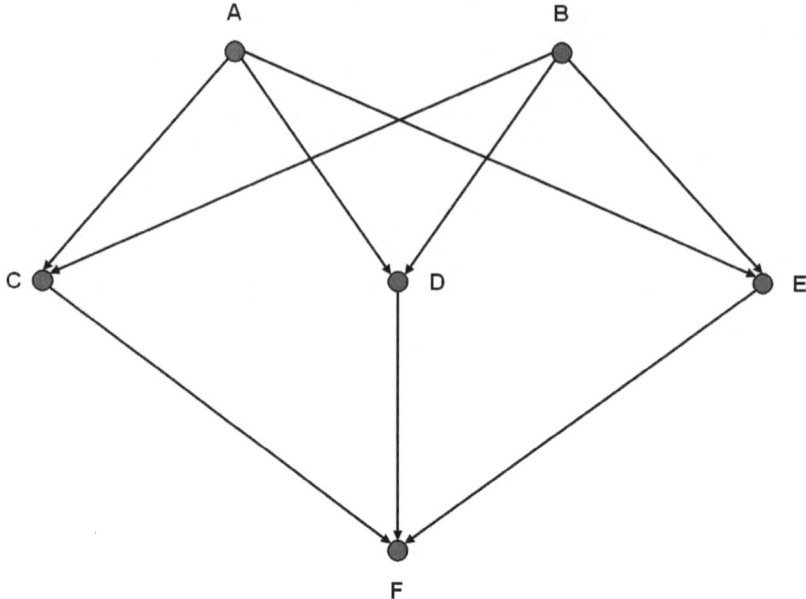

Fig. 10.15 The structure of the network required in Exercise 8.

$$C^0 = \{(A, B, C): \quad A \quad \text{AND} \quad B \quad \text{AND} \quad C \quad = \quad 0\},$$

and

$$C^1 = \{(A, B, C): \quad A \quad \text{AND} \quad B \quad \text{AND} \quad C \quad = \quad 1\}.$$

The aim of the exercise is to check if the two classes are linearly separable or not. Note that the points (A, B, C) lie on the vertices of a three-dimensional cube. Suppose that 0 stands for 'true' and that 1 stands for 'false.' From the definition of the AND operator it follows that only the point $(0, 0, 0)$ belongs to the class C^0 and all the others belong to the class C^1. Therefore, the two classes are linearly separable.

2 The classes

$$C^0 = \{(A, B, C): \quad \text{NOT} \quad A \quad \text{AND} \quad B \quad = \quad 0\}$$
$$C^1 = \{(A, B, C): \quad \text{NOT} \quad A \quad \text{AND} \quad B \quad = \quad 1\}$$

are linearly separable since they can be separated by the place having equation $B - A \geq 1$. The classes

$$C^0 = \{(A, B, C): \quad (A \quad \text{OR} \quad B) \quad \text{AND} \quad (A \quad \text{AND} \quad C) = 0\}$$
$$C^1 = \{(A, B, C): \quad (A \quad \text{OR} \quad B) \quad \text{AND} \quad (A \quad \text{AND} \quad C) = 1\}$$

are linearly separable as well. The plane $2A + B + C \geq 2$ separates the two classes.

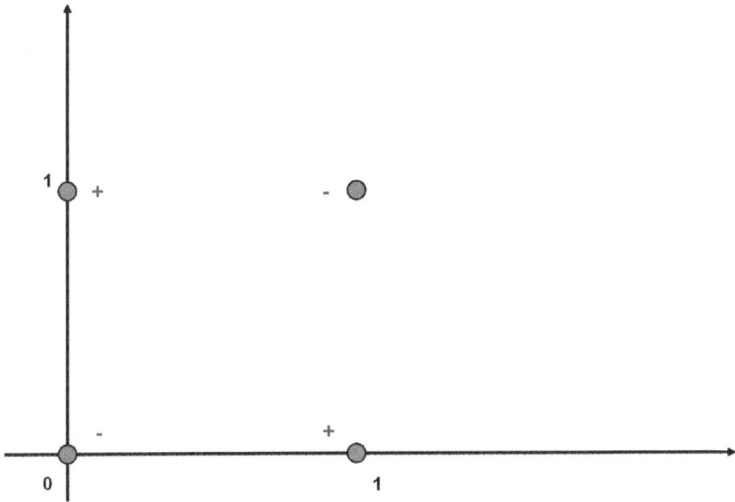

Fig. 10.16 The classes C^+ and C^- in Exercise 3.

3 A set of points and their classifications in two classes C^+ and C^- are specified as follows:

$$\big((0,0), C^-\big), \quad \big((0,1), C^+\big), \quad \big((1,0), C^+\big), \quad \big((1,1), C^-\big).$$

As it is possible to see from Figure 10.16, the classes C^+ and C^- are not linearly separable.

4 The same set of points and the same classification described in Exercise 3 are considered in this exercise. Figure 10.16 gives a geometric representation of these points. In this exercise, the transformation

$$\Phi(x_1, x_2) = \begin{pmatrix} 1 \\ \sqrt{2}x_1 \\ \sqrt{2}x_2 \\ x_1^2 \\ x_2^2 \\ \sqrt{2}x_1x_2 \end{pmatrix},$$

has to be applied in order to get the two classes C^+ and C^- linearly separable. The transformation is applied point by point:

$$\begin{aligned}
(0,0) &\implies (1,0,0,0,0,0) \\
(0,1) &\implies (1,0,\sqrt{2},0,1,0) \\
(1,0) &\implies (1,\sqrt{2},0,1,0,0) \\
(1,1) &\implies (1,\sqrt{2},\sqrt{2},1,1,\sqrt{2}).
\end{aligned}$$

Note that the first component of the transformed points is always 1, and therefore it can be discarded. Moreover, the components 2 and 3 and the components 4 and 5 of the transformed points satisfy a particular symmetry property. Indeed, the components 2 and 3 are

$$(0, 0), \quad (0, \sqrt{2}), \quad (\sqrt{2}, 0), \quad (\sqrt{2}, \sqrt{2}),$$

and the components 4 and 5 are

$$(0, 0), \quad (0, 1), \quad (1, 0), \quad (1, 1).$$

Thus, these two couples of components have the same coefficients in the separating hyperplane equation because of the symmetry. Simplifying, the given points are transformed in:

$$\left((0, 0, 0), C^-\right), \quad \left((\sqrt{2}, 1, 0), C^+\right), \quad \left((\sqrt{2}, 1, 0), C^+\right), \quad \left((2\sqrt{2}, 2, \sqrt{2}), C^-\right).$$

The second and the third point are identical, and therefore only three points are considered. There is always a plane in the three-dimensional space that can separate a point by other two different points, and therefore the obtained points belong to classes that are linearly separable.

5 The optimization problem to be solved for training a support vector machine related to the set of points in the transformed space considered in the previous exercise is

$$\min_w \frac{1}{2} \left(w_1^2 + w_2^2 + w_3^2 \right)$$

subject to

$$b \leq -1$$
$$\sqrt{2} w_1 + w_2 + b \geq 1$$
$$2\sqrt{2} w_1 + 2 w_2 + \sqrt{2} w_3 + b \leq 1.$$

6 The experiments discussed in Section 6.6 regard the use of the freeware software LIBSVM. In the quoted section, a training test and a testing set have been generated randomly by using the MATLAB function generate4libsvm. In the experiments, a support vector machine has been trained by using a sigmoidal kernel. In the following, two different support vector machines are trained by using the same training set but two different kernel functions.

```
LIBSVM>svmtrain -t 1 trainset.txt
*
optimization finished, #iter = 35
nu = 0.584346
obj = -47.033541, rho = -0.249598
nSV = 61, nBSV = 58
Total nSV = 61

LIBSVM>svmpredict testset.txt trainset.txt.model
                 testresult-polynomial-kernel.txt
Accuracy = 82.6% (826/1000) (classification)

LIBSVM>svmtrain -t 2 trainset.txt
*
```

```
optimization finished, #iter = 17
nu = 0.175650
obj = -11.319766, rho = 0.030302
nSV = 20, nBSV = 16
Total nSV = 20

LIBSVM>svmpredict testset.txt trainset.txt.model
                  testresult-polynomial-kernel.txt
Accuracy = 98.6% (986/1000) (classification)
```

These experiments show that the kernel that performs better on the considered problem is the radial basis kernel, which is specified by '2' when the option '-t' of the procedure svmtrain is used.

7 This exercise uses the same notations introduced in Section 6.1. For instance, w and b are the parameters of the general equation of the hyperplane:

$$w^T x + b = 0.$$

As known, the two parameters w and b can be normalized so that $w^T x + b = +1$ is the hyperplane that goes through the support vectors of the class C^+, and $w^T x + b = -1$ is the hyperplane that goes through the support vectors of the class C^-. If x^+ is a sample on the hyperplane C^+ and x^- is the sample closest to x^+ on the hyperplane C^-, then the margin between the two hyperplanes can be written as:

$$M = |x^+ - x^-|.$$

The aim of this exercise is to prove that the margin M between the two classes can be also written as:

$$M = \frac{2}{\sqrt{w^T w}}.$$

Since w is orthogonal to both C^+ and C^-, then

$$x^+ = x^- + \lambda w$$

for some real λ. The following system of conditions

$$\begin{cases} w^T x^+ + b = +1 \\ w^T x^- + b = -1 \\ x^+ = x^- + \lambda w \\ M = |x^+ - x^-| \end{cases}$$

implies that

$$w^T(x^- + \lambda w) = 1$$

$$\implies w^T x^- + b + \lambda w^T w = 1$$

$$\implies -1 + \lambda w^T w = 1$$

$$\implies \lambda = \frac{2}{w^T w}.$$

Therefore,

$$M = |x^+ - x^-| = |\lambda w| = \lambda |w| = \lambda \sqrt{w^T w},$$

and thus

$$M = \frac{2}{\sqrt{w^T w}},$$

and hence the proof is completed.

10.6 Problems of Chapter 7

1 The matrix

$$A = \begin{pmatrix} 1 & 2 & 3 & -4 & 5 \\ 1 & 1 & 0 & 0 & 1 \\ 0 & 1 & 2 & 2 & 0 \\ -1 & 3 & 1 & 0 & 2 \\ 3 & -1 & 1 & 2 & 1 \end{pmatrix}$$

represents a set of samples and features that can be partitioned in biclusters. Each column of the matrix represents a sample, each row of A represents instead a feature. A possible bicluster with constant row values is

$$C_A = \begin{pmatrix} 0 & 0 \\ 2 & 2 \end{pmatrix},$$

where C_A can be obtained by A by extracting its second and third rows and its third and fourth column.

2 The set of points:

$$x_1 = (7, 0, 0), \quad x_2 = (5, 0, 0), \quad x_3 = (0, 1, 0),$$
$$x_4 = (0, 3, 0), \quad x_5 = (0, 0, 1), \quad x_6 = (0, 0, 5)$$

is given and their partition is assigned as follows:

$$x_1 \in S_1, \quad x_2 \in S_1, \quad x_3 \in S_2, \quad x_4 \in S_2, \quad x_5 \in S_3, \quad x_6 \in S_3.$$

The matrix A associated to this set of data is

$$A = \begin{pmatrix} 7 & 5 & 0 & 0 & 0 & 0 \\ 0 & 0 & 1 & 3 & 0 & 0 \\ 0 & 0 & 0 & 0 & 1 & 5 \end{pmatrix}$$

and then the features are represented by the three 6-dimensional points:

$$f_1 = (7, 5, 0, 0, 0, 0)$$
$$f_2 = (0, 0, 1, 3, 0, 0)$$
$$f_3 = (0, 0, 0, 0, 1, 5).$$

Let us compute the centers of the three clusters S_1, S_2 and S_3:

$$c_1^S = \frac{x_1 + x_2}{2} = \frac{(7,0,0) + (5,0,0)}{2} = (6,0,0) = (c_{11}^S, c_{21}^S, c_{31}^S)$$

$$c_2^S = \frac{x_3 + x_4}{2} = \frac{(0,1,0) + (0,3,0)}{2} = (0,2,0) = (c_{12}^S, c_{22}^S, c_{32}^S)$$

$$c_3^S = \frac{x_5 + x_6}{2} = \frac{(0,0,1) + (0,0,5)}{2} = (0,0,3) = (c_{13}^S, c_{23}^S, c_{33}^S).$$

By applying the rule (7.2), it follows that

$$c_{11}^S > c_{12}^S \quad \text{and} \quad c_{11}^S > c_{13}^S \quad \Longrightarrow \quad f_1 \in F_1$$
$$c_{22}^S > c_{21}^S \quad \text{and} \quad c_{22}^S > c_{23}^S \quad \Longrightarrow \quad f_2 \in F_2$$
$$c_{33}^S > c_{31}^S \quad \text{and} \quad c_{33}^S > c_{32}^S \quad \Longrightarrow \quad f_3 \in F_3.$$

Thus, the partition in biclusters is

$$B = \{(x_1, x_2, f_1), (x_3, x_4, f_2), (x_5, x_6, f_3)\}.$$

3 In this exercise, the partition in biclusters obtained in the previous exercise must be checked for consistency. In such a partition, each feature is contained in a different bicluster, and therefore each center c_r^F equals the rth feature f_r:

$$c_1^F = f_1, \quad c_2^F = f_2, \quad c_3^F = f_3.$$

The rule (7.3) can be applied:

$$c_{11}^F > c_{12}^F \quad \text{and} \quad c_{11}^F > c_{13}^F \quad \Longrightarrow \quad x_1 \in \hat{S}_1$$
$$c_{21}^F > c_{22}^F \quad \text{and} \quad c_{21}^F > c_{23}^F \quad \Longrightarrow \quad x_2 \in \hat{S}_1$$
$$c_{32}^F > c_{31}^F \quad \text{and} \quad c_{32}^F > c_{33}^F \quad \Longrightarrow \quad x_3 \in \hat{S}_2$$
$$c_{42}^F > c_{41}^F \quad \text{and} \quad c_{42}^F > c_{43}^F \quad \Longrightarrow \quad x_4 \in \hat{S}_2$$
$$c_{53}^F > c_{51}^F \quad \text{and} \quad c_{53}^F > c_{52}^F \quad \Longrightarrow \quad x_5 \in \hat{S}_3$$
$$c_{63}^F > c_{61}^F \quad \text{and} \quad c_{63}^F > c_{62}^F \quad \Longrightarrow \quad x_6 \in \hat{S}_3.$$

The partition found in clusters $\hat{S}_r$ is equal to the partition in clusters S_r. Thus, the partition in biclusters is consistent.

4 The samples x_i and the features f_i related to this exercise can be summarized in the matrix

$$A = \begin{pmatrix} 1\ 2\ 3\ 4 \\ 2\ 3\ 4\ 5 \\ 3\ 4\ 2\ 1 \end{pmatrix}.$$

The columns of the matrix represent the 4 points in the three-dimensional space to which a partition in cluster is already assigned: the first two columns belong to the cluster S_1, whereas the last two columns belong to the cluster S_2. Let us compute the centers of these two clusters:

$$c_1^S = \frac{x_1 + x_2}{2} = \frac{(1, 2, 3) + (2, 3, 4)}{2} = \left(\frac{3}{2}, \frac{5}{2}, \frac{7}{2}\right)$$

$$c_2^S = \frac{x_3 + x_4}{2} = \frac{(3, 4, 2) + (4, 5, 1)}{2} = \left(\frac{7}{2}, \frac{9}{2}, \frac{3}{2}\right).$$

Let us apply the rule (7.2):

$$c_{11}^S > c_{12}^S \implies f_1 = (1, 2, 3, 4) \in F_1$$
$$c_{21}^S > c_{22}^S \implies f_2 = (2, 3, 4, 5) \in F_1$$
$$c_{31}^S < c_{32}^S \implies f_3 = (3, 4, 2, 1) \in F_2.$$

Then, the partition in biclusters is

$$B = \{(x_1, x_2, f_1, f_2), (x_3, x_4, f_3)\}.$$

Let us now check if the obtained partition B is consistent. The centers of the clusters F_r are

$$c_1^F = \frac{f_1 + f_2}{2} = \frac{(1, 2, 3, 4) + (2, 3, 4, 5)}{2} = \left(\frac{3}{2}, \frac{5}{2}, \frac{7}{2}, \frac{11}{2}\right)$$

$$c_2^F = f_3 = (3, 4, 2, 1).$$

The rule (7.3) is applied:

$$c_{11}^F < c_{12}^F \implies x_1 = (1, 2, 3) \in \hat{S}_2$$
$$c_{21}^F < c_{22}^F \implies x_2 = (2, 3, 4) \in \hat{S}_2$$
$$c_{31}^F > c_{32}^F \implies x_3 = (3, 4, 2) \in \hat{S}_1$$
$$c_{41}^F > c_{42}^F \implies x_4 = (4, 5, 1) \in \hat{S}_1.$$

The partitions in biclusters S_r and $\hat{S}_r$ are different, and therefore the obtained biclustering B is not consistent.

5 Impossible. Every α-consistent biclustering, for any α, is also consistent.

Appendix A
The MATLAB® Environment

A.1 Basic concepts

MATLAB is a numerical computing environment for scientific and numeric applications. It provides a wide variety of predefined functions that can be used for solving several problems in the field of numerical analysis. MATLAB is moreover a programming language, so that functions can be written and utilized with the ones that are predefined in the environment. The name derive from the two words **MATrix** and **LABoratory**. It indeed allows easy matrix manipulation, as they are considered as single variables. Plotting of functions and data is also simple by using MATLAB. In the following, we will pay attention to the basic concepts needed by the reader for performing the experiments discussed in this book. The reader who is interested in more details about MATLAB can make use of several tutorials on the topic.

In general, instructions that MATLAB can carry out are specified through a command window. When the symbol » is shown, MATLAB is waiting to have orders from the user. The orders can range from simple arithmetic operations such as sums and products of real numbers to the execution of complex functions. One of the easiest operations MATLAB can make is the following one:

```
>> 2 + 3

ans =

    5
```

In this case, MATLAB is used as a simple calculator. The result of the operation is stored in the auxiliary variable ans. In MATLAB, every time it is not explicitly specified, the result of an operation or function is stored in a variable called ans. The output variable can be specified as follows:

```
>> a = 2 + 3

a =

    5
```

The same result is obtained if the following is given to MATLAB:

```
>> a = 2;
>> b = 3;
>> c = a + b;
>> c

c =

   5
```

In this example, three variables are used for computing the same sum. The variable a is firstly defined and its value is set to 2. The variable b has instead value 3. The sum of the variables is this time stored in the variable called c, and the result of the operation is shown. Note that MATLAB does not produce any printed output when the given instruction ends with the symbol ;. This can be very convenient, because in many cases a lot of operations are needed, but only the last operation provides the result of interest. Every time the symbol ; is added at the end of the instruction, the instruction is executed but the result is not printed. To visualize the current value of a certain variable, for example c, it is sufficient to write its name.

Differently from other programming languages, the variables in MATLAB do not need to be declared. In other languages the declaration of a variable is needed for specifying the type of data the variable has to contain. In MATLAB, all the variables are matrices of real numbers. For instance, the instruction a = 2 implicitly declares a matrix with one row and one column and containing one real number, which corresponds to 2 in this case. Variables need to be declared in MATLAB as well if the user needs to represent different kinds of data. For simple applications, however, the explicit declaration of variables is usually not needed.

Vectors are matrices having only one row or only one column. In MATLAB, a set of sorted numbers between the symbol [and the symbol] represents a vector with such numbers as components:

```
>> v = [1 3 5 7]

v =

   1   3   5   7
```

The following ones are some of the basic operations that can be carried out on vectors:

```
>> v = [1 3 5 7];
>> w = [1 1 1 1];
>> v + w

ans =

   2   4   6   8
```

```
>> v - w

ans =

   0   2   4   5

>> 2*v

ans =

   2   6   10   14

>> v*w'

ans =

   16
```

The sum of vectors is performed component by component, as well as the difference between two vectors. A vector can also be multiplied by a number, and the result is a vector having as components the product of such number by the components of the vector v. The instruction v*w' performs the so-called *inner product* between two vectors with the same length, i.e., the same number of components. Its result is a number defined as the sum of the products of all the homologue components. In the example, the inner product v*w' is $(1 \cdot 1) + (3 \cdot 1) + (5 \cdot 1) + (7 \cdot 1)$. The symbol ' after a variable name is used for transposing the variable. The transpose of a number is the number itself, the transpose of a row vector is a column vector, the transpose of a column vector is a row vector. Before performing the inner product, the second vector, w, has to be transformed into a column vector. In fact, these vectors are actually matrices in MATLAB, and the product between two matrices can be performed only if a condition is met. From the basic mathematical theory comes that two matrices can be multiplied if and only if the number of columns of the first matrix equals the number of rows of the second matrix. In this case, the row vector v and the row vector w are two matrices with 1 row and 4 columns. The condition is then not satisfied. In order to perform the inner product, the second vector w needs to be transposed, so that it becomes a column vector, having 4 rows and 1 column. In this way the condition is satisfied, and the two vectors can be multiplied. The symbol * refers to multiplication. In general, it refers to the product between matrices. If the variables are vectors, then the inner product is performed. If the variables are just numbers, the standard arithmetic product is performed. As for the sum between vectors, if the vector having as components the products of the components in v and w is of interest, then the following instruction must be used:

```
>> v*.w

ans =

   1   3   5   7
```

The symbol . after the * specifies that the operation must be performed element by element. In the case of vectors, the operation is performed component by component. In the example, the result corresponds to the vector v because the vector w has all its components equal to 1.

The following defines a matrix in MATLAB:

```
>> A = [1 2 3; 2 3 4]

A =
     1      2      3
     2      3      4
```

Numbers separated by a space (or a comma ,) belong to the same row, whereas the symbol ; specifies that the following numbers belong to the successive row of the matrix. When this syntax is used, it is important that all the rows and all the columns of the matrix have the same number of elements, otherwise a message error is provided by MATLAB. As for the vectors, similar basic operations can be carried out by using matrices:

```
>> A = [1 2 3; 2 3 4];
>> B = [1 0 1; 0 1 2];
>> A + B

ans =
     2      2      4
     2      4      6

>> A - B

ans =
     0      2      2
     2      2      2

>> 2*A

ans =
     2      4      6
     4      6      8

>> A*B'

ans =
     4      8
     6     11
```

As before, the sum of two matrices is a matrix having as elements the sum of the homologue elements of the two matrices. If the difference is performed, the difference between the homologue elements is considered. A matrix can be multiplied by a number, and the result is a matrix having all the elements in A multiplied by that number. The symbol * refers here to the standard product between two matrices. To perform the product, the number of columns of A must equal the number of rows of B. For this reason, B is transposed before performing the product. The solution is a matrix, having a number of rows which equals the number of rows of A and a number of columns which equals the number of columns of B'. As before, the product element by element of two matrices can be carried out by using the symbol . after *.

In MATLAB, every variable is considered as a matrix. However, elements of a matrix can be considered separately, and they can define sub-matrices. The following example extracts sub-matrices, vectors and numbers from a matrix A:

```
>> A = [1 2 3 4; 2 3 4 5; 5 6 7 8]

A  =

    1    2    3    4
    2    3    4    5
    5    6    7    8

>> A(2,3)

ans =

    4

>> B = A(1:3,3:4)

B  =

    3    4
    4    5
    7    8

>> v = B(1,:)

v  =

    3    4

>> w = B(:,2)

w  =

    4
    5
    8

>> w(2)
```

```
ans =

    5
```

For referring to the element of a matrix, two indices are needed, the one related to the rows and the one related to the columns. In the example, the element with row index $i = 2$ and column index $j = 3$ is extracted. More than one element can be extracted from a matrix per time. For instance, $A(1:3,3:4)$ refers to the elements of the matrix having row indices ranging from 1 to 3 and column indices ranging from 3 to 4. If the symbol : is used instead of a number, then all the rows or columns of the matrix are considered. The symbols $1:3$ and $3:4$ define vectors by using a compact syntax. $1:3$ is actually the vector $[1\ 2\ 3]$, and $3:4$ is the vector $[3\ 4]$. In general, $x:y$ defines a vector having as first component x, having as last element y and such that the difference between any consecutive components of the vector is 1. This difference is set to 1 by default. It can be specified by using the symbology $x:d:y$, where d is the considered difference.

A.2 Graphic functions

MATLAB provides many graphic functions. They can be used for visualizing data and mathematical functions, and for building complex figures. The basic graphic function in MATLAB is `plot`. The following instructions in MATLAB draw Figure A.1.

```
>> x = [1 2 3 4];   y = [0.2 1.5 1.8 3];
>> plot(x,y,'o','MarkerSize',16)
```

The function `plot` draws on a two-dimensional Cartesian system the set of points specified by the two vectors x and y. The x coordinates of such points are in the vector x, whereas their y coordinates are in the vector y. In the example, four points are drawn, and in particular the points with coordinates $(1, 0.2)$, $(2, 1.5)$, $(3, 1.8)$ and $(4, 3)$. The third input parameter of the function `plot` specifies the symbol with which the points have to be marked. In the example, a circle (o) is used. Other symbols include stars *, crosses +, etc. The symbol o is specified between two ' symbols. Everything between two ' symbols is considered as a string of characters in MATLAB. Besides the symbol for marking the points, even the color of the points can be specified. For more details about the function `plot`, the MATLAB `help` command can be utilized. For instance, the following provides information about the `plot` function:

```
>> help plot
 PLOT   Linear plot.
    PLOT(X,Y) plots vector Y versus vector X. If X or Y is a matrix,
    then the vector is plotted versus the rows or columns of the matrix,
    whichever line up.  If X is a scalar and Y is a vector, disconnected
    line objects are created and plotted as discrete points vertically at
```

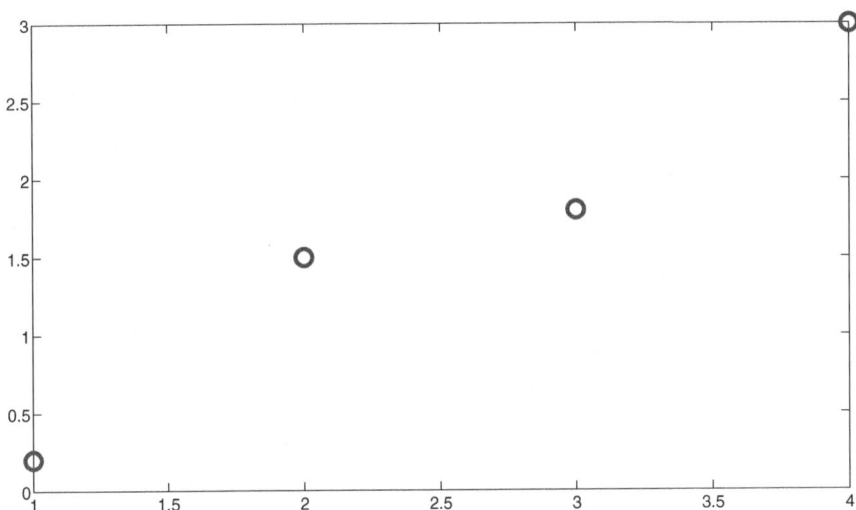

Fig. A.1 Points drawn by the MATLAB function `plot`.

```
X.
```

```
PLOT(Y) plots the columns of Y versus their index.
If Y is complex, PLOT(Y) is equivalent to PLOT(real(Y),imag(Y)).
In all other uses of PLOT, the imaginary part is ignored.
```

```
Various line types, plot symbols and colors may be obtained with
PLOT(X,Y,S) where S is a character string made from one element
from any or all the following 3 columns:
```

b	blue	.	point	-	solid
g	green	o	circle	:	dotted
r	red	x	x-mark	-.	dashdot
c	cyan	+	plus	--	dashed
m	magenta	*	star	(none)	no line
y	yellow	s	square		
k	black	d	diamond		
		v	triangle (down)		
		^	triangle (up)		
		<	triangle (left)		
		>	triangle (right)		
		p	pentagram		
		h	hexagram		

```
For example, PLOT(X,Y,'c+:') plots a cyan dotted line with a plus
at each data point; PLOT(X,Y,'bd') plots blue diamond at each data
point but does not draw any line.
```

```
PLOT(X1,Y1,S1,X2,Y2,S2,X3,Y3,S3,...) combines the plots defined by
the (X,Y,S) triples, where the X's and Y's are vectors or matrices
and the S's are strings.
```

```
For example, PLOT(X,Y,'y-',X,Y,'go') plots the data twice, with a
solid yellow line interpolating green circles at the data points.
```

```
The PLOT command, if no color is specified, makes automatic use of
the colors specified by the axes ColorOrder property.  The default
ColorOrder is listed in the table above for color systems where the
```

```
default is blue for one line, and for multiple lines, to cycle
through the first six colors in the table.  For monochrome systems,
PLOT cycles over the axes LineStyleOrder property.

If you do not specify a marker type, PLOT uses no marker.
If you do not specify a line style, PLOT uses a solid line.

PLOT(AX,...) plots into the axes with handle AX.

PLOT returns a column vector of handles to lineseries objects, one
handle per plotted line.

The X,Y pairs, or X,Y,S triples, can be followed by
parameter/value pairs to specify additional properties
of the lines. For example, PLOT(X,Y,'LineWidth',2,'Color',[.6 0 0])
will create a plot with a dark red line width of 2 points.

Example
    x = -pi:pi/10:pi;
    y = tan(sin(x)) - sin(tan(x));
    plot(x,y,'--rs','LineWidth',2,...
                    'MarkerEdgeColor','k',...
                    'MarkerFaceColor','g',...
                    'MarkerSize',10)

See also plottools, semilogx, semilogy, loglog, plotyy, plot3, grid,
title, xlabel, ylabel, axis, axes, hold, legend, subplot, scatter.

Overloaded functions or methods (ones with the same name in other
directories)
    help timeseries/plot.m
    help SimTimeseries/plot.m
    help cfit/plot.m
    help distributed/plot.m
    help fints/plot.m

Reference page in Help browser
    doc plot
```

Each function in MATLAB has a guide similar to this one, which can be accessed through the `help` command.

The `plot` function can also be used for plotting real mathematical functions defined in $\Re$. The following examples plot the function `sin` and `cos`.

```
>> x = 0:0.8:4*pi;
>> y = sin(x);
>> plot(x,y)
>> hold on
>> x = 0:0.1:4*pi;
>> y = cos(x);
>> plot(x,y,'r:')
```

Figure A.2 shows the result. The vector x defines the interval on the x axis. It is defined as the vector having as first component 0, having distance between consecutive components equal to 0.8 and having the last component smaller than 4*pi. The variable pi is predefined in MATLAB and contains an approximation of π. In the graphic of a mathematical function $f : x \in A \longrightarrow y \in B$, each point (x, y) is such that $y = f(x)$. The function sin is used in this case for computing the dependent variables y related to the independent variables x stored in x. x is a

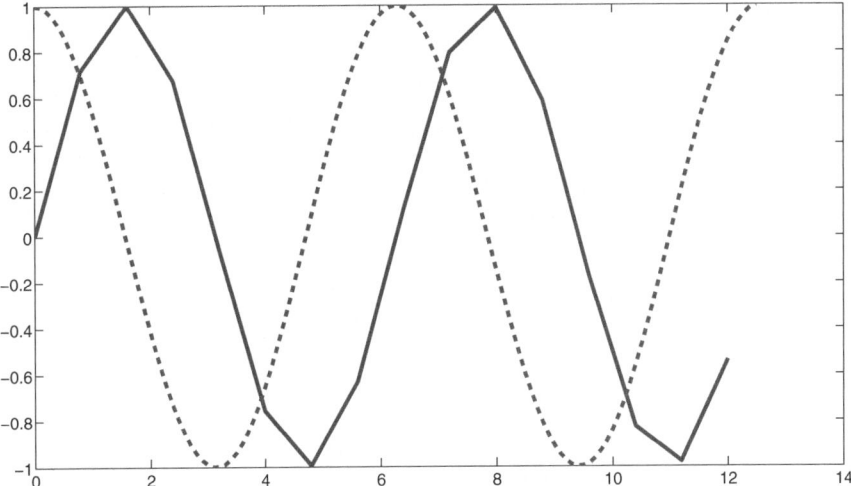

Fig. A.2 The sine and cosine functions drawn with MATLAB.

vector, and the function sin works on all its components and returns as output in y a vector containing the corresponding y variables. Note that the symbol ; is used for avoiding that the results of the instructions are printed on the screen. The function plot is then used. The third input parameter is in this case omitted, and hence the default settings are implicitly used. By default, plot draws in blue and it connects the points with straight lines. Figure A.2 shows indeed a sort of *join-the-dots* function which recalls the well-known shape of the sine function. Note that the graphic can be certainly improved if more points are used to draw it. The other instructions in the example draw the cosine function. This graphic is overlapped to the previous one in the same Figure A.2. To have that in MATLAB, the command hold on must be used. If instead the current figure is not needed anymore and it can be deleted, then the command hold off can be used. This time, the vector x is defined similarly as before, but 0.8 is substituted by 0.1. In this way, the considered interval on the x axis is always the same, but the number of points increased. This helps improve the accuracy of the graphic. The function cos is then used for computing the vector y. As x, this time vector y is longer, i.e., it contains more components. Finally, the function plot is used another time by using the newly computed x and y. The third parameter is specified, and it forces the graphic to be in red and visualized with a dashed line.

Other important graphic functions in MATLAB are fplot, axis, title, just to mention a few. The function fplot is used exclusively for drawing mathematical functions such as sin and cos. It is able to adjust by itself the number of points to use for obtaining a graphic of the function having a good quality. The function axis is used for changing the intervals of the x and y axis in a MATLAB figure. The function title adds a title to a MATLAB figure. Details about these functions can

```
%
% this function evaluates the following mathematical function
%
% f(x)=(x^2)(1.2-x)(1-e^(10(x-1)))
%
% usage:  y = fun(x);
%
% where x is a number or a vector of numbers
%

function [y] = fun(x)

    % evaluating function fun for each component of vector x
    for i=1:length(x),
      y(i)=(x(i)^2)*(1.2-x(i))*(1.-exp(10*(x(i)-1.)));
    end

end
```

Fig. A.3 The function `fun`.

be found through the `help` command. MATLAB has many other functions that can be used for drawing figures.

A.3 Writing a MATLAB function

Many built-in functions are available in the MATLAB environment. Groups of functions are collected in the so-called MATLAB *toolboxes*, where functions are grouped by specific fields of application. Other functions can be written by the user and integrated in MATLAB. In this way, MATLAB can be used as a programming language.

In Figure A.3 an example of a MATLAB function is given. In order to use it, the MATLAB code must be saved in a text file. The text file has extension `.m` and this kind of file is referred to as `m-file`. The name of the file must be the same as the function it contains. In this example, the text file must be named `fun.m`. All the rows of the `m-file` which start with the symbol `%` are considered as comments for the developer. MATLAB just ignores all such rows. In particular, the first comment rows are read by MATLAB when the `help` command is used:

```
>> help fun

this function evaluates the following mathematical function

f(x)=(x^2)(1.2-x)(1-e^(10(x-1)))

usage:  y = fun(x);

where x is a number or a vector of numbers
```

Every function in MATLAB needs to have a row in which the attributes of the function are defined. This row must have as first word the key word `function`. Then, the output parameters are specified, separated by commas and inserted between [and]. In this example, there is only one parameter, y. The list of the output parameters

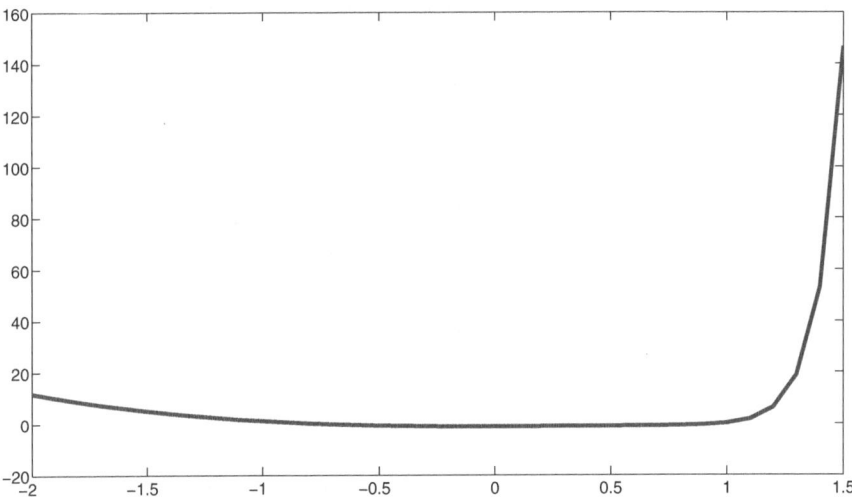

Fig. A.4 The graphic of the MATLAB function fun.

and the name of the function are separated by the rest with the symbol =. After the name of the function, the list of input parameters is specified: all the parameters are separated by commas and they are included between the symbol (and the symbol). In the example, the only input parameter x can be either a real number or a vector. All the rows in the text file between this first row and the row containing the last end represents the instructions the function carries out. In this example, just few rows are needed: one row contains a comment, three rows contain instructions. A for loop is used. The MATLAB function length counts the number of components of a vector. It returns 1 if x contains only a real number. The instruction in the for loop is repeated as many times as the number of components in x, and the index i starts from 1 and then increases its value by 1 at each iteration. The instruction in the for loop evaluates the mathematical function point by point. Note that the exponential function is used and that it is implemented in MATLAB by the function exp. The following MATLAB code:

```
x = -2:0.1:1.5;
y = fun(x);
plot(x,y)
```

exploits the function fun and creates the graphic of the mathematical function in Figure A.4.

Besides for loops, repeat..until and while..end while constructs can also be implemented in MATLAB, as well as the if construct. For other details about MATLAB, refer to the several tutorials on this topic and to the help command available in MATLAB.

Appendix B
An Application in C

B.1 h-means in C

In this section, an application in C implementing the h-means algorithm is presented. As discussed in Section 3.2, h-means is a method for clustering which is slightly different from the standard k-means algorithm or Lloyd's algorithm. We decided to implement h-means instead of k-means because it is more efficient. Moreover, as already observed, since the two algorithms are very similar, h-means can be found in the literature as the k-means algorithm.

The application presented in this section is able to partition sets of data whose samples can be represented as m-dimensional vectors. This covers a wide range of real-life applications. Sets of features are usually collected and grouped in vectors. For instance, a sound track is a vector of digital sounds, and an image is a matrix of pixels, whose rows or columns can be organized in a vector. In this application, we do not refer to a particular problem. We also try to keep the code as simple as possible. Because of this, the application may not work in particular cases. However, the reader can use this code for solving a large part of clustering problems without modifying the source code.

In C programming language, a software procedure consists of one or more functions. In the procedural approach, the tasks a procedure has to carry out are usually divided into different functions. The application we present is mainly divided into a main function, where the data are read from input files, and the h-means function, which actually performs the algorithm. As Figure 3.9 in Section 3.2 shows, the h-means algorithm can be summarized in few rows, and there are tasks that need to be repeated more than once. By using the procedural approach, every task that has to be performed more than once can be implemented in a single function, so that a function call is needed every time the task must be carried out. Different from the procedural approach is the object-oriented approach, which can be implemented by using programming languages such as C++ and Java [183]. In recent years, the object-oriented approach has been utilized more and more. However, we decided to present here a general procedural application in C, because we think it is much easier to use and modify for a user having expertise in other fields, such as agriculture. To read and understand the following, it is essential the reader has some knowledge in programming in C. C compilers are available on the Internet for free for either Windows or Linux operating systems. Any of these is good for compiling the application here presented.

In Figure B.1 the function hmeans is shown. The function returns the number of performed iterations, and it has six input parameters. The first three parameters specify the set of data to partition. n is the number of samples in the set, and m is the number of components needed for representing samples as vectors. All samples are stored in the two-dimensional array x. x is actually a matrix with n rows and m columns. Each row represents one of the samples, and each column corresponds to the values all samples have on the same component. The integer k is the number of clusters the data have to be partitioned in. iTmax specifies the maximum number of

```c
int hmeans(int n,int m,double **X,int k,int iTmax,int *clust)
{
   int i,ii;
   int iT;
   double **c,**cnew;

   // allocating memory
   c = (double**)calloc(k,sizeof(double*));
   for (ii = 0; ii < k; ii++)  c[ii] = (double*)calloc(m,sizeof(double));
   cnew = (double**)calloc(k,sizeof(double*));
   for (ii = 0; ii < k; ii++)  cnew[ii] = (double*)calloc(m,sizeof(double));

   // initializing a random partition in clusters
   rand_clust(n,k,clust);

   // computing the centers of the clusters
   compute_centers(n,m,X,k,clust,cnew);

   iT = 0;

   do
   {
      iT = iT + 1;

      // preparing for next iteration
      copy_centers(k,m,cnew,c);

      // checking the distances between samples and centers
      for (i = 0; i < n; i++)
      {
         ii = find_closest(m,X[i],k,c);
         clust[i] = ii;
      };

      // recomputing the centers
      compute_centers(n,m,X,k,clust,cnew);

   } while (isStable(k,m,c,cnew,1.e-6) == 1 && iT < iTmax);

   free(c);  free(cnew);

   return iT;
};
```

Fig. B.1 The function hmeans.

allowed iterations. Finally, the vector `clust` contains the code of the cluster each sample belongs to. The clusters are coded using an integer number, from 0 to $k - 1$. For instance, if `clust[2] = 1`, then the sample represented by vector in row 2 of matrix x belongs to cluster 1. Be aware that the indices of the vectors and of the matrices in C are counted from 0 on. Therefore, the row indexed by 2 in the matrix x is actually the third one.

Variables need to be declared in C programming language. In fact, the function `hmeans` starts declaring the local variables needed for performing the algorithm. Among the others, `c` and `cnew` are declared as pointers to pointers. Memory will be allocated later in the code for these two variables. Once the memory has been properly allocated for `c` and `cnew`, they can be considered as two matrices with k rows and m columns. Each row of these matrices represents the center of the corresponding cluster. Two matrices of this kind are needed, because the stopping criteria of the algorithm is based on the changes, iteration after iteration, of the centers of the clusters. For this reason, at each iteration, `cnew` contains the centers related to the current partition, whereas `c` contains the centers related to the previous partition. The function `calloc` (`stdlib.h`) is used for allocating the memory for `c` and `cnew`. Since they are pointers to pointers, the allocation of the memory is performed in two steps.

At the start, the function `hmeans` computes a random partition of the set of data. The function `rand_clust` is used for this purpose. It takes as inputs the set of data, through the parameters n, m and x, and the number of clusters k. The output is the vector `clust`, which provides a random division of the data in clusters. This function and all the other functions `hmeans` uses will be explained in detail below.

Once a partition in clusters has been computed, the centers of these clusters need to be computed before proceeding with the algorithm. The function `compute_centers` is called with this aim. It takes as inputs the set of data to partition (n,m,x), and a partition in clusters, specified by k and `clust`. The output of this function is stored in the matrix `cnew`, where the centers of the clusters are stored row by row. After this start-up phase where some variables are set up, the main loop of the algorithm can be implemented.

The main loop of the function is a `repeat..until` loop. Note that a `while` loop is instead used in the algorithm in Figure 3.9. The `repeat..until` is used here because the stopping criteria cannot be evaluated until one iteration of the loop is at least performed. Indeed, the matrix `cnew` must contain the centers of the current partition in clusters, where `c` must contain the centers of the previous partition. After the first iteration of the loop, `c` contains the centers of the random partition computed at the start of the function, `cnew` the newly generated partition, and then the stopping criteria can be evaluated. The stopping criteria is implemented by the function `isStable`, which checks if the centers of the clusters are stable or not, iteration after iteration. The centers are considered stable if the maximum difference upon all the matrices `c` and `cnew` elements is smaller than a given threshold. When the function `isStable` returns zero, the condition in the `repeat..until` results true, and then the loop stops.

In the `repeat..until` loop, the matrix `cnew` is soon copied in the matrix `c`. Indeed, a new iteration is starting, and therefore `cnew` contains now the centers of the previous partition. They are then moved to `c`, so that the new centers can be stored in `cnew`. The copy of the centers is obtained through the function `copy_centers`.

As already explained, the main idea in the h-means algorithm is to move samples to clusters whose center is closest to the sample. At each step of the algorithm, this condition has to be checked sample by sample, and eventually samples need to be moved from a cluster to another. The `for` loop in the `repeat..until` loop carries this task out. Sample by sample, the closest center to a sample `x[i]` is located through the function `find_closest`, and then `x[i]` is assigned to the corresponding cluster. Note that, since `x` is a matrix, `x[i]` corresponds to the ith row of the matrix, i.e., it is the ith sample. The centers used are those stored in `c`, which are related to the previous partition. Sample after sample, the vector `clust` is updated, and it provides a new partition at the end of the `for` loop. Hence, the new centers have to be computed. The function `compute_centers` is used for updating `cnew` and the stopping criteria is evaluated using the function `isStable`. When the stopping criteria is satisfied, the vector `clust` contains an optimal partition of the data. The function can stop, before freeing the allocated memory. Note that the `repeat..until` loop can also stop when the maximum number of iterations is reached.

The prototypes of the functions used in `hmeans` are shown in Figure B.2. Functions contained in standard C libraries are not included, because their prototypes can be found in the corresponding header files. All the standard functions used in the function presented in this chapter can be found in the standard input/output library (`stdio.h`), in the standard C library (`stdlib.h`), in the library for string management (`string.h`) or in the library for basic mathematics (`math.h`). The prototypes of the functions which are used need to be placed at the top of the text files where the function sources are written. If library functions are used, then the corresponding header file needs to be specified.

The source of the function `rand_clust` is given in Figure B.3. The function does not have any returning value and it expects to receive 3 input parameters. The integer n represents the number of samples in the set, and the integer k represents the desired number of clusters of the random partition. The random partition is given as output by the function through the vector `clust`. For instance, if `clust[i]` = 2, then the ith sample in the matrix x belongs to the cluster 2 (which is the third one). In order to find a random partition, a random integer number from 1 to k is assigned to each component of the vector `clust`. The standard function `rand` (`stdlib.h`) is used, and it provides an integer random number. Note that the key words `double` and `int`

```
void rand_clust(int n,int k,int *clust);
void compute_centers(int n,int m,double **X,int k,int *clust,double **c);
int find_closest(int m,double *x,int k,double **c);
int compare_centers(int k,int m,double **c1,double **c2,double tol);
void copy_centers(int k,int m,double **c1,double **c2);
```

Fig. B.2 The prototypes of the functions called by hmeans.

```
void rand_clust(int n,int k,int *clust)
{
    int i;
    double aux;

    for (i = 0; i < n; i++)
    {
        aux = (double)(rand());
        aux = k*(aux/RAND_MAX);
        clust[i] = (int)(aux);
    };
};
```

Fig. B.3 The function `rand_clust`.

between parentheses forces the following variables to be converted in the desired type. For instance, `clust` is a vector of integers, and hence the value in `aux` has to be converted from `double` to `int` before assigning it to any `clust[i]`. This kind of conversion just cuts all the decimal values of a real number. For instance, 1.3 and 1.9 are both converted to the integer 1.

The source of the function `compute_centers` is given in Figure B.4. It takes as inputs the set of data (n,m,x), the number of clusters `k` and a partition through the vector `clust`. The output is the matrix `c` that contains the centers of the clusters row by row. Therefore, `c` is a matrix with `k` rows (the number of centers) and `m` columns (the dimension of the space where the samples are represented). In this function, the vector `cclust` is used as local variable. It is a counter of the samples that are present in each cluster. The function starts initializing all the components of the vector `cclust` and all the elements of the matrix `c` to zero. After that, sample by sample, the following steps are performed. First, the cluster to which sample `x[i]` belongs is checked using `clust[i]`. The code of the cluster is stored in the auxiliary variable `ii`. Then, the counter regarding the cluster coded by `ii` is incremented by 1. Finally, the vector `x[i]` is added to `c[ii]`, because `x[i]` belongs to cluster `ii`. At the end of the `for` loop, each row in `c` is the sum of all the samples belonging to the corresponding cluster. For computing the mean among all such samples, each `c[ii]` has to be divided by the number of samples the cluster has. This information is stored in `cclust[ii]`. The third and last part of the algorithm computes these divisions. However, if there are empty clusters, then the division in correspondence of such clusters cannot be computed, because the division by 0 is not allowed. The center of an empty cluster actually does not exist, but the corresponding row in the matrix `c` cannot be left empty. In this case, the function just assigns 0 to all the components of the center. This does not affect the convergence of the *h*-means algorithm.

The function `find_closest` is shown in Figure B.5. Given a target vector and a set of vectors, this function computes the closest vector in the set to the target. In the function `hmeans`, where this function is used, the target vector is a sample and the set of vectors corresponds to the set of centers of a partition in clusters. The function has as parameters the dimension `m` of the vectors, the target vector `x`, the number `k` of the vectors in the set and the set itself. The whole set of vectors is stored in a matrix `c` with `k` rows and `m` columns. In the matrix, each row corresponds to a different

```
void compute_centers(int n,int m,double **X,int k,int *clust,double **c)
{
    int i,ii,jj;
    int *cclust;
    cclust = (int*)calloc(k,sizeof(int));

    for (ii = 0; ii < k; ii++)
    {
        cclust[ii] = 0;
        for (jj = 0; jj < m; jj++)
        {
            c[ii][jj] = 0.0;
        };
    };

    for (i = 0; i < n; i++)
    {
        ii = clust[i];
        cclust[ii] = cclust[ii] + 1;
        for (jj = 0; jj < m; jj++)  c[ii][jj] = c[ii][jj] + X[i][jj];
    };

    for (ii = 0; ii < k; ii++)
    {
        for (jj = 0; jj < m; jj++)
        {
            if (cclust[ii] != 0.0)
            {
                c[ii][jj] = c[ii][jj]/cclust[ii];
            }
            else
            {
                c[ii][jj] = 0.0;
            };
        };
    };
    free(cclust);
};
```

Fig. B.4 The function compute_centers.

vector. The returning value of the function is the row index in c of the vector which
is closest to the target x. In the function, all the distances between target and each
vector in the set are computed step by step. Every time a new distance is computed,
it is compared to mindist, which contains the minimum distance found so far. If
the new distance dist is smaller than mindist, then mindist and minindex are
updated. The variable mindist is just set to dist, while minindex is set to the
index of the row in c corresponding to the current vector. At the end of the two
nested loops, mindist contains the value of the minimum distance, and minindex
contains the corresponding row index. The information of interest is the index and
not the distance value. In fact, minindex is the function returning value. Note that
the Euclidean distance is used in this function and that it might be substituted by
other distances in particular applications.

The function isStable implements the stopping criteria of the h-means algorithm. The source code is given in Figure B.6. In general, the function compares two
matrices c1 and c2 having the same dimensions, k and m. All the homologue ele-

```
int find_closest(int m,double *x,int k,double **c)
{
    int ii,jj;
    int minindex;
    double dist,mindist;
    double aux;

    minindex = 0;
    mindist = 1.e+100;

    for (ii = 0; ii < k; ii++)
    {
        dist = 0.0;
        for (jj = 0; jj < m; jj++)
        {
            aux = x[jj] - c[ii][jj];
            dist = dist + aux*aux;
        };
        dist = sqrt(dist);
        if (dist < mindist)
        {
            mindist = dist;
            minindex = ii;
        };
    };

    return minindex;
};
```

Fig. B.5 The function find_closest.

ments of the two matrices are compared, and the difference between the two matrices is defined as the largest difference between their homologue elements. Therefore, in practice, the function is an implementation of a method for finding a maximum value in a given set of values. In this case, the values are the differences in absolute values between elements of the matrix c1 and the matrix c2 having the same row

```
int isStable(int k,int m,double **c1,double **c2,double tol)
{
    int ii,jj;
    int stable;
    double diff,max;

    max = 0.0;
    for (ii = 0; ii < k; ii++)
    {
        for (jj = 0; jj < m; jj++)
        {
            diff = fabs(c1[ii][jj] - c2[ii][jj]);
            if (max < diff)   max = diff;
        };
    };

    stable = 1;
    if (max < tol)   stable = 0;

    return stable;
};
```

Fig. B.6 The function isStable.

and column indices. The maximum value is stored in the local variable max. Such variable is initially set to 0, because this is the smallest value it can have, being all the differences in absolute values. Once max has been found, its value represents the difference between c1 and c2. If such value is greater than the threshold tol given as input, then the two matrices are considered to be different. Otherwise, if max < tol, then the matrices are considered similar. This is reflected on the h-means stopping criteria: if the matrices are different, the centers are not stable, and other iterations of the algorithm need to be performed; if the matrices are similar, instead, the centers converged, and the algorithm can stop. The C programming language does not provide a boolean data type. Therefore, the integer variable stable is used for storing the following information: matrices are different / matrices are similar. In particular, when the functions returns 0, the h-means algorithm can stop, otherwise it returns 1, and another iteration of the algorithm is needed to be performed.

The last function which is used by the function hmeans is copy_centers (Figure B.7). It is simply used for copying the centers of a cluster partition from one variable to another. Since the centers are stored in matrices, the function performs the copy of two matrices in practice. k and m represent the dimensions of the matrices: the number of rows and the number of columns. c1 and c2 are the two variables containing the matrices. c1 is the matrix to be copied in c2. Hence, c2 is the only output parameter of the function copy_centers. The function does not return any value.

B.2 Reading data from a file

In the previous section, the function hmeans is presented for the partition of a given set of data in clusters. A detailed description of the source codes is provided, for the function hmeans itself and all the functions it uses. As already pointed out, the set of data is considered in the function through three variables: the number n of samples in the set of data, the number m of components each vector representing a sample has, and a matrix x containing all the samples row by row. In order to use the function hmeans, these variables need to be defined. In the experiments in the MATLAB® environment shown in Section 3.6, the data have been randomly

```
void copy_centers(int k,int m,double **c1,double **c2)
{
    int ii,jj;

    for (ii = 0; ii < k; ii++)
    {
        for (jj = 0; jj < m; jj++)
        {
            c2[ii][jj] = c1[ii][jj];
        };
    };
};
```

Fig. B.7 The function copy_centers.

```
4  3
12  23  34
45  56  67
78  89  90
13  46  79
```

Fig. B.8 An example of input text file.

generated. Something similar will be done also in this case. However, the aim of this chapter is to provide to the reader an application which can be used for personal purposes. For this reason, two functions are introduced in this section. They allow one to read the set of data from an input file, and store them under the format (n,m,x).

Data files can have different formats. Just to quote some example, files in WAV or MP3 format are used for storing digital audio tracks, whereas BMP and JPEG formats are used for digital images. In this case, a set of vectors needs to be stored. Following the same logical organization utilized when storing the data in x, the vectors can be placed in a text file line by line. On the same line of the text file, the numeric values related to each component of the vector can be saved. Then, a matrix structure is built inside the file. The other two pieces of information that are needed are the length of the vectors (the number of their components) and the number of lines in the file. Knowing this information helps in reading the data. The n and m values can be placed at the top of the text file. An example of text file formatted in this way is given in Figure B.8.

In the following, two functions for reading these input text files are presented. The reading task is split into two phases: in the first one, the variables n and m are read, whereas the whole matrix x is read in the second one. This is done because the memory for the matrix x can be allocated dynamically during the execution. In the function main of the application, x can be declared as a pointer to pointers to double variables. The first function can then be called, and the variables n and m can be read. After that, the memory for x can be dynamically allocated, and the second function can be called for transferring the data from the text file to the matrix x.

In Figure B.9 the function dimfile is reported. It takes a string of characters

```
int dimfile(char *filename,int *n,int *m)
{
    FILE *input;

    input = fopen(filename,"r");
    if (!input)   return 1;

    if (fscanf(input,"%d %d",n,m) == EOF)   return 1;

    if (*n <= 0 || *m <= 0)   return 1;

    fclose(input);

    return 0;
};
```

Fig. B.9 The function dimfile.

filename as input and it gives as outputs the two variable values n and m. The values n and m are specified as pointers, because they represent the output of the function. The returning value for the function dimfile is an error variable: it is 0 if the two variables n and m are read correctly, or 1 if some problem occurs.

The only local variable declared in the function dimfile is a pointer to a FILE type. It is needed for opening a file and read or write it. The file named as specified in filename is opened by calling the function fopen (stdio.h). This function has two input parameters, a string of characters containing the name of the file, and another string of characters containing the options to use when opening the file. In this case, only one option is specified: r. By using this option, the file is opened in reading mode only. Specifying w or rw, the file is opened in writing mode or both reading and writing mode. By default, fopen opens file in text mode. If a binary file is instead needed to be opened, the option b must specified. After fopen has been called, the variable input contains a pointer to a file. If its value is NULL, then some problem occurred when fopen tried to open the file. NULL is a variable declared and defined in stdio.h. Reasons for having the input variable equal to NULL might be the attempt of opening a file which does not exist or a damaged file. In that case, the function dimfile stops and returns 1.

If the input file has been correctly opened, then the two parameters n and m can be read. The function fscanf is used for this purpose (stdio.h). Four parameters are specified when the function is called. The first one is the pointer to the file to read. Then, the format of the text to read from the file is specified. The symbol %d indicates that an integer value is expected in the text. n and m are the two variables where the integer values have to be stored. Note that, in this function, n and m are pointers, as the fscanf function requires. The returning value of the function fscanf is EOF if the pointer to the input file reached the end of the file. In that case, the function stops and returns 1. Just like NULL, EOF is a predefined variable, whose declaration and definition can be found in the header file stdio.h. After that n and m have been read, their value can be checked. They represent the dimensions of the matrix x, and so they need to be positive numbers. Therefore, if n or m has as value zero or a negative value, then the function stops and returns 1. If the n and m values are acceptable, instead, the function closes the text file and returns 0.

The function that actually reads the text file is the one in Figure B.10. The function readfile has the same parameters of the function dimfile, plus the matrix x, where the set of data is stored. Moreover, in this function, the variables n and m are not pointers. It is supposed indeed that the dimensions n and m are already known. At the start, the source code of the function readfile is similar to the one of the function dimfile. Besides the pointer to the file, two integer variables are declared, the two indices i and j. The file named as specified in the variable filename is opened for reading and its pointer is assigned to the pointer input. If no errors occurred when opening the file, the n and m values are read from the file. This would not be needed, because the n and m values are already known, as it is supposed that the function dimfile already read it. However, the pointer input to the text file moves sequentially over the file. Then, it finds n and m at the top of the file and the remaining data after them. Therefore, n and m are read another time, and their values are stored

```
int readfile(char *filename,int n,int m,double **X)
{
    int i,j;
    FILE *input;

    input = fopen(filename,"r");
    if (!input)   return 1;

    if (fscanf(input,"%d %d\n",&i,&j) == EOF)   return 1;

    for (i = 0; i < n; i++)
    {
        for (j = 0; j < m; j++)
        {
            if (fscanf(input,"%lf",&X[i][j]) == EOF)   return 1;
        };
    };

    fclose(input);

    return 0;
};
```

Fig. B.10 The function readfile.

in auxiliary variables. Since they are integer numbers, the variables i and j are used. At this point, the data can be read. Two for loops start. The one on i counts the number of vectors, whereas the one on j counts the components of each vector. For each i and for each j, the corresponding value is stored in X[i][j]. As before, if the file should be damaged or not well-formatted, then the file may end earlier than expected. For controlling this, the returning value of the function fscanf is checked and compared to EOF. Note that the data are saved in a matrix of double variables, and hence in the format specified in the function fscanf the symbol used is %lf. If no errors occurred when reading all the data, then the function readfile closes the text file and returns 0.

B.3 An example of main function

In this section an example of function main in C is presented. The function is the main one of an application which reads a set of data from a file and then it partitions the set by using the function hmeans. This is just an example. The aim of this section is to provide a function main for exploiting the function hmeans. Note that the function hmeans might even be included in a general C function in which the h-means algorithm is only a sub-procedure of a more complex procedure.

An example of function main is presented in Figure B.11. The standard function main in C has as parameters the integer argc and the array of strings of characters argv. In this implementation, the final user is supposed to provide, as input arguments, the name of the file containing the data, and the number k of clusters in which the data need to be partitioned. The arguments of this application are provided through the parameters argc and argv. In practice, argc contains the number of

```
int main(int argc,char *argv[])
{
    int i;
    int n,m;
    int len,iT;
    int k,*clust;
    double **X;
    char *outfile;
    FILE *output;

    // checking the number of arguments
    if (argc < 3)
    {
        fprintf(stderr,"%s: too few arguments\n",argv[0]);
        fprintf(stderr,"%s: usage: %s nomefile.txt k\n",argv[0],argv[0]);
        return 1;
    };

    // checking the k value
    k = atoi(argv[2]);
    if (k < 2)
    {
        fprintf(stderr,"%s: invalid k value (%d)\n",argv[0]);
        return 1;
    };

    // checking the input file
    if (dimfile(argv[1],&n,&m) == 1)
    {
        fprintf(stderr,"%s: error while opening file '%s' or
                        invalid dimensions (n=%d,m=%d)\n",argv[0],argv[1],n,m);
        return 1;
    };

    fprintf(stderr,"%s: input file = '%s', k = %d\n",argv[0],argv[1],k);
    fprintf(stderr,"%s: n = %d, m = %d\n",argv[0],n,m);

    // memory allocation
    X = (double**)calloc(n,sizeof(double*));
    for (i = 0; i < n; i++)  X[i] = (double*)calloc(m,sizeof(double));
    clust = (int*)calloc(n,sizeof(int));

    // reading the input file
    if (readfile(argv[1],n,m,X) == 1)
    {
        fprintf(stderr,"%s: input file '%s' is not well formatted\n",argv[0],argv[1]);
        return 1;
    };

    // applying the hmeans algorithm
    iT = hmeans(n,m,X,k,1000,clust);

    // checking the number of iterations
    fprintf(stderr,"%s: algorithm terminated; %d iterations performed\n",argv[0],iT);
    if (iT >= 1000)
        fprintf(stderr,"%s: maximum number of iterations (1000) reached\n",argv[0]);

    // saving the partition on an output file

    len = strlen(argv[1]);
    outfile = (char*)calloc(len+5,sizeof(char));
    sprintf(outfile,"out.%s",argv[1]);

    output = fopen(outfile,"w");
    if (!output)
    {
        fprintf(stderr,"%s: error while writing output file\n",argv[0]);
        return 1;
    };

    for (i = 0; i < n; i++)  fprintf(output,"%d\n",clust[i]+1);

    fclose(output);
    fprintf(stderr,"%s: file '%s' saved\n",argv[0],outfile);

    free(X);  free(clust);  free(outfile);

    return 0;
};
```

Fig. B.11 The function main.

arguments, and `argv` contains the arguments, stored as strings of characters. In this example, the arguments must be 2. Since the name of the application is always given as first argument in C, they are actually 3. The array `argv` therefore contains: the name of the application, the name of the input text file, and the k value.

Before that the algorithm implemented in the function starts, all the local variables need to be declared. `i` is an integer variable used as counter in the `for` loops. n and m contain the dimensions of the clustering problem: the number of samples in set of data and the length of the vectors representing such samples. The variable `len` is used for storing the length of the specified name of the input text file: it is needed because another file name is defined in the algorithm taking characters from the input file name. `iT` is used for monitoring the number of iterations the *h*-means algorithm performs. k is the number of clusters, and `clust` is the vector containing the partition in clusters of the data. The matrix x contains all the data, organized row by row. `outfile` is a string of characters, where the name of the output file is defined. Note that `clust`, x and `outfile` are declared as pointers. Therefore, memory is supposed to be allocated for them before their use. Finally, `output` is a pointer to FILE.

Good applications should be able to manage exceptions. If the arguments the application is expecting are not provided, or if they are not well-defined, the application should stop and provide an error message. In this function `main`, only some basic exceptions are checked. First of all, the number of arguments is checked. It should be 3. If instead `argc` has a value smaller than 3, an error message is printed on the screen. The function `fprintf` (`stdio.h`) is used for this purpose. Its first parameter specifies where to print the message: `stderr` refers to the standard error stream of the prompt command. Every time an error occurs, the function `main` stops, prints an error message and returns 1. An application returning 1 is an application which stopped its execution because of some error. After the number of arguments, the k value is checked. It is stored as string of characters in `argv[2]`. The function `atoi` (`stdlib.h`) is therefore used for converting the string of characters in an integer value. After that, the obtained integer value is checked: it should be at least 2, otherwise the *h*-means algorithm cannot be carried out. The last thing that is controlled is that the file specified through `argv[1]` can be opened and that the values n and m can be read from it.

At this point, it seems that there are not problems related to bad arguments. However, other possible errors are checked later in the code. The function `main` starts printing on the screen information regarding the clustering problem to solve. Then, the memory for x and `clust` is dynamically allocated, because n and m are now known. Since x is a pointer to pointers, the allocation of the memory requires two steps. Once the memory has been allocated for x, the function `readfile` can be called for transferring the data from the text file `argv[1]` to x. If some error occurs during the process, the function `readfile` returns 1 and the function `main` stops after printing an error message.

The function `hmeans` can be called at this point. The set of data to partition is known and specified through the variables n, m and x. The number k of clusters has been provided from the arguments of the application. The maximum number of iterations the algorithm is allowed to perform is set to 1000. The vector `clust` is used for storing the final partition provided by the function `hmeans`. The returning value of this function provides the number of iterations performed. If it corresponds to 1000, then the number of allowed iterations has been reached. A message is then printed on the screen, since the algorithm did not stop because the centers of the clusters were stable: the found partition may not be optimal.

The partition in clusters of the data is stored in a text file. The name of the text file is defined as follows. The characters "`out.`" are inserted before the ones defining the input text file. For instance, if the input file is named "`apples.txt`", then the output file is named "`out.apples.txt`". This name is stored in the string of characters `outfile`. At the start, `outfile` is declared as a pointer, and so memory needs to be allocated before using it. The length of the string is equal to the length of the string `argv[1]` plus the other characters needed for "`out.`" The function `strlen` (`string.h`) is used for computing the length of `argv[1]`. The function `calloc` (`stdlib.h`) is used for allocating the memory, and `sprintf` is used for defining the string of characters `outfile`.

The file named as specified in `outfile` is then opened for writing. In this case, if the file already exists, it is deleted and re-created empty; if the file does not exist, it is simply created. If some problem occurs when opening the file, an error message is printed and the function `main` stops returning 1. If no errors occurred, the result of the clustering algorithm can be stored in the output file. In the input file, the samples are stored row by row. Therefore, the cluster to which they belong are stored row by row as well in the output file. In C, indices start from 0. In general, they are considered starting from 1, and this is why the values `clust[i] + 1` are printed. When the found partition is saved in the output file, the file can be closed, a message can notify that the output file has been saved, the memory can be freed, and the function `main` can finally stop.

B.4 Generating random data

In this section another application in C is presented. This application generates a random set of data that can be used as input in the application implementing the h-means algorithm. The application presented here can be exploited for testing the h-means algorithm if other data are not available. This application consists of a single main function only. It takes as input arguments the number of samples to generate, the dimension of the vectors representing the samples, the number of clusters in which the samples will be partitioned, and an output file to store the data. The data might be generated in a completely random way, but then the effectiveness of the clustering algorithm could not be validated. Therefore, this application generates data containing intrinsic patterns. This is why the application requires as input argument the number of clusters in which the data will be partitioned by h-means.

The algorithm applied for generating the data is the following one. If the data will be partitioned in k clusters, then k random samples c_i are generated at the start of the algorithm. After that, the distances between each couple of generated samples are computed, and the smallest one (`mindist` in the code) is located. These samples are not inserted into the final set of data, but they rather represent the centers of possible clusters a clustering algorithm might find. Indeed, the data of the set are generated as follows. One of the k firstly generated samples c_i is randomly chosen, and a new sample is generated such that it has a distance from c_i smaller than a certain value. In this implementation, this value is set to half of the value `mindist` previously computed. In this way, all the samples associated randomly to a certain c_i are generated so that they are much closer to c_i than to any other c_j, with $j \neq i$. A clustering algorithm, such as the h-means algorithm, should be able to find these patterns.

The function `main` of the application for the generation of a random set of data is given in Figures B.12 and B.13. These are the local variables used in the function: `i`, `i1` and `j` are used as counters in the `for` loops. The integer variables `n`, `m` and `k` represent, respectively, the number of samples, the components of the vectors representing the samples, and the number of clusters. `ci` contains the information regarding the random c_i chosen during the algorithm. The matrix `c` contains the samples that act like centers. The two `double` variables `dist` and `mindist` are used for storing distances. `aux` is an auxiliary variable used in the computation of the distances. Finally, `output` is the pointer to the output file providing the generated random set of data, and `outputp` is the pointer to the output file providing the corresponding codes of the clusters.

As any standard function `main` in C, it has as parameters the two variables `argc` and `argv`. This application requires as input arguments three integer numbers, `n`, `m` and `k`, and the name of the file in which the data must be stored. The arguments are checked soon in the function. If `argc` is smaller than 5, then not all the required arguments have been specified, and the function stops, after that an error message has been printed on the screen. If the number of arguments is right, then their values are checked one by one. The first three arguments represent integer numbers, but they are passed to the application through strings of characters. The function `atoi` (`stdlib.h`) is therefore used for converting these values. Then, the obtained values are checked, and if they are negative numbers, the function stops giving an error. Even the `k` value is required to be non-negative. If `k` is 1, then the application generates a set of data where there are not intrinsic patterns. Finally, the output file whose name is given in `argv[4]` is opened by the function `fopen` in writing mode. The function `main` checks if some problem occurred when opening the file.

The matrix `c` has been declared as a pointer to pointers to `double` variables. Therefore, before using it, memory needs to be dynamically allocated. `c` is used for storing the first `k` randomly generated samples that are utilized as reference for generating the others. At this point, the algorithm for generating the set of data can start. The procedure described above is then implemented. The samples are stored in the output file as soon as they are generated. In fact, only memory for `c` has been allocated. Simultaneously, in another file all the codes of the clusters are stored. When all the set has been generated, the output file is closed, the memory is freed, and the application stops returning 0.

```c
int main(int argc,char *argv[])
{
    int i,i1,j;
    int n,m,k;
    int len,ci;
    double **c;
    double dist,mindist;
    double aux;
    char *filename;
    FILE *output;
    FILE *outputp;

    // checking the number of arguments
    if (argc < 5)
    {
        fprintf(stderr,"%s: too few arguments\n",argv[0]);
        fprintf(stderr,"%s: usage: %s n m k nomefile.txt\n",argv[0],argv[0]);
        return 1;
    };

    // checking the arguments

    n = atoi(argv[1]);
    if (n <= 0)
    {
        fprintf(stderr,"%s: invalid n value (%d)\n",argv[0],n);
        return 1;
    };

    m = atoi(argv[2]);
    if (m <= 0)
    {
        fprintf(stderr,"%s: invalid m value (%d)\n",argv[0],m);
        return 1;
    };

    k = atoi(argv[3]);
    if (k <= 0)
    {
        fprintf(stderr,"%s: invalid k value (%d)\n",argv[0],k);
        return 1;
    };

    output = fopen(argv[4],"w");
    if (!output)
    {
        fprintf(stderr,"%s: error while opening file '%s'\n",argv[0],argv[4]);
        return 1;
    };

    len = strlen(argv[4]);
    filename = (char*)calloc(len+5,sizeof(char));
    sprintf(filename,"cls.%s",argv[4]);
    outputp = fopen(filename,"w");
    if (!outputp)
    {
        fprintf(stderr,"%s: error while opening file '%s'\n",argv[0],filename);
        return 1;
    };
```

Fig. B.12 The function main of the application for generating random sets of data. Part 1.

```
// memory allocation
c = (double**)calloc(k,sizeof(double*));
for (i = 0; i < k; i++)   c[i] = (double*)calloc(m,sizeof(double));

// generating the random set of data

fprintf(output,"%d %d\n",n,m);

for (i = 0; i < k; i++)
{
    for (j = 0; j < m; j++)   c[i][j] = (double)rand()/(double)RAND_MAX;
};

mindist = 1.e+100;
for (i = 0; i < k; i++)
{
    for (i1 = i + 1; i1 < k; i1++)
    {
        dist = 0.0;
        for (j = 0; j < m; j++)
        {
            aux = c[i][j] - c[i1][j];
            dist = dist + aux*aux;
        };
        dist = sqrt(dist);
        if (dist < mindist)   mindist = dist;
    };
};
mindist = mindist/2.0;

for (i = 0; i < n; i++)
{
    ci = (int)(k*(double)rand()/(double)RAND_MAX);
    fprintf(outputp,"%d\n",ci+1);
    for (j = 0; j < m; j++)
    {
        fprintf(output,"%lf ",c[ci][j]
                + mindist*((double)rand()/(double)RAND_MAX - 0.5) );
    };
    fprintf(output,"\n");
};

fclose(output);

free(c);

return 0;
};
```

Fig. B.13 The function main of the application for generating random sets of data. Part 2.

B.5 Running the applications

All the source codes described above, the ones for the application implementing the h-means algorithm and the one for generating a random set of data, have been compiled by using the MinGW free compiler under Windows XP [166]. In this section, some examples of execution of these applications is provided. For clearness, let us refer to the application implementing h-means with hmeans, and let us refer to the application generating the data with generate. In the compiling phase, the executable file hmeans.exe can be created in correspondence with the application

```
10  4
-10 -12 -15 -12
-11 -12 -13 -11
 11  12  13  11
 15  16  12  18
-15 -16 -12 -12
-11 -11 -11 -11
 11  11  12  12
 18  20  11  11
-10 -15 -13 -14
 10  10  10  10
```

Fig. B.14 An example of input text file for the application hmeans.

hmeans, and the executable file generate.exe can be created in correspondence with the application generate.

Let us start by showing a very simple example in which the input text file required by the application hmeans is not randomly generated. For simplicity, let us suppose that the following set of data is available: vectors in a four-dimensional space having either all the components negative or all the components positive. The h-means algorithm is able to correctly partition these data into two clusters. One cluster contains all the vectors having all its components negative, and the other one all the vectors having only positive components. Let us consider only few samples in this first example, so that the input text file can be edited by hand. Figure B.14 shows an example of an input text file. In the file, 10 samples represented by 10 vectors having 4 components are listed. All the vectors satisfy a property: all the components of a vector have the same sign. Five vectors have all the components negative, and 5 vectors have all the components positive. The execution of the application hmeans on this set of data and with k set to 2 gives as a result the text file shown in Figure B.15. It is easy to note that the application hmeans partitioned the data so that each vector with negative components is in cluster 1, and each vector with positive components is in cluster 2.

Let us consider now a more complex example. The application generate can create random sets of data having predetermined features. Let us generate a set of data containing 100 samples with dimension 6. The application generate can then be run by setting as arguments n to 100, m to 6 and k to 4. The application creates the set of data which is grouped in four different categories. The aim is to present these data

```
1
1
2
2
1
1
2
2
1
2
```

Fig. B.15 The output file provided by the application hmeans when the input is the file in Figure B.14 and $k = 2$.

```
100 6
-0.058106 0.578975 0.227904 0.858204 0.637098 0.317643
0.686582 0.441658 0.088690 0.067326 0.229113 0.511537
0.614237 0.696280 0.414173 0.236579 0.298327 0.383580
-0.017180 0.434939 0.369374 0.667589 0.720939 0.646909
0.390194 0.220094 0.936243 0.332058 0.020550 0.169999
-0.059085 0.365576 0.279437 0.593035 0.346252 0.683430
0.239906 0.938691 0.915766 0.910704 0.284197 0.851623
0.119721 0.548255 0.172871 1.027057 0.703821 0.289474
0.654719 0.627763 0.356959 0.050178 0.024006 0.195087
-0.035129 0.710803 0.201619 1.046905 0.707277 0.404807
0.077712 0.559647 -0.018841 0.905835 0.587346 0.308535
-0.026903 0.362810 1.082282 0.349962 0.083386 -0.204153
0.236353 -0.002650 1.171893 0.602048 0.158964 -0.145352
0.155426 0.551563 0.025883 0.810647 0.697783 0.433985
0.383705 0.984542 0.947199 0.854469 0.162100 0.675709
0.512966 0.669992 0.831110 0.825825 0.138247 0.666794
0.352131 0.189981 0.913666 0.431949 0.058317 0.173322
0.328953 0.785087 1.057592 0.648323 0.290368 0.891645
0.128266 0.728574 0.143559 0.809178 0.774652 0.250194
0.182595 -0.052581 1.003752 0.296991 0.285823 0.066282
0.563639 0.679844 0.120909 -0.174614 0.209576 0.274077
0.043335 0.086338 0.813672 0.558985 0.281728 0.105769
0.830663 0.393344 0.131189 -0.227259 -0.121974 0.513302
0.006636 -0.020941 1.014388 0.209590 -0.068646 -0.017440
0.237792 0.187015 0.781408 0.414934 -0.025791 0.100043
0.319519 0.765907 0.860688 0.762440 0.236655 0.693776
0.935567 0.608686 0.510474 -0.135719 0.011664 0.208111
0.126627 0.407964 0.792889 0.506533 -0.078065 -0.025599
-0.072175 0.357707 0.886461 0.313145 0.249865 -0.038727
0.063501 0.614206 0.169785 0.792386 0.632559 0.545357
0.307125 0.226547 1.095900 0.477652 0.058391 -0.148794
0.737299 0.710431 0.179013 -0.136209 0.142305 0.460938
0.638224 0.510843 0.100038 0.132001 0.145865 0.422964
0.858520 0.342983 0.340953 0.193724 0.291444 0.420205
0.488477 0.595084 0.535617 -0.074930 0.125039 0.270072
-0.189014 0.742963 0.364019 0.927492 0.417261 0.395743
-0.204545 0.632110 0.348844 0.830657 0.559859 0.435631
0.333552 0.896548 0.654009 0.660591 0.289789 0.768480
0.370450 0.347368 1.097902 0.348509 0.334893 -0.176518
0.139762 0.575534 0.246595 1.030439 0.377106 0.662455
0.624043 0.321445 0.306450 -0.117667 -0.010528 0.568587
0.789632 0.326177 0.300368 -0.043765 0.089853 0.507042
0.653325 0.604993 0.319622 0.066643 0.040456 0.124285
0.516379 0.573352 0.480554 0.010853 0.211623 0.362946
0.488995 0.921632 0.753381 0.968111 0.237708 0.701742
0.171240 1.014210 1.042403 0.801037 -0.055393 0.776238
0.068916 0.384534 0.362002 0.985819 0.632099 0.587604
-0.088582 -0.015660 1.089580 0.502632 -0.018196 -0.206304
0.061573 0.486294 0.105525 0.744369 0.663265 0.626854
0.160211 0.219901 1.133459 0.640408 -0.053560 0.067943
0.230575 0.719299 0.995483 0.535735 0.139019 0.894596
0.373647 0.762569 0.731145 0.557348 0.323833 0.681390
0.490330 1.088068 0.963620 0.865105 0.230113 0.791442
0.015344 0.342621 0.857016 0.323959 0.139695 0.240327
...
```

Fig. B.16 An output file containing a set of data generated by the application `generate`.

to the application `hmeans` and check if it is able to recognize these inherent patterns in the data. If the three arguments n, m and k are specified, and the name of a file is given to the application `generate`, two output files are generated, one containing the set of data, and another one containing their intrinsic classification. Figure B.16 shows the first samples of the set of data. The list of cluster codes (contained in the second file `generate` creates), is given in Figure B.17 together with the partition the application `hmeans` finds.

As the spreadsheet shows, the h-means algorithm has been able to recognize all the inherent patterns that the application `generate` inserted into the data. On the

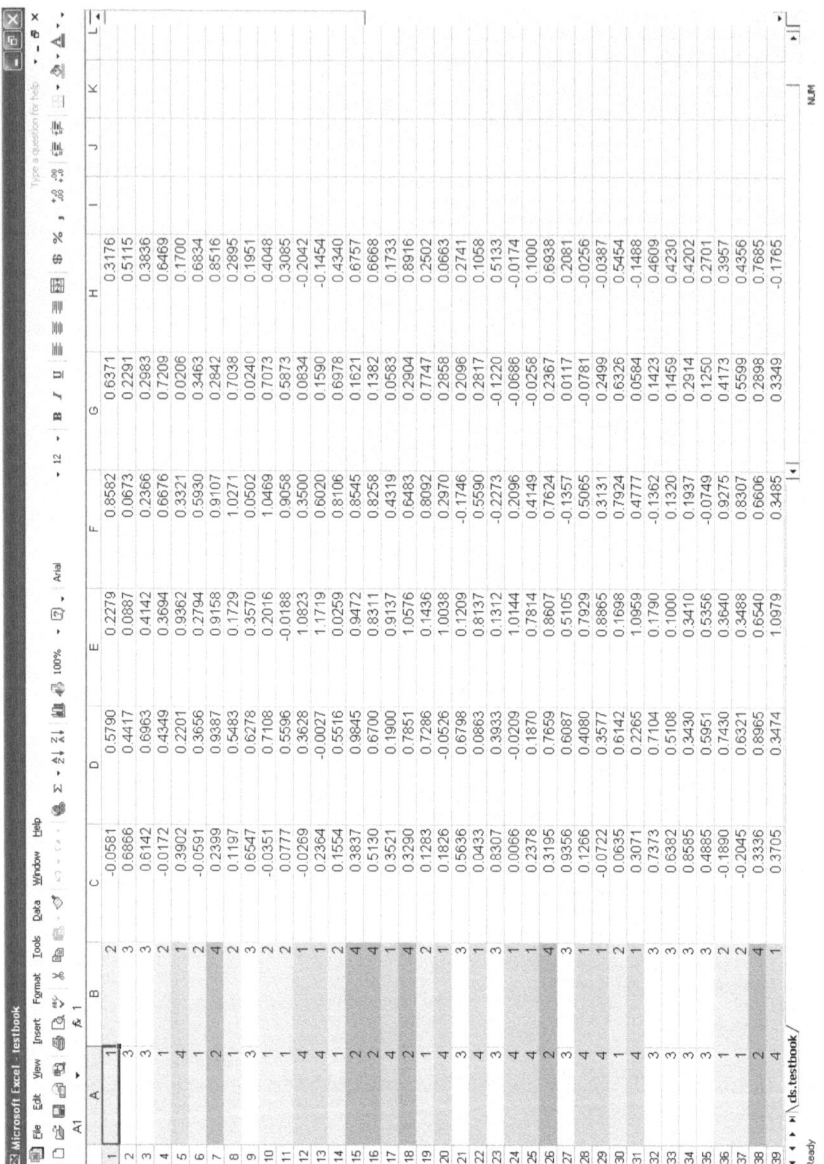

Fig. B.17 The partition provided by the application `generate` (column A), the partition found by hmeans (column B) and the components of the samples (following columns) in an Excel spreadsheet.

first column of the spreadsheet, the partition randomly generated by the application `generate` is shown. On the second column, instead, the partition found by the application `hmeans` is shown. The figure shows 39 samples on 100, but the same result is obtained on the remaining samples. The codes of the clusters in these two columns are often different, and only the code 3 can be found on the same row on both columns. This does not mean that only the cluster coded by 3 has been located correctly. Indeed, even all the other clusters have been identified, but the application hmeans assigned to them a different code. Colors or gray scales show the correspondence between a code in a column and another code in the other column. For instance, the cluster coded by 1 by generate corresponds to the cluster coded by 2 by hmeans. This result shows a very good performance of the h-means algorithm.

The reader can deeply analyze the converge properties of the algorithm by creating other sets of data (having different features) and checking the ability of the application hmeans in partitioning the data. Moreover, real data can also be used, if they are stored in a text file formatted as described in Section B.2.

References

1. C.S. Adjiman, S. Dallwig, C.A. Floudas, and A. Neumaier, *A Global Optimization Method, αBB, for General Twice-Differentiable Constrained NLPs - I. Theoretical Advances*, Computers and Chemical Engineering **22**, 1137–1158, 1998.
2. C.S. Adjiman, I.P. Androulakis, and C.A. Floudas, *A Global Optimization Method, αBB, for General Twice-Differentiable Constrained NLPs - II. Implementation and Computational Results*, Computers and Chemical Engineering **22**, 1159–1179, 1998.
3. J.-M. Aerts, P. Jans, D. Halloy, P. Gustin, D. Berckmans, *Labeling of Cough Data from Pigs for On-Line Disease Monitoring by Sound Analysis*, American Society of Agricultural and Biological Engineers **48** (1), 351–354, 2004.
4. A. Andoni and P. Indyk, *Near-Optimal Hashing Algorithms for Approximate Nearest Neighbor in High Dimensions*, Communications of the ACM **51** (1), 117–122, 2008.
5. I.P. Androulakis and C.A. Floudas, *Distributed Branch and Bound Algorithms for Global Optimization*, In: Parallel Processing of Discrete Problems, P.M. Pardalos (Ed.), volume 106 of IMA Volumes in Mathematics and Its Applications, Springer, 1–36, 1998.
6. F. Angiulli and G. Folino, *Efficient Distributed Data Condensation for Nearest Neighbor Classification*, A.-M. Kermarrec, L. Bouge, and T. Priol (Eds.), Lecture Notes on Computer Science **4641**, Springer, New York, 338–347, 2007.
7. F. Angiulli and G. Folino, *Distributed Nearest Neighbor Based Condensation of Very Large Datasets*, IEEE Transactions on Knowledge and Data Engineering **19** (12), 1593–1606, 2007.
8. H. Apaydin, F.K. Sonmez, Y.E. Yildirim, *Spatial Interpolation Techniques for Climate Data in the GAP Region in Turkey*, Climate Research **28**, 31–40, 2004.
9. C. Arima, T. Hanai, M. Okamoto, *Gene Expression Analysis using Fuzzy K-Means Clustering*, Genome Informatics **14**, 334–335, 2003.
10. A. Arulselvan, G. Baourakis, V. Boginski, E. Korchina, P.M. Pardalos, *Analysis of Food Industry Market using Network Approaches*, British Food Journal **110**(9), 916–928, 2008.
11. AveSound Project Web site,
    http://www.acoustics.hut.fi/research/avesound/avesound.html
12. M. Aznar, R. Lopez, J. Cacho, and V. Ferreira, *Prediction of Aged Red Wine Aroma Properties from Aroma Chemical Composition. Partial Least Squares Regression Models*, Journal of Agriculture and Food Chemistry **51**, 2700–2707, 2003.
13. G.A. Baigorria, J.W. Jones, J.J. O'Brien, *Potential Predictability of Crop Yield using an Ensemble Climate Forecast by a Regional Circulation Model*, Agricultural and Forest Meteorology **148**, 1353–1361, 2008.
14. L. Baoli, Y. Shiwen, and L. Qin, *An Improved k-Nearest Neighbor Algorithm for Text Categorization*, ArXiv Computer Science e-prints, 2003.
15. M.E., Bauer, T.E. Burk, A.R. Ek, P.R. Coppin, S.D. Lime, T.A. Walsh, D.K. Walters, W. Befort, and D.F. Heinzen, *Satellite Inventory of Minnesota's Forest Resources*, Photogrammetric Engineering and Remote Sensing **60** (3), 287–298, 1994.

16. R. Benetis, C.S. Jensen, G. Karciauskas, S. Saltenis, *Nearest and Reverse Nearest Neighbor Queries for Moving Objects*, The International Journal on Very Large Data Bases **15** (3), 229–250, 2006.

17. P. Berkhin, *Survey Of Clustering Data Mining Techniques*, Tech. Report, Accrue Software, San Jose, CA, 2002.

18. H.M. Berman, J. Westbrook, Z. Feng, G. Gilliland, T.N. Bhat, H. Weissig, I.N. Shindyalov, P.E. Bourne, *The Protein Data Bank*, Nucleic Acids Research **28**, 235–242, 2000.

19. J.L. Bentley, *Multidimensional Binary Search Trees Used for Associative Searching*, Communications of the ACM **18** (9), 509–517, 1975.

20. M. Bewernitz, G. Ghacibeh, O. Seref, P.M. Pardalos, C.-C. Liu and B. Uthman, *Quantification of the Impact of Vagus Nerve Stimulation Parameters on Electroencephalographic Measures*, AIP Conference Proceedings **953**, Data Mining, System Analysis and Optimization in Biomedicine, 206–218, 2007.

21. J. Bezdek, *Pattern Recognition with Fuzzy Objective Function Algorithms*, Plenum, New York, 1981.

22. B. Bhattacharya, R.K. Price, D.P. Solomatine, *A Machine Learning Approach to Modeling Sediment Transport*, ASCE Journal of Hydraulic Engineering **133**(4), 440–450, 2007.

23. L. Boruvka, O. Vacek, J. Jehlicka, *Principal Component Analysis as Tool to Indicate the Origin of Potentially Toxic Elements in Soils*, Geoderma **128**, 289–300, 2005.

24. O. Bousquet and B. Scholkopf, *Comment*, Statistical Science **21** (3), 337–340, 2006.

25. V. Bovinski, S. Butenko, P.M. Pardalos, *Mining Market Data: a Network Approach*, Computers and Operation Research **33**, 3171–3184, 2006.

26. S. Bradley, M. Fayyad, *Refining Initial Points for k-means Clustering*, In: J. Shavlik (Ed.), Proceedings of the 15th International Conference on Machine Learning (ICML98). Morgan Kaufmann, San Francisco, 91–99, 1998.

27. S.E. Brossette, and P.A. Hymel Jr., *Data Mining and Infection Control*, Clinics in Laboratory Medicine **28** (8), 119–126, 2008.

28. R.L. Brown, *Accelerated Template Matching using Template Trees Grown by Condensation*, IEEE Transactions on Systems, Man and Cybernet **25** (3), 523–528, 1995.

29. K. Brudzewski, S. Osowski, T. Markiewicz, *Classification of Milk by Means of an Electronic Nose and SVM Neural Network*, Sensors and Actuators **B 98**, 291–298, 2004.

30. C.J.C. Burges, *A Tutorial on Support Vector Machines for Pattern Recognition*, Data Mining and Knowledge Discovery 2 (2), 955–974, 1998.

31. S. Busygin, N. Boyko, P.M. Pardalos, M. Bewernitz and G. Ghacibeh, *Biclustering EEG Data from Epileptic Patients Treated with Vagus Nerve Stimulation*, AIP Conference Proceedings **953**, Data Mining, System Analysis and Optimization in Biomedicine, 162–173, 2007.

32. S. Busygin, O.A. Prokopyev, P.M. Pardalos, *Feature Selection for Consistent Biclustering via Fractional 0–1 Programming*, Journal of Combinatorial Optimization **10**, 7–21, 2005.

33. S. Cafieri, M. D'Apuzzo, M. Marino, A. Mucherino, G. Toraldo, *Interior Point Solver for Large-Scale Quadratic Programming Problems with Bound Constraints*, Journal of Optimization Theory and Applications **129** (1), 55–75, 2006.

34. C.Z. Cai, W.L. Wang, L.Z. Sun, Y.Z. Chen, *Protein Function Classification via Support Vector Machine Approach*, Mathematical Biosciences **185**, 111–122, 2003.

35. W.M. Campbell, J.P. Campbell, D.A. Reynolds, E. Singer, P.A. Torres-Carrasquillo, *Support Vector Machines for Speaker and Language Recognition*, Computer Speech and Language **20**, 210–229, 2006.

36. G. Camps-Valls, L. Gomez-Chova, J. Calpe-Maravilla, E. Soria-Olivas, J.D. Martin-Guerrero, J. Moreno, *Support Vector Machines for Crop Classification using Hyperspectral Data*, Lectures Notes on Computer Science **2652**, Springer, New York, 134–141, 2003.

37. G. Castellano, A.M. Fanelli, and M. Pelillo, *An Iterative Pruning Algorithm for Feedforward Neural Networks*, IEEE Transactions on Neural Networks **8** (3), 1997.

38. G. Ceci, A. Mucherino, M. D'Apuzzo, D. di Serafino, S. Costantini, A. Facchiano, G. Colonna, *Computational Methods for Protein Fold Prediction: an Ab-Initio Topological Approach*, In: Data Mining in Biomedicine, Springer Optimization and Its Applications **7**, P.M. Pardalos et al (Eds.), Springer, 2007.

39. S. Cocke and T.E. LaRow, *Seasonal Predictions using a Regional Spectral Model Embedded with a Coupled Ocean - Atmosphere Model*, Monthly Weather Review **128**, 689–708, 2000.
40. A.R. Conn, N.I.M. Gould, *Trust-Region Methods*, SIAM Mathematical Optimization, 2000.
41. C. Cortes and V. Vapnik, *Support Vector Networks*, Machine Learning **20**, 273–297, 1995.
42. T.M. Cover and P.E. Hart, *Nearest Neighbor Pattern Classification*, IEEE Transactions on Information Theory **IT-13** (1), 1967.
43. C.-C. Chang and C.-J. Lin, *LIBSVM: a Library for Support Vector Machines*, manual available at http://www.csie.ntu.edu.tw/~cjlin/libsvm/, 2001.
44. C.L. Chang, *Finding Prototypes for Nearest Neighbor Classifiers*, IEEE Transactions on Computers **23** (11), 1179–1184, 1974.
45. A. Chedad, D. Moshou, J.M. Aerts, A. Van Hirtum, H. Ramon, D. Berckmans, *Recognition System for Pig Cough based on Probabilistic Neural Networks*, Journal of Agricultural Enginnering Research **79** (4), 449–457, 2001.
46. M.-S. Chen, J. Han, P.S. Yu, *Data Mining: an Overview from a Database Perspective*, IEEE Transactions on Knowledge And Data Engineering **8**, 866–883, 1996.
47. J. Cheng, M.J. Sweredoski, P. Baldi, *Accurate Prediction of Protein Disordered Regions by Mining Protein Structure Data*, Data Mining and Knowledge Discovery **11**, 213–222, 2005.
48. R. Chinchuluun, W.S. Lee, J. Bhorania, P.M. Pardalos, *Clustering and Classification Algorithms in Food and Agricultural Applications: A Survey*, in Advances in Modeling Agricultural Systems, Springer Optimization and Its Applications Series, P.J. Papajorgji, P.M. Pardalos (Eds.), Springer, 433–454, 2008.
49. M.L. Chiusano, T. Gojobori, G. Toraldo, *A C++ Computational Environment for Biomolecular Sequence Management*, Computational Management Science **2**, 165–180, 2005.
50. K.L. Chung, and K.S. Lin, *An Efficient Line Symmetry-Based K-means algorithm*, Pattern Recognition **27** (7), 765–772, 2006.
51. K.L. Chung, J.S. Lin, *Faster and More Robust Point Symmetry-Based k-means Algorithm*, Pattern Recognition **40** (2), 410–422, 2007.
52. K.L. Clarkson, *Nearest-Neighbor Searching and Metric Space Dimensions*, In: Nearest-Neighbor Methods for Learning and Vision: Theory and Practice, MIT Press, Cambridge, MA, 2005.
53. K.C. Das, M.D. Evans, *Detecting Fertility of Hatching Eggs using Machine Vision II: Neural Network classifiers*, Transactions of the American Society of Agricultural Engineers **35** (6), 2035–2041, 1992.
54. S.B. Davis and P. Mermelstein, *Comparison of Parametric Representations for Monosyllabic Word Recognition in Continuously Spoken Sentences*, IEEE Transactions on Acoustic, Speech, and Signal Processing **28** (4), 357–366, 1980.
55. P.A.D. de Castro, F.O. de Franca, H.M. Ferreira and F.J. Von Zuben, *Applying Biclustering to Perform Collaborative Filtering*, Seventh International Conference on Intelligent Systems Design and Applications, 421–426, 2007.
56. D. Delen, G. Walker, A. Kadam, *Predicting Breast Cancer Survivability: a Comparison of three Data Mining Methods*, Artificial Intelligence in Medicine **34**, 113—127, 2005.
57. W. De Neve, P. Lambert, S. Lerouge, and R.V. de Walle, *Assessment of the Compression Efficiency of the MPEG-4 AVC Specification*, Proceedings of SPIE **5308**, Visual Communications and Image Processing, 1082–1093, 2004.
58. G. Destefanis, M.T. Barge, A. Brugiapaglia, S. Tassone, *The Use of Principal Component Analysis (PCA) to Characterize Beef*, Meat Science **56**, 255–259, 2000.
59. V.S. Devi, M.N. Murty, *An Incremental Prototype Set Building Technique*, Pattern Recognition **35**, 505–513, 2002.
60. I.S. Dhillon, S. Mallela, and D.S. Modha, *Information-Theoretical Coclustering*, Proceedings of The Ninth ACM SIGKDD International Conference on Knowledge Discovery and Data Mining, 89–98, 2003.
61. I.S. Dhillon and D.M. Modha, *Concept Decompositions for Large Sparse Text Data using Clustering*, Machine Learning **42** (1), 143–175, 2001.

62. Q. Ding and N. Zhang, *Classification of Recorded Musical Instruments Sounds Based on Neural Networks*, Proceedings of the 2007 IEEE Symposium on Computational Intelligence in Image and Signal Processing (CIISP 2007), 157–162, 2007.

63. J.-x. Dong, A. Krzyzak, C.Y. Suen, *An Improved Handwritten Chinese Character Recognition System using Support Vector Machine*, Pattern Recognition Letters **26**, 1849–1856, 2005.

64. M. Dorigo, G. Di Caro, *Ant Colony Optimization: A New Meta-Heuristic*, In: New Ideas in Optimization, D. Corne, M. Dorigo and F. Glover (Eds.), McGraw-Hill, London, UK, 11-32, 1999.

65. C.-J. Du, D.-W. Sun, *Pizza Sauce Spread Classification using Colour Vision and Support Vector Machines*, Journal of Food Engineering **66**, 137–145, 2005.

66. D. Duffy and A. Quiroz, *A Permutation Based Algorithm for Block Clustering*, Journal of Classification **8**, 65–91, 1991.

67. D.A. Elizondo, R.W. McClendon, G. Hoogenboom, *Neural Network Models for Predicting Flowering and Physiological Maturity of Soybean*, Transactions of the American Society of Agricultural Engineers **37** (3), 981–988, 1994.

68. C. Elkan, *Using the Triangle Inequality to Accelerate k-means*, Proceedings of the Twentieth International Conference on Machine Learning (ICML-2003), Washington, DC, 2003.

69. P.A. Estevez, C.M. Held, C.A. Perez, *Subscription Fraud Prevention in Telecommunications using Fuzzy Rules and Neural Networks*, Expert Systems with Applications **31**, 337–344, 2006.

70. I. Etikan, M.Z. Caglar, *Prediction Methods for Babies' Birth Weight using Linear and Nonlinear Regression Analysis*, Technology and Health Care **13**(2), 131–135, 2005.

71. S. Fagerlund, *Bird Species Recognition Using Support Vector Machines*, EURASIP Journal on Advances in Signal Processing **2007**, Article ID 38637, 1–8, 2007.

72. U. Fayyad, G. Piatetsky-Shapiro, and P. Smyth, *From Data Mining to Knowledge Discovery in Databases*, Artificial Intelligence Magazine **17**, 37–54, 1996.

73. I. Ferreras, A. Pasquali, R.R. de Carvalho, I.G. de la Rosa, and O. Lahav, *A Principal Component Analysis Approach to the Star Formation History of Elliptical Galaxies in Compact Groups*, Journal of Monthly Notices of the Royal Astronomical Society **370**, 828–836, 2006.

74. A. Flexer, *Connectionists and Statisticians, Friends or Foes?*, In: From Natural to Artificial Neural Computation, Proceedings International Workshop Artificial Neural Networks, J. Mira and F. Sandoval (Eds.), Lecture Notes in Computer Science **930**, Springer, New York, 454–461, 1995.

75. A. Flexer, *Statistical Evaluation od Neural Network Experiments: Minimum Requirements and Current Practice*, Proceedings of the 13th European Meeting on Cybernetics and systems research, EMCSR 96, Vienna, 1996.

76. R. Fletcher, *Practical Methods of Optimization*, Wiley, New York, Second Edition, 1987.

77. M. Flynn, *Some Computer Organizations and Their Effectiveness*, IEEE Trans. Comput. **C-21**, 948–960, 1972.

78. N. Fnaiech, S. Abid, F. Fnaiech and M. Cheriet, *A Modified Version of a Formal Pruning Algorithm Based on Local Relative Variance Analysis*, First International Symposium on Control, Communications and Signal Processing, 2004.

79. I. Foster, C. Kesselman, *The Grid: Blueprint for a New Computing Infrastructure*, Morgan Kaufmann, 2nd edition, 2004.

80. W. Fx, W.J. Zhang, and A.J. Kusalik, *A Genetic K-means Clustering Algorithm Applied to Gene Expression Data*, Advances in Artificial Intelligence, Lecture Notes in Artificial Intelligence **2671**, Springer, New York, 520–526, 2003.

81. G.W. Gates, *The Reduced Nearest Neighbor Rule*, IEEE Transactions in Information Theory **18**, 431–433, 1972.

82. Z. W. Geem, J. H. Kim, G. V. Loganathan, *A New Heuristic Optimization Algorithm: Harmony Search*, Simulations **76** (2), 60–68, 2001.

83. G. Getz, E. Levine, and E. Domany, *Coupled Two-Way Clustering Analysis of Gene Microarray Data*, Proceedings of the National Academy of Sciences of the United States of America, 12079–12084, 2000.

84. R. Gil-Garcia, J.M. Badia-Contelles, and A. Pons-Porrata, *Parallel Nearest Neighbour Algorithms for Text Categorization*, Lecture Notes in Computer Science **4641**, Springer, New York, 328–337, 2007.
85. R. Gil-Garcia and A. Pons-Porrata, *A New Nearest Neighbor Rule for Text Categorization*, Lecture Notes in Computer Science **4225**, Springer, New York, 814–823, 2006.
86. M.K. Gill, T. Asefa, M.W. Kemblowski, and M. McKee, *Soil Moisture Prediction using Support Vector Machines*, Journal of the American Water Resources Association **42** (4), 1033–1046, 2006.
87. L. Goddard, S.J. Mason, S.E. Zebiak, C.F. Ropelewski, R. Basher and M.A. Cane, *Current Approaches to Seasonal-to-Interannual Climate Predictions*, International Journal of Climatology **21**: 1111-1152, 2001.
88. D. E. Goldberg, *Genetic Algorithms in Search, Optimization & Machine Learning*, Addison-Wesley Longman Publishing Co., Inc., 1989.
89. T.R. Golub, D.K. Slonim, P. Tamayo, C. Huard, M. Gaasenbeek, J.P. Mesirov, H. Coller, M.L. Loh, J.R. Downing, M.A. Caligiuri, C.D. Bloomfield, and E.S. Lander, *Molecular Classification of Cancer: Class Discovery and Class Prediction by Gene Expression Monitoring*, Science **286**, 531–537, 1999.
90. K.C. Gowda and G. Krishna, *The Condensed Nearest Neighbor Rule using the Concept of Mutual Nearest Neighbothood*, IEEE Transactions on Information Theory **IT-25** (4), 488–490, 1979.
91. H.P. Graf, E. Cosatto, L. Bottou, I. Dourdanovic, and V. Vapnik, *Parallel Support Vector Machines: The Cascade SVM*, In: Advances in Neural Information Processing Systems, Lawrence Saul, Bernhard Scholkopf, and Leon Bottou (Eds.), volume 17, MIT Press, 2005.
92. R.M. Gray, *Vector Quantization*, IEEE ASSP Magazine, 4–28, 1984.
93. T. H. Grubesic, *On The Application of Fuzzy Clustering for Crime Hot Spot Detection*, Journal of Quantitative Criminology **22** (1), 2006.
94. Y. Guan, A. A. Ghorbani, and N. Belacel. *Y-means: A Clustering Method for Intrusion Detection*, IEEE Canadian Conference on Electrical and Computer Engineering Proceedings, 1083–1086, 2003.
95. G. Guo, H. Wang, D. Bell, Y. Bi and K. Greer, *Using kNN Model for Automatic Text Categorization*, Soft Computing - A Fusion of Foundations, Methodologies and Applications **10** (5), 423–430, 2006.
96. I. Guyon, J. Weston, S. Barnhill, V. Vapnik, *Gene Selection for Cancer Classification using Support Vector Machines*, Machine Learning **46**, 389–422, 2002.
97. R. Haapanen, A.R. Ek, M.E. Bauer, A.O. Finley, *Delineation of Forest/Nonforest Land Use Classes using Nearest Neighbor Methods*, Remote Sensing of Environment **89**, 265–271, 2004.
98. M. Halkidi, Y. Batistakis, M. Vazirgiannis, *On Clustering Validation Techniques*, Journal of Intelligent Information Systems, **17** (2/3), 107–145, 2001.
99. D. Hammerstrom, *Neural Networks at Work*, IEEE Spectrum, 26–32, 1993.
100. C.G. Han, P.M. Pardalos and Y. Ye, *Computational Aspects of an Interior Point Algorithm for Quadratic Programming problems with Box Constraints*, Large-Scale Numerical Optimization, T. Coleman and Y. Li (Eds.), SIAM, 1990.
101. P. Hansen and N. Mladenovic. *J-means: a New Local Search Heuristic for Minimum Sum-of-Squares Clustering*, Pattern Recognition, **34** (2): 405–413, 2002.
102. P.E. Hart, *The Condensed Nearest Neighbor Rule*, IEEE Transactions on Information Theory **IT-14**, 515–516, 1968.
103. J. Hartigan, *Clustering Algorithms*, John Wiles & Sons, New York, 1975.
104. R. Hathaway, J. Bezdek, Y. Hu, *Generalized Fuzzy c-means Clustering Strategies using Lp Norm Distances*, IEEE Transactions on Fuzzy Systems **8** (5), 576–582, 2000.
105. R. Hathaway, J. Bezdek, *Fuzzy c-Means Clustering of Incomplete Data*, IEEE Transactions on Systems, Man and Cybernetics - Part B, Cybernetics **31** (5), 735–744, 2001.
106. J.C. Hemphill III, C.W. Barton, D. Morabito and G.T. Manley, *Influence of Data Resolution and Interpolation Method on Assessment of Secondary Brain Insults in Neurocritical Care*, Physiological Measurement **26**, 373–386, 2005.

107. K. Hiroaki, *Three-Dimensional Protein Structural Data Mining Based on the Glycine Filter Reduced Representation*, Journal of Computer Chemistry **4** (2), 33–42, 2005.
108. P. Holmgren, and T. Thuresson, *Satellite Remote Sensing for Forestry Planning: a Review*, Scandinavian Journal of Forest Research **13** (1), 90–110, 1998.
109. P.W. Holland, R.E. Welsch, *Robust Regression using Iteratively Reweighted Least-Squares*, Communications in Statistics, Theory and Methods **6** (9), 813–827, 1977.
110. R. Howard, *Classifying a Population into Homogeneous Groups*, In: J.R. Lawerence (Ed.), Operational Research in the Social Sciences, Tavistock Publ., London. 1966.
111. L.-L. Hsiao, F. Dangond, T. Yoshida, R. Hong, R.V. Jensen, J. Misra, W. Dillon, K.F. Lee, KE. Clark, P. Haverty, Z.Weng, G. Mutter, M.P. Frosch, M.E. MacDonald, E.L. Milford, C.P. Crum, R. Bueno, R.E. Pratt, M. Mahadevappa, J.A. Warrington, G. Stephanopoulos, G. Stephanopoulos, and S.R. Gullans, *A Compendium of Gene Expression in Normal Human Tissues*, Physiological Genomics **7**, 97–104, 2001.
112. HuGE Index.org Web site: http://www.hugeindex.org
113. L.C.K. Hui, K.-Y. Lam and C.W. Chea, *Global Optimisation in Neural Network Training*, Neural Computing & Applications **5**, 58–64, 1997.
114. ILOG Inc. CPLEX 9.0 User's Manual, 2004.
115. L.S. Itzhaki and P.G. Wolynes, *The Quest to Understand Protein Folding*, Current Opinion in Structural Biology **18**, 1–3, 2008.
116. A.K. Jain, M.N. Murty, P.J. Flynn, *Data Clustering: a Review*, ACM Computing Surveys **31**(3), 264–323, 1999.
117. S.S. Jagtap, J.W. Jones, T. LaRow, A. Ajayan, and J.J. O'Brien, *Statistical Recalibration of Precipitation Outputs from Coupled Climate Models*, submitted to Journal of Applied Meteorology.
118. S.S. Jagtap, U. Lall, J.W. Jones, A.J. Gijsman, J.T. Ritchie, *Dynamic Nearest-Neighbor Method for Estimating Soil Water Parameters*, Transactions of the American Society of Agricultural Engineers **47** (5), 1437–1444, 2004.
119. T. Jinlan, Z. Lin, Z. Suqin, L. Lu, *Improvement and Parallelism of k-Means Clustering Algorithm*, Tsinghua Science and Technology **10** (3), 277–281, 2005.
120. I.T. Jolliffe, *Discarding Variables in a Principal Component Analysis. I: Artificial Data*, Applied Statistics **21** (2), 160–173, 1972.
121. J.W. Jones, J.W. Hansen, F.S. Royce, C.D. Messina, *Potential Benefits of Climate Forecasting to Agriculture*, Agriculture, Ecosystems and Environment **82**, 169–184, 2000.
122. J.W. Jones, G.Y. Tsuji, G. Hoogenboom, L.A. Hunt, P.K. Thornton, P.W. Wilkens, D.T. Imamura, W.T. Bowen, and U. Singh. *Decision Support System for Agrotechnology Transfer: DSSAT v3*, In: Understanding Options For Agricultural Production, 157–177, G. Y. Tsuji, G. Hoogenboom, and P. K. Thornton (Eds.), Dordrecht, The Netherlands: Kluwer Academic Publishers, 1998.
123. H. Jorquera, R. Perez, A. Cipriano, and G. Acuna, *Short Term Forecasting of Air Pollution Episodes*, In: Environmental Modeling **4**, P. Zannetti (Ed.), WIT Press, UK, 2001.
124. Y. Karimi, S.O. Prasher, R.M. Patel, S.H. Kim, *Application of Support Vector Machine Technology for Weed and Nitrogen Stress Detection in Corn*, Computers and Electronics in Agriculture **51**, 99–109, 2006.
125. O. Karkacier, Z.G. Goktolga, A. Cicek, *A Regression Analysis of the Effect of Energy Use in Agriculture*, Energy Police **34**, 3796–3800, 2006.
126. J. Kennedy, R. Eberhart, *Particle Swarm Optimization*, Proceedings IEEE International Conference on Neural Networks **4**, Perth, WA, Australia, 1942–1948, 1995.
127. Kernel-Machines Web site: http://www.kernel-machines.org/
128. S. Kirkpatrick, C. D. Jr. Gelatt and M. P. Vecchi, *Optimization by Simulated Annealing*, Science **220** (4598), 671–680, 1983.
129. D. Kim, H. Kim, and D. Chung, *A Modified Genetic Algorithm for Fast Training Neural Networks*, Lecture Notes in Computer Science **3496**, Springer, New York, 660–665, 2005.
130. J.L. Klepeis and C.A. Floudas, *ASTRO-FOLD: Ab Initio Secondary and Tertiary Structure Prediction in Protein Folding*, European Symposium on Computer Aided Process Engineering **12**, Elsevier, 2002.

131. J.L. Klepeis and C.A. Floudas, *ASTRO-FOLD: a Combinatorial and Global Optimization Framework for Ab Initio Prediction of Three-Dimensional Structures of Proteins from the Amino Acid Sequence*, Biophysical Journal **85**, 2119–2146, 2003.

132. K.A. Klise and S.A. McKenna, *Water Quality Change Detection: Multivariate Algorithms*, Proceedings of SPIE **6203**, Optics and Photonics in Global Homeland Security II, T.T. Saito, D. Lehrfeld (Eds.), 2006.

133. K. Krishna, K.R. Ramakrishnan, M.A.L. Thathachar, *Vector Quantization using Genetic K-Means Algorithm for Image Compression*, International Conference on Information, Communications and Signal Processing, ICICS '97, Singapore, 1997.

134. K. Krishna, M. Murty, *Genetic k-means Algorithm*, IEEE Transactions on Systems, Man and Cybernetics - Part B, Cybernetics, **29** (3), 433–439, 1999.

135. N. Kondo, U. Ahmad, M. Monta, H. Murase, *Machine Vision based Quality Evaluation of Iyokan Orange Fruit using Neural Networks*, Computers and Electronics in Agriculture **29**, 135–147, 2000.

136. S.R. Kulkarni, G. Lugosi, and S.S. Venkatesh, *Learning Pattern Classification - A Survey*, IEEE Transactions on Information Theory **44** (6), 1998.

137. S.-Y. Lai, W.-J. Chang, and P.-S. Lin, *Logistic Regression Model for Evaluating Soil Liquefaction Probability Using CPT Data*, Journal of Geotechnical and Geoenvironmental Engineering **132**(6), 694–704, 2006.

138. C. Lavor, L. Liberti, and N. Maculan, *Computational Experience with the Molecular Distance Geometry Problem*, In: Global Optimization: Scientific and Engineering Case Studies, J. Pintér (Ed.), 213–225. Springer, Berlin, 2006.

139. C. Lavor, L. Liberti, and N. Maculan, *Molecular distance geometry problem*, In: Encyclopedia of Optimization, C. Floudas and P. Pardalos (Eds.), 2nd edition, Springer, New York, 2305–2311, 2009.

140. C. Lavor, L. Liberti, A. Mucherino and N. Maculan, *Recent Results on the Discretizable Molecular Distance Geometry Problem*, Proceedings of the conference ROADEF09, Nancy, France, Febraury 10/12 2009.

141. C. Lavor, L. Liberti, A. Mucherino and N. Maculan, *On a Discretizable Subclass of Instances of the Molecular Distance Geometry Problem*, Proceedings of the Conference SAC09, Honolulu, Hawaii, March 8/12, 2009.

142. L. Lazzeroni and A. Owen, *Plaid Models for Gene Expression Data*, technical report, Stanford Univ., 2000.

143. J.R. Leathwick, D. Rowe, J. Richardson, J. Elith and T. Hastie, *Using Multivariate Adaptive Regression Splines to Predict the Distributions of New Zealand's Freshwater Diadromous Fish*, Freshwater Biology **50**(12), 2034–2052, 2005.

144. V. Leemans, H. Magein, and M.-F. Destain, *Defect Segmentation on 'Jonagold' Apples using Colour Vision and Bayesian Method*, Computers and Electronics in Agriculture **23**, 43–53, 1999.

145. V. Leemans, H. Magein, and M.-F. Destain, *On-line Apple Grading According to European Standards using Machine Vision*, Biosystem Engineering, **83** (4), 397–404, 2002

146. V. Leemans, M.F. Destain, *A Real Time Grading Method of Apples based on Features Extracted from Defects*, Journal of Food Engineering **61**, 83–89, 2004.

147. R.A. Leonard, W.G. Knisel, and D.A. Still. *GLEAMS: Groundwater-Loading Effects of Agricultural Management Systems*, Transactions of American Society of Agricultural Engineers **30** (5), 1403–1418, 1987.

148. L. Lhotska, M. Macas, and M. Bursa, *PSO and ACO in Optimization Problems*, E. Corchado et al. (Eds.), Intelligent Data Engineering and Automated Learning 2006, Lecture Notes in Computer Science **4224**, Springer, New York, 1390–1398, 2006.

149. L. Li, D.M. Umbach, P. Terry and J.A. Taylor, *Application of the GA/KNN Method to SELDI Proteomics Data*, Bioinformatics **20** (10), 1638–1640, 2004.

150. Y. Liao, V.R. Vemuri, *Use of K-Nearest Neighbor Classifier for Intrusion Detection*, Computers & Security **21** (5), 439–448, 2002.

151. L. Liberti, S. Cafieri, F. Tarissan, *Reformulations in Mathematical Programming: a Computational Approach*, In: Foundations on Computational Intelligence, volume 3, A.-E. Hassanien, A. Abraham, F. Herrera, W. Pedrycz, A. Carvalho, P. Siarry, A. Engelbrecht (Eds.), Studies on Computational Intelligence **203**, Springer, New York, 153–234, 2009.

152. L. Liberti, C. Lavor, and N. Maculan, *Discretizable Molecular Distance Geometry Problem*, Tech. Rep. q-bio.BM/0608012, arXiv, 2006.

153. L. Liberti, C. Lavor, and N. Maculan, *A Branch-and-Prune Algorithm for the Molecular Distance Geometry Problem*, International Transactions in Operational Research **15** (1): 1–17, 2008.

154. A. Likasa, N. Vlassis, J.J. Verbeek, *The Global k-means Clustering Algorithm*, Pattern Recognition **36** (2), 451–461, 2003.

155. P.L. Lisboa, A.F.G. Taktak, *The Use of Artificial Neural Networks in Decision Support in Cancer: A Systematic Review*, Neural Networks **19**, 408–415, 2006.

156. Y. Liu, B.G. Lyon, W.R. Windham, C.E. Lyon, and E.M. Savage, *Principal Component Analysis of Physical, Color, and Sensory Characteristics of Chicken Breasts Deboned at Two, Four, Six, and Twenty-Four Hours Postmortem*, Poultry Science **83**, 101–108, 2004.

157. Y. Lu, S. Lu, F. Fotouhi, Y. Deng, S. Brown, *Fast Genetic K-means Algorithm and Its application in Gene Expression Data Analysis*, Detroit, Wayne State University, 2003.

158. Y. Lu, S. Lu, F. Fotouhi, Y. Deng, and S. J. Brown, *Incremental Genetic K-means Algorithm and Its Application in Gene Expression Data Analysis*, BMC Bioinformatics **5**, 172, 2004.

159. S.C. Madeira and A.L. Oliveira, *Biclustering Algorithms for Biological Data Analysis: a Survey*, IEEE Transactions on Computational Biology and Bioinformatiocs **1** (1), 24–44, 2004.

160. H.R. Maier, G.C. Dandy, *Neural Networks for the Prediction and Forecasting of Water Resources Variables: a Review of Modelling Issues and Applications*, Environmental Modelling & Software **15**, 101–124, 2000.

161. U.K. Mandal, D.N. Warrington, A.K. Bhardwaj, A. Bar-Tal, L. Kautsky, D. Minz, G.J. Levy, *Evaluating Impact of Irrigation Water Quality on a Calcareous Clay Soil using Principal Component Analysis*, Geoderma **144**, 189–197, 2008.

162. R.T. Marler, J.S. Arora, *Survey of Multi-Objective Optimization Methods for Engineering*, Structural and Multidisciplinary Optimization **26**(6), 369–395, 2004.

163. H. Martens, T. Naes, *Multivariate Calibration*, John Wiley & Sons, 1989.

164. N. Metropolis, A.W. Rosenbluth, M.N. Rosenbluth. A.H. Teller, and E. Teller. *Equation of State Calculations by Fast Computing Machines*, Journal of Chemical Physics **21**, 1087–1092, 1953.

165. G. E. Meyer, J. C. Neto, D. D. Jones, T. W. Hindman, *Intensified Fuzzy Clusters for Classifying Plant, Soil, and Residue Regions of Interest from Color Images*, Computers and Electronics in Agriculture **42**, 161–180, 2004.

166. MinGW: gnu C compiler: http://www.mingw.org

167. A. Moore, *Lecture on Validation Techniques*, available on the Internet at the address: http://www.autonlab.org/tutorials/overfit.html

168. MPI - Message Passing Interface: http://www-unix.mcs.anl.gov/mpi/

169. B. Moreaux, D. Beerens and P. Gustin, *Development of a Cough Induction Test in Pigs: Effects of SR 48968 and Enalapril*, Journal of Veterinary Pharmacology and Therapeutics **22**, 387–389, 1999.

170. D. Moshou, A. Chedad, A. Van Hirtum, J. De Baerdemaeker, D. Berckmans, H. Ramon, *An Intelligent Alarm for Early Detection of Swine Epidemics based on Neural Networks*, American Society of Agricultural Engineers **44** (1), 167–174, 2001.

171. D. Moshou, A. Chedad, A. Van Hirtum, J. De Baerdemaeker, D. Berckmans, H. Ramon, *Neural Recognition System for Swine Cough*, Mathematics and Computers in Simulation **56**, 475–487, 2001.

172. A. Mucherino, S. Costantini, D. di Serafino, M. D'Apuzzo, A. Facchiano and G. Colonna, *Towards a Computational Description of the Structure of all-alpha Proteins as Emergent Behaviour of a Complex System*, Computational Biology and Chemistry **32** (4), 233–239, 2008.

173. A. Mucherino and O. Seref, *Monkey Search: A Novel Meta-Heuristic Search for Global Optimization*, "Data Mining, System Analysis and Optimization in Biomedicine", AIP Conference Proceedings **953**, O. Seref, O.E. Kundakcioglu, P.M. Pardalos (Eds.), 162–173, 2007.

174. A. Mucherino and O. Seref, *Modeling and Solving Real Life Global Optimization Problems with Meta-Heuristic Methods*, Advances in Modeling Agricultural Systems, Springer Optimization and Its Applications Series, P.J. Papajorgji, P.M. Pardalos (Eds.), Springer, 403–420, 2008.

175. A. Mucherino, O. Seref, P.M. Pardalos, *Simulating Protein Conformations through Global Optimization*, arXiv e-print, arXiv:0811.3094v1, November 2008.

176. A. Nahapatyan, S. Busygin, and P. Pardalos, *An Improved Heuristic for Consistent Biclustering Problems*, In: Mathematical Modelling of Biosystems, R.P. Mondaini and P.M. Pardalos (Eds.), Applied Optimization **102**, Springer, 185–198, 2008.

177. K. Nakano, *Application of Neural Networks to the Color Grading of Apples*, Computers and Electronics in Agriculture **18**, 105–116, 1997.

178. A.J. Nebro, E. Alba and F. Luna, *Multi-Objective Optimization using Grid Computing*, Soft Computing - A Fusion of Foundations, Methodologies and Applications **11** (6), 531–540, 2007.

179. J. Ni, Q. Song, *Dynamic Pruning Algorithm for Multilayer Perceptron Based Neural Control Systems*, Neurocomputing **69**, 2097–2111, 2006.

180. A. Nurnberger, W. Pedrycz and R. Kruse, *Neural Network Approaches*, In: Handbook of Data Mining and Knowledge Discovery, W. Klosgen and J.M. Zytkow (Eds.), Oxford University Press, 2002.

181. N.R. Pal, J.C. Bezdek, *On Cluster Validity for the Fuzzy c-means Model*, IEEE Transactions on Fuzzy Systems **3** (3), 370–379, 1995.

182. T.N. Pappas, *An Adaptive Clustering Algorithm for Image Segmentation*, IEEE Transactions on Signal Processing **40** (4), 1992.

183. P.J. Papajorgji, P.M. Pardalos, *Software Engineering Techniques Applied to Agricultural Systems An Object-Oriented and UML Approach*, Applied Optimization Springer Series **100**, 2006.

184. P.M. Pardalos, and H.E. Romeijn (eds.), *Handbook of Global Optimization*, Vol. 2, Kluwer Academic, Norwell, MA, 2002.

185. V.C. Patel, R.W. McClendon, J.W. Goodrum, *Crack Detection in Eggs using Computer Vision and Neural Networks*, Artificial Intelligence Applications **8** (2), 21–31, 1994.

186. PDB - Protein Data Bank, ftp://ftp.wwpdb.org/

187. J.A. Fernandez Pierna, V. Baeten, A. Michotte Renier, R.P. Cogdill and P. Dardenne, *Combination of Support Vector Machines (SVM) and Near-Infrared (NIR) Imaging Spectroscopy for the Detection of Meat and Bone Meal (MBM) in Compound Feeds*, Journal of Chemometrics **18**, 341–349, 2004.

188. J.C. Platt, *Fast Training of Support Vector Machines using Sequential Minimal Optimization*, In: Advances in Kernel Methods - Support Vector Learning, B. Schölkopf, C. Burges, and A. Smola (Eds.), MIT Press, 185–208, 1999.

189. L. Prechelt, *A Quantitative Study of Experimental Evaluations of Neural Network Learning Algorithms: Current Research Practice*, Neural Networks **9** (3), 457–462, 1996.

190. J. Puchinger, G.R. Raidl, *Combining Metaheuristics and Exact Algorithms in Combinatorial Optimization: A Survey and Classification*, Lecture Notes in Computer Science **3562**, Springer, New York, 41–53, 2005.

191. S. Rahimi, M. Zargham, A. Thakre, D. Chhillar, *A Parallel Fuzzy C-Mean Algorithm for Image Segmentation*, IEEE Annual Meeting of the Fuzzy Information, Processing NAFIPS '04, **1**, 234–237, 2004.

192. B. Rajagopalan, U. Lall, *A k Nearest Neighbor Simulator for Daily Precipitation and Other Weather Variables*, Water Resources Research **35** (10), 3089–3101, 1999.

193. R. Reed, *Pruning Algorithms - A Survey*, IEEE Transactions on Neural Networks **4** (5), 1993.

194. M. Reyes-Sierra and C.A.C. Coello, *Multi-Objective Particle Swarm Optimizers: A Survey of the State-of-the-Art*, International Journal of Computational Intelligence Research **2** (3), 287–308, 2006.

195. G.L. Ritter, H.B. Woodruff, S.R. Lowry, T.L. Isenhour, *An Algorithm for a Selective Nearest Neighbor Decision Rule*, IEEE Transactions on Information Theory **21**, 665–669, 1975.

196. A. Riul Jr., H.C. de Sousa, R.R. Malmegrim, D.S. dos Santos Jr., A.C.P.L.F. Carvalho, F.J. Fonseca, O.N. Oliveira Jr., L.H.C. Mattoso, *Wine Classification by Taste Sensors Made from Ultra-Thin Films and using Neural Networks*, Sensors and Actuators **B98**, 77–82, 2004.

197. L.E. Rocha-Mier, L. Sheremetov and I. Batyrshin, *Intelligent Agents for Real Time Data Mining in Telecommunications Networks*, Lecture Notes in Computer Science **4476**, Springer, New York, 138–152, 2007.

198. F. Rosenblatt, *The Percentron: a Probabilistic Model for Information Storage and Organization in the Propagation*, Physichological Review **65**, 386–408, 1958.

199. M. Rova, R. Haataja, R. Marttila, V. Ollikainen, O. Tammela and M. Hallman, *Data Mining and Multiparameter Analysis of Lung Surfactant Protein Genes in Bronchopulmonary Dysplasia*, Human Molecular Genetics **13** (11), 1095–1104, 2004.

200. H.A. Rowley, S. Baluja, and T. Kanade, *Neural Network-Based Face Detection*, IEEE Transations on Patterns Analysis and Machine Intelligence **20** (1), 1998.

201. D. Salomon, *Data Compression: The Complete Reference*, Springer 2004.

202. C. Saunders, M.O. Stitson, J. Weston, L. Bottou, B. Scholkopf, and A. Smola, *Support Vector Machine Reference Manual*, Royal Holloway Technical Report CSD-TR-98-03, 1998.

203. T. F. Schatzki, R. P. Haff, R. Young, I. Can, L-C. Le, N. Toyofuku, *Defect Detection in Apples by Means of X-ray Imaging*, Transactions of the American Society of Agricultural Engineers **40** (5), 1407–1415, 1997.

204. R.B. Schnabel, J.E. Jr. Dennis, *Numerical Methods for Unconstrained Optimization and Nonlinear Equations*, Prentice Hall, 1983.

205. F. Schwenker, *Hierarchical Support Vector Machines for Multi-Class Pattern Recognition*, Proceedings of the 4th International Conference on Knowledge-Based Intelligent Engineering Systems and Allied Technologies (KES '00), vol. 2, 561–565, Brighton, UK, 2000.

206. U. Seiffert, *Artificial Neural Networks on Massively Parallel Computer Hardware*, European Symposium on Artificial Networks proceedings, Bruges (Belgium), 319–330, 2002.

207. G. Serban, A. Campan, *Hierarchical Adaptive Clustering*, Informatica **19**(1), 101–112, 2008.

208. O. Seref, O.E. Kundakcioglu, and P.M. Pardalos, *Selective Linear and Nonlinear Classification*, In: Data Mining and Mathematical Programming, P.M. Pardalos, P. Hansen (Eds.), CRM Proceedings and Lecture Notes **45**, American Mathematical Society, Providence, RI, 2008.

209. M.A. Shahin, E.W. Tollner, M.D. Evans, H.R. Arabnia, *Watercore Features for Sorting Red Delicious Apples: a Statistical Approach*, Transactions of the American Society of Agricultural Engineers **42** (6), 1889–1896, 1999.

210. M.A. Shahin, E.W. Tollner, R.W. McClendon, *Artificial Intelligence Classifiers for Sorting Apples based on Watercore*, Journal of Agricultural Engineering Research **79** (3), 265–274, 2001.

211. J.Shawe-Taylor, N. Cristianini, *Kernel Methods for Pattern Analysis*, Cambridge University Press, 2004.

212. Y. Shen, H. Shi, and J.Q. Zhang, *Improvement and Optimization of a Fuzzy C-Means Clustering Algorithm*, IEEE Instrumentation and Measurement Technology Conference, Budapest, Hungary, 2001.

213. H. Sherali, L. Liberti, *Reformulation-Linearization Technique for Global Optimization*, In: Encyclopedia of Optimization, P.M. Pardalos and C. Floudas (Eds.), 2nd Edition, 3263–3268, Springer, 2008.

214. K. Shin , A. Abraham and S.Y. Han, *Improving kNN Text Categorization by Removing Outliers from Training Set*, Lecture Notes in Computer Science **3878**, Springer, New York, 563–566, 2006.

215. J. Si, B.J. Nelson and G.C. Runger, *Artificial Neural Network Models for Data Mining*, In: The Handbook of Data Mining, N. Ye (Eds.), Lawrence Erlbaum Associates Publishers, 2003.

216. J. Sim, S.-Y. Kim and J. Lee, *Prediction of Protein Solvent Accessibility using Fuzzy k-Nearest Neighbor Method*, Bioinformatics **21** (12), 2844–2849, 2005.

217. L.C. Sim, H. Schroder, G. Leedham, *MIMD–SIMD Hybrid System - Towards a New Low Cost Parallel System*, Parallel Computing **29**, 21–36, 2003.

218. A.N. Skodras, T. Ebrahimi, *JPEG2000 Image Coding System Theory and Applications*, Proceedings of the IEEE International Symposium on Circuits and Systems, 3866–3869, 2006.
219. H. Spath, *Cluster Analysis Algorithms for Data Reduction and Classification of Objects*, Ellis Horwood, Chichester, 1980.
220. I. Steinwart, *Consistency of Support Vector Machines and Other Regularized Kernel Classifiers*, IEEE Transactions on Information Theory **51**, 128–142, 2005.
221. C.O. Stockle, S.A. Martin, and G.S. Campbell. *CropSyst, a Cropping Systems Model: Water/Nitrogen Budgets and Crop Yield*, Agricultural Systems **46** (3), 335–359, 1994.
222. K. Stoffel, A. Belkoniene, *Parallel k/h-Means Clustering for Large Data Sets*, Lecture Notes in Computer Science **1685**, Proceedings of the 5th International Euro-Par Conference on Parallel Processing, Springer, New York, 1451–1454, 1999.
223. M. Su, C. Chou, *A Modified Version of the K-means Algorithm with a Distance Based on Cluster Symmetry*, IEEE Transactions on Pattern Analysis and Machine Intelligence **23** (6), 674–680, 2001.
224. R. Sulej, K. Zaremba, K. Kurek and E. Rondio, *Application of the Neural Networks in Events Classification in the Measurement of the Spin Structure of the Deuteron*, Measurement Science and Technology **18**, 2486–2490, 2007.
225. A. Tellaeche, X.-P. Burgos-Artizzu, G. Pajares and A. Ribeiro, *A Vision-Based Hybrid Classifier for Weeds Detection in Precision Agriculture Through the Bayesian and Fuzzy k-Means Paradigms*, Advances in Soft Computing **44**, 72–79, 2008.
226. S. Tripathi, V.V. Srinivas, R.S. Nanjundiah, *Downscaling of Precipitation for Climate Change Scenarios: A Support Vector Machine Approach*, Journal of Hydrology **330**, 621–640, 2006.
227. Y.-H. Tseng, C.-J. Lin, and Y.-I Lin, *Text Mining Techniques for Patent Analysis*, Information Processing & Management **43** (5), 1216–1247, 2007.
228. L. Ungar and D.P. Foster, *A Formal Statistical Approach to Collaborative Filtering*, Proceedings of the Conference on Automated Learning and Discovery (CONALD '98), 1998.
229. A. Urtubia, J.R. Perez-Correa, M. Meurens, E. Agosin, *Monitoring Large Scale Wine Fermentations with Infrared Spectroscopy*, Talanta **64** (3), 778–784, 2004.
230. A. Urtubia, J. R. Perez-Correa, A. Soto, P. Pszczolkowski, *Using Data Mining Techniques to Predict Industrial Wine Problem Fermentations*, Food Control **18**, 1512–1517, 2007.
231. A. Van Hirtum and D. Berckmans, *Fuzzy Approach for Improved Recognition of Citric Acid Induced Piglet Coughing from Continuous Registration*, Journal of Sound and Vibration **266** (3), 667–686, 2003.
232. V.N. Vapnik, *Statistical Learning Theory*, John Wiley & Sons, 1998.
233. K. Verheyen, D. Adriaens, M. Hermy, S. Deckers, *High-Resolution Continuous Soil Classification using Morphological Soil Profile Descriptions*, Geoderma **101**, 31–48, 2001.
234. K.N. Vikram, V. Vasudevan and S. Srinivasan, *Rate-Distortion Estimation for Fast JPEG2000 Compression at Low Bit-Rates*, Electronic Letters **41** (1), 16–18, 2005.
235. H.D. Vinod, *A Survey of Ridge Regression and Related Techniques for Improvements over Ordinary Least Squares*, The Review of Economics and Statistics **60** (1), 121–131, 1978.
236. Y. Wu, K. Ianakiev, V. Govindaraju, *Improved k-Nearest Neighbor Classification*, Pattern Recognition **35**, 2311–2318, 2002.
237. X. Wu, V. Kumar, J.R. Quinlan, J. Ghosh, Q. Yang, H. Motoda, G.J. McLachlan, A. Ng, B. Liu, P.S. Yu, Z.-H. Zhou, M. Steinbach, D.J. Hand, D. Steinberg, *Top 10 Algorithms in Data Mining*, Knowledge and Information Systems **14**, 1–37, 2008.
238. J. Xu, D.W.C. Ho, *A New Training and Pruning Algorithm Based on Node Dependence and Jacobian Rank Deficiency*, Neurocomputing **70**, 544–558, 2006.
239. R. Xu, D. Wunsch II, *Survey of Clustering Algorithms*, IEEE Transactions on Neural Networks **16** (3), 645–678, 2005.
240. S. Xu and M. Zhang, *A New Adaptive Neural Network Model for Financial Data Mining*, Lecture Notes in Computer Science **4491**, Springer, New York, 1265–1273, 2007.
241. Q. Yang, *An Approach to Apple Surface Feature Detection by Machine Vision*, Computers and Electronics in Agriculture **11**, 249–264, 1994.
242. Z.R. Yang, R. Hamer, *Bio-Basis Function Neural Networks in Protein Data Mining*, Current Pharmaceutical Design **13** (14), 1403–1413, 2007.

243. Q. Yang, and J.A. Marchant, *Accurate Blemish Detection with Active Contour Models*, Computers and Electronics in Agriculture **14**, 77–89, 1996.

244. J. Yang, W. Wang, H. Wang, and P. Yu, *Enhanced Biclustering on Expression Data*, Proceedings of the Third IEEE Conference in Bioinformatics and Bioengineering, 321–327, 2003.

245. N. Yano, M. Kotani, *Clustering Gene Expression Data using Self-Organizing Maps and k-means Clustering*, SICE Annual Conference in Fukui, Japan, 2003.

246. K.Y. Yeung and W.L. Ruzzo, *Principal Component Analysis for Clustering Gene Expression Data*, Bioinformatics **17** (9), 763–774, 2001.

247. S. Ying, Y. Zheng, G. Kanglin, *Mining Stock Market Tendency by RS-Based Support Vector Machines*, IEEE International Conference on Granular Computing, 659–659, 2007.

248. R. Yu, P.S. Leung, P. Bienfang, *Predicting Shrimp Growth: Artificial Neural Network versus Nonlinear Regression Models*, Aquacultural Engineering **34**, 26–32, 2006.

249. X. Zeng, D.S. Yeung, *Hidden Neuron Pruning of Multilayer Perceptrons using a Quantified Sensitivity Measure*, Neurocomputing **69**, 825–837, 2006.

250. Y. Zhang, Z. Xiong, J. Mao, L. Ou, *The Study of Parallel k-means Algorithm*, Proceedings of the 6th World Congress on Intelligent Control and Automation **2**, 5868–5871, 2006.

251. S. Zhong, *Efficient Online Spherical K-means Clustering*, Proceedings of International Joint Conference on Neural Networks **5**, 3180–3185, 2005.

Glossary

agriculture The science, art, or occupation concerned with cultivating land, raising crops, and feeding, breeding, and raising livestock. Data mining techniques applied to agriculture are discussed in this book.

algorithm A set of unambiguous rules or instructions for solving a given problem in a finite number of steps.

center of a cluster The mean among all the vectors representing the samples in a unique cluster.

class A subset of samples having the same classification. The word "class" is used when classification methods are used.

classification The problem of dividing a given set of data into different classes.

cluster A subset of samples having some common properties. The word "cluster" is used when clustering methods are employed.

clustering The problem of dividing a given set of data in different clusters into which samples having some common properties are grouped.

covariance A statistical measure of the variance of two variables. It corresponds to the product of the deviations of the corresponding values of the two variables from their respective means.

covariance matrix A matrix of covariances between elements of a vector.

cubic spline A spline in which all the polynomial pieces have degree 3.

data mining The nontrivial extraction of previously unknown, potentially useful and reliable patterns from a set of data; it is the process of analyzing data from different perspectives and summarizing it into useful information; it is also known as "knowledge discovery."

dependent variable A mathematical variable whose value is determined by the value other variables have. For example, if f is a function in $\Re$ and x is a real variable, then $y = f(x)$ is a dependent variable.

deterministic method A method that is able to provide the solution to the problem to be solved if specific hypotheses are met.

eigenvalues and eigenvectors Given a square matrix Σ, if there is a vector x such that

$$(\Sigma - \lambda I)x = 0,$$

where I is the identity matrix having the same dimensions of Σ, then x is an eigenvector of Σ and the real number λ is the corresponding eigenvalue. Usually, the above linear system is solved in order to obtain all the eigenvalues and all the eigenvectors of Σ.

Euclidean plane A Euclidean space in dimension 2.

Euclidean space The space of all possible n-tuples $(x_1, x_2, \ldots, x_n)$ of real numbers. It is denoted by the symbol $\Re^n$.

exact method See "deterministic method."

function A rule or law that associates uniquely an element of a set A to one and only one element of another set B.

heuristic method Heuristic methods are used to rapidly come to a solution that is reasonably close to the best possible answer, or "optimal solution." They do not guarantee that the solution found is the optimal one. However, they are used when any deterministic method for solving the same problem cannot be applied or it is too computationally expensive.

independent variable A variable whose value determines the value of other variables. For example, if f is a function in $\Re$ and x is an independent variable, then the value of x influences the value of the variable $y = f(x)$.

interpolating function A function that interpolates a given set of points

$$\{(x_1, y_1), (x_2, y_2), \ldots, (x_n, y_n)\}$$

in a Euclidean space. In other words, f is an interpolating function if

$$f(x_i) = y_i \quad \forall i = 1, 2, \ldots, n.$$

learning phase The process in which a given system learns how to perform a certain task. The learning phase is employed by artificial neural networks and support vector machines.

logarithmic function The logarithmic function in base b of the real number x is the exponent to give to the base for obtaining x.

multilayer perceptron A type of artificial neural network in which the neurons of the network are organized in layers.

natural logarithm The logarithmic function having as base the Nepero number $(e = 2.71\ldots)$.

Newton polynomial A polynomial interpolating a given set of points in the two-dimensional space if its coefficients correspond to the "divided differences," obtained from the points to interpolate.

objective function The function to be optimized in an optimization problem; depending on the problem at hand, it can be required that the function is minimized or maximized.

optimization problem The problem of optimizing (minimizing or maximizing) a given objective function, subject to certain constraints.

outliers Any sample which is uniquely different from a given subgroup of samples (a cluster or a class).

parabola A polynomial of degree 2.

parallel computing Parallel computing is a form of computation in which several calculations are carried out simultaneously.

pattern A distinctive style, model, or form.

polynomial Any function having equation

$$p(x) = a_n x^n + a_{n-1} x^{n-1} + \cdots + a_2 x^2 + a_1 x + a_0,$$

in the Euclidean two-dimensional space, is a polynomial of degree n.

pruning process The process of removing useless or redundant objects or information from a given system. For example, artificial neural networks can be pruned after the learning phase.

regression function A function which approximates a given set of points in a Euclidean space. The coefficients of such a function are usually identified by solving a certain optimization problem.

sample One that is representative of a group or class or cluster.

software Software is a general term used to describe a collection of computer programs, procedures and documentation that perform some tasks on an operating system.

spline A function

$$S : [a, b] \subset \Re \longrightarrow \Re$$

formed by polynomial pieces

$$P_i : [t_i, t_{i+1}) \in [a, b] \longrightarrow \Re \quad \forall i \in \{1, 2, \ldots, K\},$$

where $a = t_1 < t_2 < \cdots < t_K < t_{K+1} = b$. Each polynomial piece usually has a predetermined degree.

testing set A set of samples with known classification used for testing a data mining technique.

training phase See "learning phase."

training set A set of samples with known classification used for tuning the parameters of a given classification technique.

unsupervised classification See "clustering."

validation set A set of samples with known classification used for validating the results obtained by certain classification technique.

variance The variability of a given variable. It can be obtained by locating the minimum and the maximum value of the variable.

vector A sorted set of a variables which are called components.

Index